AF527814

FOSSIL REMAINS

ENCYCLOPAEDIA OF LIFE SCIENCE SERIES

FOSSIL REMAINS

BARON CUVIER

ANMOL PUBLICATIONS PVT. LTD.
NEW DELHI-110 002 (INDIA)

ANMOL PUBLICATIONS PVT. LTD.
4374/4B, Ansari Road, Daryaganj
New Delhi - 110 002

Fossil Remains
First Edition, 1998
ISBN 81-261-0114-8

PRINTED IN INDIA

Published by J.L. Kumar for Anmol Publications Pvt. Ltd., New Delhi and Printed at Tarun Offset Press, Delhi.

LIST OF PLATES.

LIST OF PLATES.

FOSSIL MAMMALIA.

Researches into Fossil Osteology are comparatively of very recent date, and almost all that they possess of a scientific form is owing to the exertions of the illustrious naturalist whose steps we have thus far pursued, though at a very humble distance. It was reserved for him to ascertain, for the most part, the genera and species to which the osseous remains of terrestrial animals, so abundantly discovered in the superficial strata of this planet, are attached. Prepared for the execution of this Herculean task by the profoundest study of comparative anatomy, and the natural history of existing, he was enabled to characterize with precision the fragments of extinct species, to reconstruct those ancient animals, and present to our astonished view the wonders of former creations. More certain data have been thus obtained for the revolutions and duration of the globe; geology has ceased to be a romance, and a solid basis is at length established for a rational theory of the earth. Our intention is to present to our readers, in an abridged form, the result of such researches, not only of the Baron, but of every other modern naturalist who has investigated the subject; but, before we enter on any specific details, it will be necessary to take a brief general view of the revolutions which the surface of this globe has undergone, and the consequent alterations which have taken place in animal existence. Our limits will prevent us from entering very deeply into the purely geological portion of the subject, or, in fact, of considering it at all, except in relation to its connection with the organic fossils.

An inspection of the various strata in which fossil remains have been deposited, serves to prove that, in general, a constant order has been observed in their formation.

The sea, by which the entire earth appears to have been covered, having rested in certain situations a sufficient length of time to collect particular substances, and sustain the life of certain genera and species of animals, has been afterwards replaced by another sea, which has collected other substances, and nourished other animals.

It may be believed that the primitive strata, which contain no organic remains, had all of them one contemporaneous origin. But, with respect to the strata which cover them, the study of fossil osteology has clearly proved that they were formed at different eras of time, during each of which animals existed distinct from those which lived in other eras, and distinct from almost all the known species which exist at the present day. It is true, that those causes to which the production of mountains is owing, have, in the countries which are intersected by primitive chains, or which border on them, disturbed the original established order of the strata. But, in level countries, it is perfectly obvious that they have been formed by a long and tranquil sojournment of the waters, in the same manner as are formed, at the present day, those depositions which cover the bottom of the seas.

Vegetable and animal remains are sometimes found at a depth of three or four thousand feet, and even below the sea, as in the instance of the coal-pits of Whitehaven. In all parts of the world, marine productions are to be found in a fossil state. They are found at very considerable degrees of elevation, on mountains far remote from the neighbourhood of any sea. So numerous, indeed, are they in certain places, that they constitute, to a very great extent, the aggregate of the soil. These remains of organised bodies were formerly considered as mere *lusus naturæ,* generated in the bosom of the earth by its creative powers. But this absurdity has been completely refuted, by a thorough examination of their forms and composition. It has been clearly demonstrated that there is no difference of texture between the bodies of which we now speak, and those which exist in our present seas.

The marine genera, found in the most ancient, do not appear to be as numerous as those contained in the more recent strata; and it is worthy of remark that the fossil organic bodies of every description, differ more from existing species in proportion to the antiquity of the strata in which they are found. Those very ancient formations, to which the name of *transition-strata* has been given, rest upon the granite or other primitive rocks, which, as far as we can tell, form the substratum of the globe, and in which no organic remains have ever been discovered. We are thus led to the knowledge of a fact equally astonishing and certain, namely, that there was a period when life did not exist upon this earth; the era, indeed, of its commencement is clearly observable. This evidently proves the doctrine of a creation, and utterly confounds the absurd speculations of atheism respecting the eternity of the world, and the generative powers of inanimate matter.

The mode in which these primitive strata were formed is a mooted question. Some are of opinion that the ancient granite owed its origin to a fluid which once held every thing in solution; and others, that it was the first substance that became fixed, on the cooling of a mass of matter in a state of fusion. The Marquess de La Place has conjectured, that the materials of which the earth is composed were at first of an elastic form, and became, in cooling, of a liquid, and, finally, of a solid consistence. The recent experiments of M. Mitscherlich, says the Baron Cuvier, go far in support of this opinion. That gentleman has completely succeeded in composing and crystallizing several of the mineral species which enter into the composition of primitive rocks. Be this, however, as it may, we find the shelvy summits of all the grand mountain-chains which intersect our continents composed of these primitive strata. The granite almost invariably constitutes the central ridges of these mighty chains, and, singular to relate, it occupies the highest and the lowest position in their stratification. That it was forced upwards by some tremendous convulsions of nature, which have shaken this globe to its very centre, is indubitable. The indented ridges, the ragged precipices, the bristling peaks, by which these primitive chains are always characterized, prove to demonstration the violence which was exerted in their production. In this respect, they exhibit a decided contrast to those more convex mountains, and undulating ranges of hills, whose mass was quietly deposited by the last retiring sea, and has since remained undisturbed by any violent revolution.

The lateral ridges of these chains are formed of schistus, porphyry, talc-rocks, &c. which rest on the sides of the granite. Finally, the external ranges are composed of granular marble, and other calcareous strata, but devoid of shells, which rest upon the schistus, and form the last boundary of the empire of mere inanimate matter. We now begin to find, but few in number, and, at intervals, in the transition strata, the earliest animal productions. We find the larger orthoceræ, those singular crustacea, the tribolites, the calymenes, the ogygiæ. We find encrinites, numerous species of cornua ammonis, and of terebratulæ, belemnites, trigoniæ, and other genera, most of which are no longer found in less ancient strata. Terebratulæ are found in these ancient strata, in the chalk formations above them, in the shelly limestone above those, and in the living state; but the number of species, and even of individuals of this genus, are found to diminish in an inverse ratio to the antiquity of the periods in which they existed.

We find shelly strata occasionally interposed between beds of granite and other primitive substances, which must have occupied their present situation at a more recent period. That these primitive masses experienced changes and convulsions even previously to the appearance of life on this globe is evident. These masses indicate violent removals of position, some of which must have taken place before they were covered by the strata of shells. The disruptions we observe among them are sufficient proof of this. But since the formation of the secondary strata those same primitive masses have undergone similar convulsions. They have, not improbably, caused, most certainly they have shared in, the violent changes which have as evidently taken place in the secondary strata. How, if it were otherwise, could it happen that we find immense portions of those primitive rocks uncovered, though not situated so high as the secondary strata? We find numerous blocks of granite, &c. scattered over the secondary strata, even in situations where deep vallies, where portions of the sea intervene between them and those mountainous ridges from which they must have been transported. They must have been driven thither by tremendous eruptions or violent inundations, far exceeding in force and velocity any impelling cause with which we are now acquainted capable of changing the face of nature.

To enter very deeply into an examination of the acting causes which contribute at present, and have contributed ever since the era at which authentic history dates its commencement, to change the earth's surface, would be foreign to our present purpose. It will be sufficient to remark, that these causes are rains and thaws, which bring down portions of mountains; streams which carry these on, and form what are termed alluvial depositions, in places where their course is slackened; the sea, which gradually changes the outline of the land, by undermining the more elevated coasts and forming precipitous cliffs, and throwing in heaps of sand upon the more level shores, thus gradually overspreading a considerable extent of terra firma; and finally, volcanoes, piercing through the solid strata, and throwing heaps of matter around them to certain degrees of extent and elevation.

Now, the action of the waters, whether in rains, thaws, or running streams, which extend the land by the eventual deposition of debris, presupposes the existence of mountains, vallies, plains, and other inequalities, and consequently could not have produced them. The action of the sea is still more limited, and its phenomena have no affinity with the immense masses of whose revolutions we have been speaking; and volcanoes, though they have formed both mountains and islands, formed them of nothing but lava, i. e. of substances modified by volcanic action, which is never the case with the substances to which we have above alluded, nor do volcanoes ever disturb the strata which traverse their apertures. In a word, none of the agents acting on the earth's surface with which we are now acquainted, are capable of producing those tremendous revolutions which have left their traces so indelibly marked on the external covering of the globe. Neither will the circular motion of the pole of the earth, nor the gradual inclination of its axis on the plane of the ecliptic, better serve to explain such phenomena. The slowness and limited direction of these motions bear no proportion to the extent and overwhelming rapidity of such catastrophes. It follows, then, that they must have been occasioned by causes whose operation has long ceased, which were external to this planet, and most probably totally out of the course of things in existing nature. Many conjectures have been made by naturalists respecting the character of these causes, some eminent for absurdity, and all resting on hypothesis; into any of which it would be as wide of our design to enter as to propound any new solution of our own. It is enough to repeat that nothing in the agency of nature, as it has operated for ages in relation to this earth, could have produced the grand revolutions which this earth has evidently undergone; nor is it any absurdity to suppose that the agency which did produce them was preternatural.

To return to our more immediate subject; it is comparatively but a short time since the study of marine fossils has been pursued with that degree of attention which it deserves. Involving infinitely greater difficulties than that of the conchology of existing species, much fewer of the former than of the latter have been discovered. It is yet the opinion of some eminent naturalists that the number of ancient species may equal, if not exceed, that of the modern. They are led to this conclusion by considering that the latter appertain but to a single era, while the former are attached to many successive periods in which animals of different descriptions have been abundantly produced.

It is but seldom that we meet with shells in the fossil state of species perfectly analogous to those which now exist. There is scarcely any exception to this but in the case of the fossils imbedded in the low hills of the Apennine range, of which a considerable number are found in a living state in the neighbouring Mediterranean. It is, however, remarkable that a number of mollusca and marine polypi found in this sea in abundance are not discovered in the fossil state, as, in like manner, fossil species are found in the Apennines that no longer exist. This want of perfect similarity is by no means surprising when we find that even species of the same strata, and of the existing seas, do not perfectly resemble when the habitat differs.

The remains of mollusca and zoophytes are much more numerous than other fossils, and the strata in which they are found are sometimes changed into calcareous stone. They are found in falun, in marl, in clay, and in grès or granulated brownish quartz. Shells, nearly resembling those of our marshes and streams, are found in the more recent strata.

Between the strata composed of marine fossils, we sometimes meet others containing terrestrial remains of animals or vegetables, which prove the settlement and return, at different periods, of the sea and fresh water, and even between these periods, the absence for a time of both, as it appears that the terrestrial animals must have lived in the place where their remains are found.

The circumstance of finding, amidst the ice of the north, the carcasses of elephants and rhinoceroses with flesh and hair, proves that the retreat of the waters at the era of their destruction must have been prompt. A sudden change must have also taken place in the temperature of these countries; for these carcasses were found in places to which they could not have been transported at the present day, and were besides so frozen up, that in the instance of the elephant found by a Tungoose in 1799 (as will be subsequently seen), several years elapsed before all the parts of the body could be extricated.

Had the waters retired slowly, the entire surface of the earth so abandoned would have been like the shore of the sea; ancient cliffs would have been found wherever elevations existed, and the fossil shells would have been defaced like those now found on the sea coast. But nothing of all this occurs. Many fossil shells are found broken, but not worn: their angular points are not blunted. This is declared by M. Defrance to be invariably the case with those of France, Italy, England, and North America, which he has examined, with the exception of such as are found in the falun of Touraine, which in all respects resembles the shelly sand of the sea-shore. The shells found there are almost all broken; their angles are blunted, and in the apertures of the univalves stones or other shells are repeatedly found, which are difficult to be got out, just exactly as we find it to be with those on the sea-coasts. Terrestrial helices are even found there of a species unknown in the country, filled with the debris of marine polypi and shells. It is natural to imagine that the soil of Touraine where the falun is found was exposed to the dashing of the waves which covered those parts of France where the bed of coarse, shelly limestone is found, and with which the falun of Touraine has the strongest affinity.

Fossil fish are found in the ancient marine strata as well as in the more recent. So are the crustacea which frequently accompany them. There is reason to believe, that a sudden revolution like that which a volcano might occasion, may have overwhelmed such of them as are found in the greatest abundance in certain places. The debris of osseous fishes are often found; but of the cartilaginous we find nothing but the vertebræ and teeth of squali. The coarse, shelly limestone, as well as the more recent strata, contains an immense quantity of debris of the claws of crustacea, and of the auricular bones of different sorts of fish.

The remains of terrestrial animals found in the fossil state consist of bones, the antlers of certain species of cervus, and teeth. It may be noticed, that such remains are rarely in a state of petrifaction. The horns of other ruminants, hoofs, claws, &c. are never found.

Oviparous quadrupeds, such as the crocodiles of Honfleur, of England, and the monitors of Thuringia, are found in very ancient strata. The saurians and tortoises of Maestricht are met with in the more recent chalk formation. The bones of lamantins and phocæ are found in a coarse, shelly limestone, very analogous to that which covers the chalk formation near Paris. The Baron has observed, in his great work, that up to this point no remains of mammiferous land animals have been found. Professor Buckland, however, to whose researches fossil osteology is so much indebted, has, in the first volume of the "Transactions of the Geological Society," given an account of a mammiferous quadruped occurring in an ancient secondary rock. In the calcareous slate of Stonesfield, in Oxfordshire, which lies in the upper part of the lowest division of oolitic rocks, have been found, says the Doctor, "two portions of the jaw of the didelphis, or opossum, being of the size of a small kangaroo rat, and belonging to a family which now exists chiefly in America, Southern Asia, and New Holland." The Doctor refers this fossil to didelphis on the authority of the Baron himself, who has examined it twice, and the second time pronounced it to have been mammiferous, like an opossum, but of an extinct genus, and differing from all carnivorous mammalia in having ten teeth in a series in the lower jaw.

It is right, however, to remark here, that some controversy has arisen respecting the exact position of this calcareous slate in the minor subdivisions of the oolitic series at Stonesfield. To enter into the merits of this controversy would be quite beside our purpose; and, though we most strongly lean to the belief that the Doctor is justified in his conclusions, it would be presumptuous in humble compilers like ourselves, to pronounce decisively on so important a question.

Waiving this exception, if it be one, we find no bones of terrestrial mammalia until we come to the strata deposited above the last-mentioned formation of shelly limestone. There we first discover them, and there is a remarkable succession among the species. The debris of genera unknown at the present day, of anoplotheria, of palæotheria, found in the fresh-water formation, are the first which exhibit themselves above the shelly limestone. With those we find some lost species of known genera, oviparous quadrupeds and fishes. The beds in which they are found are covered by other strata, filled with marine fossils.

The fossil elephant, the rhinoceros, the hippopotamus, and the mastodon, are not found with those more ancient genera. They are found in the ancient alluvial strata, sometimes with marine and sometimes with fresh-water productions; but never in the regular rocky strata. The species of these animals, and every relic found with them, are either unknown or doubtful: and it is only in the latest alluvial depositions that species which appear similar to those now existing are to be found.

Among the most astonishing phenomena which fossil osteology unfolds to our view, are those *osseous breccie*, which, though removed from each other the distance of many hundred leagues, do yet present analogous peculiarities. Scattered rocks composed of the same stone are divided in different directions. Their fissures are filled with a calcareous concretion, very hard, and forming a sort of ochreous cement, in which bones mixed with terrestrial shells are found imbedded. These bones, which are not petrified, have been almost all of them broken previously to their incrustation. These breccie are found in the rock of Gibraltar, at Cette, Nice, Antibes, in Corsica, in Dalmatia, and in the island of Cerigo; depositions nearly similar are found at Corcud, near Terruel in Arragon, in the Vicentine territory, and in the Veronese.

In the rock of Gibraltar have been found the bones of a ruminant animal, which the Baron thinks may appertain to antelope, and the teeth of one species of the genus lepus.

In the deposition at Cette, the bones of rabbits, of the size and form of such as now exist, have been found; others of the same genus, but one-third smaller; rodentia, similar to the campagnol; birds of the size of the wagtail, and snakes.

In the osseous breccia of Nice and Antibes are found the bones of horses, and ruminants' teeth of the latter order, of species about the magnitude of cervus.

The breccia of Corsica contains debris of lagomys, existing at present only in Siberia, and bones of a rodens resembling perfectly the water-rat, except that it is smaller.

We find in those of Dalmatia the bones of ruminants of the size of dama.

Of the bones in the breccia of the island of Cerigo, we have no account, except from Spallanzani, who imagined that human bones existed among them. This opinion, however, appears totally destitute of foundation.

In the deposition of Corcud, the bones of asses and oxen have been found, resembling those of the present day, and of sheep of a very diminutive size.

In the Vicentine and the Veronese breccia, the antlers and bones of cervus have been found, together with the bones of oxen and elephants. A tusk of one of these last was nearly twelve feet in length.

Similar discoveries have also been made in the fissures of Sicily and Sardinia, and in different parts of Germany. But it is impossible to afford in this place any further detail concerning them.

In the plaster-quarries in the neighbourhood of Paris are found skeletons of genera for ages extinct, such as the anoplotherium and palæotherium: also bones of an animal bearing affinity to the sarigue, of four species of carnivora, with debris of tortoises, birds, and fishes.

The loose strata exhibit bones, teeth, and tusks of elephants, mingled with bones of horses, in almost every country; of mastodons, in America, in Little Tartary, in Siberia, in Italy, in France; of the rhinoceros, in France, in England, in Italy, in Germany, and Siberia; of the hippopotamus, near Montpellier, in Italy, and England, &c. &c.; of an animal resembling the tapir in the south of France; of a gigantic species of cervus, resembling the elk in Ireland and England; of the Indian musk-ox in Siberia; of fallow-deer of an unknown species in Scania; of hyænas, near Eichstadt; of balænæ, in the Plaisantin; and of an immense animal of the family tardigrada, called the megatherium, a species unknown in the living state, near Buenos Ayres.

In the *turbaries* of the department of the Somme in France, have been found debris of the aurochs, of oxen, far surpassing in magnitude our domestic races; of beavers, of cervi, of unknown species of horses, of roebucks, and of wild boars.

We are far, indeed, from having enumerated all the discoveries of this description that have been made, nor will our limits permit us to do so. Even since the publication of the last edition of the "Ossemens Fossiles," thirty species have been found in volcanic tufa in the strata of Mount Perrier, near the Issoire, in France;—namely, nine ruminants, six pachydermata, one edentatum, twelve carnivora, and two rodentia. In the calcareous fresh-water formation of Volvic, ten species;—one ruminant, two anoplotheria, one palæotherium, two rodentia, two carnivora, and two reptiles. In the similar formation of Gergovia, four species;—one anoplotherium, one reptile, and two ornitholites. Nay, even as we write, these discoveries are being prosecuted on the Continent and in America, with a zeal, assiduity, and success, unexampled in any former era in the annals of science; nor can it be expected, by any possibility, that a sketch like the present should embrace them all.

Phenomena not less astonishing than those on which we have been hitherto commenting, are exhibited in certain ancient caverns which have been discovered in Germany, in Hungary, and in England. They equally surprise and interest us by the immense quantity of debris of fossil animals which they contain, and the remarkable analogy that exists among them all in a geological point of view. To attempt any thing even approaching to a complete account of them here would be impossible. We shall, however, notice some of their most striking peculiarities; and for a fuller description refer our readers to the "Reliquiæ Diluvianæ" of Professor Buckland, a work equally admirable for deep research, luminous exposition of facts, and sound deduction.

The most anciently celebrated of these caverns, according to the Baron, is that of Bauman, near the city of Brunswick. The entrance faces the north, but the entire direction is from east to west. The entrance is very narrow. The first chamber is the largest. Into the second it is necessary to descend by a passage, first creeping, and then with the assistance of a ladder. The difference of level is thirty feet. This second chamber most abounds in stalactite, of a variety of forms. The passage to the third chamber is at first the most difficult of all. It is necessary to climb with hands and feet, but it gradually enlarges, and the stalactites upon its roof and sides exhibit an astonishing variety of fantastic and beautiful figures. There are in this passage two lateral dilatations, constituting a third and fourth chamber in the map of the *Acta Erud:* At its extremity it is necessary again to re-ascend to arrive at the entrance of the third chamber, which forms a sort of portico. Behrens, in his *Hercynia Curiosa*, says that there is no penetrating there, as it would be necessary to descend more than sixty feet. But the map above-mentioned and the description of Van der Hardt which accompanies it, characterize this third chamber as the fifth, and place beyond it another tunnel or passage terminated by two small caverns. Silbersschlag, in his *Geogenie*, adds that one of them leads into a final tunnel, which, descending considerably, leads under the other chambers, and is terminated by a place filled with water. There are abundance of fossil remains in this remote and unfrequented part.

The principal portion of the bones discovered in this cavern belong to the genus of the bear.

Other caverns very nearly similar are found in the chain of the Hartz mountains. Many are also found in Hungary on the southern declivities of the Krapach mountains. But the most celebrated of all is that of Gaylenreuth, situated on the left bank of the Wiesent. It is composed of six grottoes, which form an extent of more than two hundred feet. These caverns are strewed with bones, great and small, which are all of the same description as are to be found over an extent of more than two hundred leagues. More than three-fourths of these bones belong to a species of bear as large as our horses, and which is no longer found in the living state. The half or two-thirds of the remaining bones belong to an hyæna, of the size of the living bear. There were also the remains of a tiger, wolf, fox, glutton, and pole-cat, or some species approximating to it. The bones of herbivora are also found there, particularly of cervi, but in smaller number. Soemmerring has also mentioned that a portion of the cranium of an elephant was extracted from this cavern.

It is the opinion of the Baron that the remains in question belonged to animals which lived and died in the caves in which their debris are found, and that the period of their establishment there was considerably posterior to the era in which the extensive rocky strata were formed. In his first edition, he expressed his opinion that it was subsequent to the formation of the loose strata in which the bones of the elephant, rhinoceros, and hippopotamus have been discovered. But he has since altered this opinion, and fully coincides with Dr. Buckland that the bones of the caverns and the osseous breccia, are of the same antiquity with those of the loose strata, and that all were prior to the last general catastrophe which overwhelmed this globe.

Of the *Megalonix*, an extinct animal of the sloth genus, the remains have been found in a cavern in Western Virginia.

Of the caves of this country the most remarkable is that of Kirkdale, in Yorkshire, visited and first described by Dr. Buckland. The generality of educated readers must be so well acquainted, through the medium of various publications, with the researches of the learned professor, that we shall be excused from following him through his very minute and lucid description of the geological position and internal peculiarities of this cavern : for our present purpose it will be sufficient to observe, that the teeth and bones discovered in the cave of Kirkdale are referable to twenty-three different species of animals ;—six carnivora, four pachydermata, four ruminantia, four rodentia, and five birds. Among the carnivora, the most numerous by far appear to have been hyænas of a larger size than any known at present. Their teeth were so very abundant, that the professor does not calculate the number of animals to which they belonged at less than two or three hundred. Two large canine teeth of the tiger were found four inches in length, and a few molars exceeding in size those of the largest lion or Bengal tiger. There was one tusk of a bear, which appears to have been specifically identical with the *ursus spelæus* of the Germanic caves, and which, as we have already observed, equalled the horse in magnitude. The bones of the elephant, rhinoceros, and hippopotamus were found co-extensively with all the rest, even in the inmost and smallest recesses. The teeth of deer of two or three species are also numerous, but the most abundant of all are those of the water-rat.

The conclusion of the professor respecting this cave is, that it was inhabited during a long succession of years previous to the last general deluge, by hyænas, and that they dragged into it the other animal bodies whose remains are found there. The bones are all comminuted and broken; and many of them distinctly bear the impress of the canine fangs of the hyæna, an animal whose appetite for bones and tremendous power in fracturing them is well known. The professor considers, that, at the period of the last general inundation, the floor of the cave was covered with the diluvial loam and pebbles under which these bones were found, and had been so long preserved from decomposition by this covering of mud, and the coating of stalagmite above it. Several other caverns and fissures have been discovered in this country and in Wales, containing osseous remains, the greater portion of which are referable to the antediluvian era. Near Wirksworth, in Derbyshire, in a cave called Dream Cave, was found the skeleton of a rhinoceros, nearly entire. At Oreston, near Plymouth, three deposits were found of a similar nature, containing great quantities of bones. In the cave of Paviland, in Glamorganshire, were found remains of elephant, rhinoceros, horse, hog, bear, hyæna, &c. In this cave, a human skeleton was also found; but its circumstances, position, state of preservation, &c., prove it to be most clearly postdiluvian.

From all the facts of this description which have been ascertained, Dr. Buckland concludes, that previously to the last general catastrophe, the extinct species of the hyæna, tiger, bear, elephant, rhinoceros and hippopotamus, as also wolves, foxes, oxen, deer, horses, and other animals not distinguishable from existing species, existed contemporaneously in this country; and the Baron, from similar inductions, has drawn a similar conclusion respecting the continent of Europe.

The general circumstances of all these caverns are, as we have observed, extremely similar. The hills in which they are excavated resemble each other in their composition. They are all calcareous, and produce stalactite in abundance. The roofs, sides, and passages in the caverns are ornamented and contracted by it in all its boundless variety of configuration. The bones are nearly in a similar state in all these deposits. Detached, scattered, partly broken, but never rolled, as would be the case had they been brought from a distance by the force of inundations. Somewhat specifically lighter, and less solid than the recent bones, they yet preserve their genuine animal nature, are not much decomposed, still contain plenty of gelatine, and are never in a state of petrifaction. A hardened earth, but still liable to break or pulverize, impregnated with animal matter, and sometimes of a blackish colour, constitutes their natural envelope. This is, in many instances, interpenetrated and covered by a crust of stalactite of the finest alabaster. The bones themselves are sometimes clothed with the same material which enters their natural cavities, and occasionally attaches them to the walls of the cavern. From the admixture of animal matter, this stalactite often exhibits a reddish hue. At other times its surface is tinted with black. But these are accidents of recent occurrence, and independent of the cause which introduced the bones into their present *locale*. It is easy to observe, that this same stalactite is daily making a rapid progress, and invading those groups of osseous remains which it had hitherto left untouched.

This mass of earth, intermixed with animal matter, envelopes without distinction the bones of all the species, and if we except a few on the surface of the soil, and which, from their comparative freshness, we must conclude to have been transported thither at a much later era, all were evidently interred in the same manner, and by the same agents. In a great many of these caverns, especially in that of Gaylenreuth, are found pieces of bluish marble, the angles of which are rounded and blunted, clearly testifying the influence of that diluvial action which hurried them along. A similar phenomenon is observable in the osseous breccia of Gibraltar and Dalmatia.

The pachydermatous remains, so common in the loose or ancient alluvial strata, are very rarely found among the fossil carnivora of the Germanic caves: nor are the bones of the latter very frequent in the alluvial strata. This circumstance, at first, led M. Cuvier to assign different eras to those respective remains. But, independently of the fact that the reverse is sometimes the case in these situations, the discoveries made in our British caves have clearly demonstrated the contemporaneous existence of the animals in question; and the Baron, with that single-minded devotion to the cause of science invariably characteristic of the highest order of philosophical genius, has subsequently avowed that this important fact has been completely established by Dr. Buckland.

There are but three imaginable causes by which such quantities of bones could have been accumulated in those vast subterraneous repositories. Either they are the remains of animals which lived and died there undisturbed, or they were carried thither by inundations, or some other violent cause, or they were originally enveloped in the stony strata whose dissolution produced those excavations, and were not dissolved by the agent which removed the material of the excavated strata. The last supposition is refuted by the fact of the same strata containing no bones; and the second, by the integrity of their smallest prominences, which will not permit us to suppose that they have been rolled, or have suffered any violent change of place. Some of these bones, indeed, are a little worn, but, as Dr. Buckland remarks, on one side alone; which only proves that some transient current has passed over them in the deposit where they are found. The first supposition, then, which, as we observed, the Professor adopted in explanation of the phenomena of Kirkdale-cave, is the only one that can be admitted in reference to all the rest which exhibit phenomena precisely similar. There are certainly some cases of caverns in which we may otherwise account for the presence of bones. Animals may have retired there to die, fallen accidentally into their fissures, or been washed in by diluvial waters. But such hypotheses will only explain the cases where the bones are few and not gnawed, the caverns large, and their fissures extending upwards. In all other instances we can only account for the vast accumulation of bones by the agency of beasts of prey.

It is, however, quite certain that the period in which the animals lived in these caves was considerably subsequent, not only to the formation of the extended rocky strata of which the mountains where they are excavated are composed, but to that of similar strata of more recent date. No permanent inundation penetrated into these subterraneous recesses, or formed there any regular stony deposition. The rolled pebbles that are found there, and any traces of detrition on the bones, indicate nothing but the transitory passage of evanescent waters.

"How, then," exclaims the Baron, "have the ferocious beings which once peopled our ancient forests been extirpated from the surface of the earth? The only reply that can be given is, that they were destroyed at the same time, and by the same agency, as the large herbivora, which were fellow-inhabitants with themselves, and of which no traces in existing nature are any longer to be found."

The debris of birds have been found in the fossil state, the genera of which involve some difficulty in the determination. Of these we treat in the proper place.

The genera of fossil reptiles are well characterized; such are the tortoises, the crocodiles or saurians, the monitors, the salamanders, the protei, the frogs, and a lizard with the wings of a bat, called pterodactylus, on which the present is no place to extend our remarks.

Insects are found in the fossil state in calcareous, foliated stones, and in amber, where they are preserved without any alteration. In Prussia, where this resinous fossil production is most usually found, the insects exhibited there are all of them foreign to the climate.

The debris of vegetable fossils are found in the ancient as well as in the recent strata; but they are more common in the latter, and even on the surface of the earth. They consist for the most part in ligneous trunks, almost always changed into silex, in kernels, seeds, and the impressions of leaves, disposed between the veins of fissile stones. Those found in the mines of pit-coal belong for the most part to the family of fern, bamboo, casuarinas, and other plants, foreign to the climate in which they are thus found. These mines, situated between the granitic or porphyritic schists, are very ancient, and contain no marine shells. It is not thus with similar mines which occur in the calcareous formation. They do not appear to be equally ancient; and instead of recognising in them the impressions of fern, &c., we find in some of them succinum, and shells of the genus ampularia, which appear to appertain to marine depositions. The palm-tree of different kinds has been found in many situations in the fossil state, in countries to which the particular species were not native. Near Canstadt, in the duchy of Wirtemberg, an entire forest was discovered of palm-trees in a horizontal position, each two feet in diameter. In Cologne, from Bruhl, Liblar, Kierdorf, Bruggen, and Balkhausen, as far as Watterberg, are found over many leagues of country, immense depositions of wood changed almost entirely into mould, and covered with a bed of rolled flints, from ten to twenty feet in height. This deposit, which exceeds fifty feet in thickness, also contains trunks of trees, and nuts, which exhibit a strong analogy to the *areka* which grows in India. In the midst of quartzose sands of the most arid kind in the African deserts, and on the surface of a soil for ages under the curse of sterility, are found considerable quantities of the trunks of trees changed into silex. Buried in the peat, on a mountain in the department of the Isère, in France, fossil-wood was found not less than 2000 feet above the most elevated line where trees can grow at the present day*.

We have now to notice a fact connected with fossil osteology of the most singular and striking kind. We find, as has been seen, quadrupeds of different genera, cetacea, birds, reptiles, fishes, insects, mollusca, and vegetables, in the fossil state. But to the present moment no human remains have been found, nor any traces of the works of man in those particular formations where these different organic fossils have been discovered. What is meant by this assertion is, that no human bones have been found in the regular strata of the surface of the globe. In turf-bogs, alluvial beds, and ancient burying-grounds, they are disinterred as abundantly as the bones of other living species. Similar remains are found in the clefts of rocks, and sometimes in caves, where stalactite is accumulated upon them; and the stage of decomposition in which they are found, and other circumstances, prove the comparative recentness of their deposition; but not a fragment of human bone has been found in such situations as can lead us to suppose that our species was contemporary with the more ancient races,—with the palæotheria, the anoplotheria, or even with the elephants and rhinoceroses of comparatively a later date. Many authors, indeed, have asserted, that debris of the human species have been found among the fossils, properly so called; but a careful examination of the facts on which

* In speaking of fossil vegetables, we should not omit to mention the name of our countryman, Mr. Parkinson, who, in his "Organic Remains," was one of the first writers who threw considerable light on the subject of fossils in general. Though erroneous in some of his speculative notions, his work contains a summary of facts of the utmost importance to this branch of science.

such assertions were founded, have proved that these authors were utterly mistaken. We refer our readers to the preliminary discourse of the Baron to his "Ossemens Fossiles," for the most complete satisfaction on this question. The same may be asserted of all articles of human fabrication. Nothing of that description has ever been found indicating the existence of the human race at an era antecedent to the last general catastrophe of this globe, in those countries where the strata have been examined, and the fossil discoveries we are treating of been made. Yet there is nothing in the composition of human bones that should prevent their being preserved as well as any others. There is no principle of premature decomposition in their texture. They are found in ancient fields of battle equally well preserved with those of horses, whose bones we know are found abundantly in the proper fossil state. Neither can it be said that the comparative smallness of human bones has any thing to do with the question, when it is recollected, that fossil remains of some of the smallest of the rodentia are to be found in a state of preservation.

The result, then, of all our investigations serves to prove that the human race was not coeval with the fossil genera and species: for no reason can be assigned why man should have escaped from the revolutions which destroyed those other beings, nor, if he did not escape, why his remains should not be found intermingled with theirs. His bones are found occasionally in sufficient abundance in the latest and most superficial depositions of our globe, where their bones are never found: their bones are in immense quantities in some of the ancient strata of the earth, where no traces of him exist. Human remains in caverns and fissures, along with some of those more ancient debris, prove nothing for the affirmative of man's coeval existence with the lost species. Their freshness proves the lateness of their origin; their fewness, the impossibility that mankind could have been established in the adjacent regions at the period when those other animals lived there; and their situation and general circumstances, the accident of their introduction.

It is a fact not less remarkable, that no remains of the quadrumanous races, which occupy the next rank in creation to man, at least in physical conformation, are to be found in the strata of which we have been speaking. Nor will this fact be deemed less remarkable when it is considered that the majority of the mammifera there found have their congeners at present in the warmest regions of the globe, in the intertropical climates where those anthropomorphous animals are almost exclusively located.

Where, then, was the human species during the periods in question? Where was this most perfect work of the Creator, this self-styled image of the divinity? If he existed any where, was he surrounded by such animals as now surround him, and of which no traces are discoverable among the organic fossils? Were the countries which he and they inhabited overwhelmed by some desolating inundation, at a time when his present abodes had been left dry by the retreating waters? These are questions, says the Baron, to which the study of the extraneous fossils enables us to give no reply.

It is not meant, however, to deny that man did not exist at all in the eras alluded to—he might have inhabited a limited portion of the earth, and commenced to extend his race over the rest of its surface, after the terrible convulsions which had devastated it were passed away. His ancient country, however, remains as yet undiscovered. It may, for aught we know, lie buried, and his bones along with it, under the existing ocean, and but a remnant of his race have escaped to continue the human population of the globe. All this, however probable, is but conjecture. But one thing is certain, that in a great part of Europe, Asia, and America, countries where the organic fossils have been found, man did not exist previously to the revolutions which overwhelmed these remains, nor even previously to those by which the strata containing such remains have been denudated, and which were the latest by which this earth has been convulsed.

It only remains for us now to give a summary view of the succession of strata, and an enumeration of the different fossil genera and species in the respective strata, by the order of which we are enabled to calculate to a certain extent the number of revolutions the globe has undergone. In doing this, we shall pursue the order observed by the Baron in his great work.

In speaking of the strata of which this globe is composed, we must be understood to mean here nothing more recent than that formation which is proved to have resulted from the last grand catastrophe by which the earth was overwhelmed. The strata then formed, the most superficial of the regular strata, consisting of beds of loam and argillaceous sand, mixed with rolled pebbles from remote regions, and filled with debris of land animals unknown, or foreign to the places in which they are found, appear to have covered all the plains, and the floors of the caverns, and choked up the fissures of the rocks within their reach. To such formations, Dr. Buckland has given the name of *diluvium*, and described them with his usual clearness and accuracy. They must be considered as totally distinct from the other strata, which, like them, are equally loose, but have been continually deposited, by streams and rivers, in the usual course of nature, since the last great convulsion of the globe, and which contain no fossil remains, but such as are indigenous to the country where they are found. These last depositions Dr. Buckland distinguishes by the term *alluvium*, and they must be considered as entering for nothing into the question of the grand revolutions of the earth. But in the *diluvial* strata, all modern geologists have discovered the clearest evidence of that tremendous inundation, which constituted the last general catastrophe by which the surface of our planet has been modified. It may not be amiss to inform our readers here, that both these formations agreeing in their character of uncompactness, but differing in their antiquity, are alike termed loose, or *alluvial*, by Cuvier and other geologists. We do not altogether deviate from this usage in our subsequent account of the fossil species; but when we use the term alluvial, in relation to organic debris, we must be understood to mean the diluvial formations.

Between this diluvium and the chalk formation are strata alternately filled with fresh and salt water productions. These mark the irruptions and retreats of the sea to which our portion of the globe has been subjected, subsequently to the formation of the chalk. First come marly beds, and cavernose silex, similar to those of our ponds and morasses. Under these are marle again, sandstone, and limestone, containing nothing but marine productions.

At a greater depth we find fresh-water strata, of an era more remote. Among these are reckoned the celebrated plaster-quarries in the neighbourhood of Paris, where the remains of entire genera of terrestrial animals have been found, which exist no longer.

These last-mentioned strata rest on beds of calcareous stone, in which an immense number of sea-water shells have been collected, the great majority of which belong to species unknown in the existing seas. In this formation are also found the bones of fishes, cetacea, and other marine mammalia.

Under this marine limestone we have again another fresh-water stratum, composed of argilla, in which are interposed considerable beds of lignite, or that species of coal which is of a more recent origin than our pit-coal. Here are found shells only of the fresh water, and bones among them, not of mammiferous animals, but of reptiles. It is filled with crocodiles and tortoises, &c., whereas the mammiferous genera contained in the gypsum are not seen there. They did not yet exist in the country when the argilla and lignites were in a course of formation.

This last fresh-water formation, which supports all the

strata just enumerated, and appears the most ancient of the Parisian depositions, is itself supported by the chalk. This formation, of immense thickness and extent, appears in countries as remote from us as Pomerania and Poland. But in the neighbourhood of Paris, in Berri, in Champagne, in Picardy, and a considerable part of England, it predominates uninterruptedly, and forms a most extensive circle, or basin, in which all the strata we have mentioned are contained, and its edges are barely covered by them in those places where the superstrata are least elevated.

Such superstrata are not confined to the countries just instanced, or to the basin in question. Depositions, more or less similar, and containing organic remains, are found in other regions, wherever the surface of the chalk affords similar cavities for their reception. They are found even where no chalk formation exists, and where the most ancient strata constitute their only support. The two distinct formations with fresh-water shells have been found in England, Spain, and even on the confines of Poland; the marine beds interposed between them exist along the entire range of the Apennines. Some of the quadrupeds of the Parisian plaster-stones have been found elsewhere, as, for instance, in the gypseous strata of Valai, and in the molasse quarries in the South of France.

Thus it appears, that the partial revolutions which took place between the era of the chalk-formation, and that of the last great inundation, and which consisted in the alternate invasion and retreat of the sea, occurred in many countries. This globe has undergone a long series of agitations and changes, which appear to have been rapid in their operation, from the comparative slightness of the depositions they have left behind. The chalk has evidently been the production of a more tranquil and extensive sea. It contains marine productions alone, but among them the most remarkable remains of vertebrated animals, all of the fish or reptile class—tortoises and lizards of colossal size and extinguished genera.

A very considerable portion of Germany and England is composed of strata anterior to the chalk, in the hollows of which the chalk reposes, just as the intermediate strata before mentioned rests in its own cavities. Immediately under the chalk, and indeed partially intermingled in its lowest strata, are depositions of green sand, and ferruginous sand below it. In many countries both are found condensed into banks of sandstone, in which are seen lignites, succinum, and debris of reptiles.

After these come the immense accumulation of strata composing the mountain-chain of Jura, which extends into Suabia and Franconia, the chief summits of the Apennines, and many similar formations in England and France. They consist in calcareous schistus, abounding in fish and crustacea, immense banks of oolite, marly and pyriteous grey limestone, containing ammonites, oysters with curved valves, and reptiles of more extraordinary character and conformation than any of their predecessors.

These, which we shall take leave to call *Jurassic* strata, are supported by extensive beds of sand and sandstone, in which the impressions of vegetables are frequently found, and which rest upon a limestone which has been termed *coquillaceous*, from the immense quantities of shells and zoophytes with which it abounds. It is separated by other strata of sandstone of the variegated kind, from a limestone still more ancient, called *Alpine*, because it composes the loftier range of the Tyrol Alps; but, in fact, it appears also continually in the east of France, and the entire south of Germany.

In the limestone called *coquillaceous*, are deposited considerable accumulations of gypsum and rich beds of salt. Below it we find slender strata of coppery schistus, with abundant remains of fish, and some fresh-water reptiles. The coppery schistus rests upon a red sandstone, of the same age as the pit-coal, which we have before alluded to as bearing the impressions of the earliest vegetable productions which adorned the surface of the globe.

We now come rapidly to the transition strata, where matter lifeless and unorganised appears to have made its last stand against the vivifying and organising principle of nature. Here we find black limestone and schistus, with crustacea and shells of unknown genera, alternating with the latest of the primitive strata. We finally arrive at the most ancient formations which we are permitted to discover,—the marble, the primitive schistus, the gneiss, and the granite, the ancient foundations of the earth, and which are themselves, in all probability, the result of the united action of fire and water, after myriads of revolving ages.

Such is the exact enumeration of the series of strata which compose our globe; such is the order of facts which geology has been enabled to establish, by calling in the aid of mineralogy, and the sciences of organization, by abandoning the reveries of arbitrary hypothesis, and steadily pursuing the safer path of observation and induction.

We shall now rapidly enumerate the fossils in those various allocations, beginning with the earliest, and ending with the latest formations.

We have observed that zoophytes, mollusca, and certain crustacea, begin to appear in the transition strata. Bones and skeletons of fish may, perhaps, be also found there. But we are far from discovering, among those early formations, the remains of land animals, or any formed for the direct respiration of atmospheric air.

The great strata of pit-coal, and the trunks of palm and fern of which they bear the impression, must presuppose the existence of dry land and aërial vegetation. Yet no bones of quadrupeds are found there, not even of the oviparous species.

Their first traces are found a step higher, in the bituminous coppery schistus. There we find quadrupeds of the family of the lizards, very similar to the monitors, which are now natives of the torrid zone. Many individuals of this description are found in the mines of Thuringia in Germany, among innumerable fish, of a genus unknown at present, but which, from its analogy with some existing genera, appears to have inhabited the fresh water. The monitors we know to be inhabitants of the same element.

A little higher is the Alpine limestone, and on it the coquillaceous limestone, abounding in entrochi and encrini, and forming the basis of a great portion of Germany and Lorraine. It contains the osseous remains of a very large sea-tortoise, and another reptile of the lizard tribe, of very great length, and pointed muzzle.

Next come certain sandstone strata, having only vegetable impressions of the large reeds, bamboos, palms, &c.; and then the Jurassic limestone, where the remains of the reptile class exhibit a diversity of singular conformations, and a gigantic degree of development. Its middle portion is composed of oolites and lias, or the gray limestone, containing the recurvivalve oysters; and in it were found the debris of two most extraordinary genera, uniting the characters of oviparous quadrupeds, with locomotive organs, like those of the cetacea. Those are the *ichthyosaurus* and *plesiosaurus*, first discovered and determined here by our distinguished countrymen, Sir Everard Home and Mr. Conybeare. These, with their species, shall be described in the proper place. Their skeletons are in a state of high perfection, and their remains have been found extended through all the formations of lias.

In the same deposition were also found two species of the crocodile, amidst ammonites, terebratulæ, and other shells of the ancient sea. These are called by Cuvier the long-beaked and short-beaked gavial. Another crocodile was discovered in the oolite at Caen, and another in the same formation here.

The megalosaurus, a fossil reptile of prodigious size, has been discovered by Dr. Buckland in this country. Its remains appear to have been contemporaneous with the concretion of the lias, but are also dispersed abundantly in the oolite and higher sands. From the magnitude of a femur and other bones, found in the ferruginous sandstone of Tilgate-Forest, in Sussex, the Doctor calculates that the animal in question could not have been less than from sixty to seventy feet in length. Remains of this reptile, or at least of species referrible only to this genus, have been also discovered in France and in Germany, in the calcareous slate above the oolitic beds.

In this same slate, the long-beaked crocodiles continue to abound; but the most remarkable animals there are the pterodactyls, or flying-lizards. They appear to have been sustained in the air, on the same principle as the cheiroptera: they had long jaws, armed with trenchant teeth, hooked claws, and some species, as would seem from the fragments remaining, arrived at a considerable size.

In the nearly-homogeneous limestone of the crests of Jura, a little higher than the calcareous slate are bones, but invariably of the reptile class. There are crocodiles, but more especially fresh-water tortoises, as yet not fully determined, but many of which, by their magnitude and conformation, are strongly distinguished from all known species.

Amid those innumerable reptiles, whose varied structure and colossal dimensions rival, if not surpass, the fabled monsters of poetical antiquity, we begin, as is said for the first time, to recognise the remains of some small mammalia. Jaws and bones have been discovered in England, appertaining to the families of the didelphis and insectivora, in these situations. But if the locale of Dr. Buckland's discovery of the opossum before mentioned be completely established, it must be granted that they appear sooner. Cuvier, however, seems to think that the rocks in which the bones in question are incrusted may owe their existence to some local recomposition, posterior to the era of the original formation of these strata; and it is most certain that, even for a period considerably subsequent, the reptile class exclusively predominate. In the ferruginous sands above the chalk in England, abundance of those already enumerated occur; and an additional reptile has been discovered there by Mr. Mantell, of Lewes, which seems to have been herbivorous, and to have inhabited the fresh water. This is the *iguanodon*, and it is supposed to have been sixty feet long.

In the chalk, according to Cuvier, there are only reptiles, the remains of crocodiles and tortoises. In the tufa of Mount St. Pierre, near Maëstricht, which is of the chalk formation, has been found amidst marine tortoises, shells, and zoophytes, another gigantic member of the saurian family, a distinct genus, for which Mr. Conybeare has proposed the name of *mosasaurus*.

In the argilla and lignites, covering the superior portion of the chalk, the Baron declares that he has discovered nothing but crocodiles; and thinks that the lignites of Switzerland, in which are bones of the beaver and mastodon, must be assigned to a more recent era. He adds, that it was only in what the French call the *calcaire grossier*, surmounting the argilla, that he commenced to discover mammiferous remains, and that they belonged to marine mammalia. But Dr. Buckland mentions the occurrence of these in the stratum of Cuckfield, in Sussex, much anterior to the formation of which we now speak; and they are also declared to have been found in the calcareous slate of Stonesfield, and in the corn-brash limestone in Oxfordshire.

These marine mammalia are dolphins, lamantins, and morses, apparently of an unknown species. The lamantin is at present confined to the torrid zone, and the morse to the Icy Sea; still those two genera are found together in the coarse limestone, in the midst of France. This union of species, whose consimilars are now allocated in opposite zones, is by no means uncommon.

In the strata succeeding this coarse limestone, or in the ancient contemporaneous fresh-water depositions, the class of terrestrial mammifera first begins to appear in tolerable abundance. Belonging to the same age are the animal remains buried in the molasse and ancient gravel-beds in the south of France;

in the gypsum, mixed with limestone, in the environs of Paris and of Aix, and in the marly fresh-water formations, covered again with marine strata, in Alsace, Orleannais, and Berri.

These organic remains are singularly remarkable, as belonging to a variety and abundance of certain genera of pachydermata, now totally extinct, and approximating more or less in character to the tapir, rhinoceros, and camel. These genera, for whose discovery we are entirely indebted to the Baron, are the palæotherum, laphiodon, anoplotherium, antracotherium, cheropotamus, and adapis. Of these there are about forty species, all extinct, and to which there are none analogous in the living world, except two tapirs and a daman.

In the same formation with these pachydermata are some remains of carnivora, of rodentia, of birds, of crocodiles, and tortoises. Of the first, a bat (and, singular to relate, the only instance of the kind occurring in this or subsequent formations,) a fox, an animal approximating to the racoons and coatis, a peculiar species of genet, some other carnivora not so easily determined, and, most remarkable of all, a small sarigue, a genus now confined to America. There are two small rodentia of the dormouse kind, and a head of the genus squirrel. In the gypsum of Paris, bones of birds are very abundant, constituting the remains of at least ten species. The crocodiles approximate to those of the present age, and the tortoises are all of the fresh water. There are also remains of fish and shells, in great part unknown at present.

There can be no doubt that this immense animal population of what Cuvier calls the middle age of the earth, has been entirely destroyed. Wherever its debris have been discovered, there are vast superincumbent beds of marine formation, proving the invasion and long continuance of the sea in the countries inhabited by these races. Whether the countries subjected to such inundation at this era were of considerable extent or not, our present acquaintance with the strata in question does not enable us to decide. These formations, however, embrace the gypsum or plaster-quarries of Paris, and those of Aix in Provence, and many quarries of marle-rocks and molasse in the south of France. Certain portions of the molasse of Switzerland, and the lignites of Liguria and Alsace, are referrible to the same. As for the fossil bones of England, Italy, and Germany, they belong either to an earlier or a later era,—to the ancient reptiles of the Jurassic strata and copper-slate, or to the diluvial formations of the last universal inundation.

We may, then, fairly suppose, that, at the period when these numerous pachydermata existed, there were not many fertile plains to afford pasture for their support. These plains, too, in all probability, were insulated districts intersected by those elevated mountain-chains in which we discover no traces of those extinct animals.

In the same strata with those pachydermatous remains are found the trunks of palm, and many other relics of those magnificent vegetable productions which at present are indigenous to tropical climates alone.

The sea which covered these formations has left extensive depositions, constituting, at a moderate depth, the foundation of our present large plains. It again retired, and left open immense surfaces of soil to a new population, the debris of which abound in all the sandy and loamy strata of every region of the globe which has been subjected to examination.

To this last tranquil deposition of the sea we must refer some cetacea, very similar to the existing species. Among these an entirely new genus has been discovered, and named *ziphius* by Cuvier. It contains three species, and approximates to the cachalots and hyperoodontes.

Among the animals which lived on the surface of this deposition, when it became dry land, and whose debris now fill the loose strata of the earth, we find no palæotheria, or anoplotheria, none of the extraordinary and extinct genera contained in the gypseous formations. Still, however, the order pachydermata predominates; but in the gigantic genera of the elephant, rhinoceros, and hippopotamus, accompanied with innumerable bones of horses, and many of the larger ruminantia. This new animal kingdom was devastated by carnivora, of the magnitude and generic characters of the lion, the tiger, and the hyæna. This population, the remains of which extend to the extremity of the north, and the borders of the Frozen ocean, has, generally speaking, nothing congeneric at present, except in the torrid zone, but in all cases a specific difference is sufficiently marked.

Among these species are the *elephas premigenius*, or mammoth of the Russians, whose remains are found from Spain to the coasts of Siberia, and throughout all North America; the mastodon, with narrow teeth, common in the temperate parts of Europe and the mountains of South America; the great mastodon, in immense abundance in North America; an hippopotamus, very common in England, Germany, France, and Italy, and a smaller species; three rhinoceroses, chiefly in Germany and England; a gigantic tapir, in Germany and France, and an apparently extinct genus, resting on a single fragment, discovered in Siberia, and called *elasmotherium* by Fischer.

The bones of the horses are not so clearly determined to belong to distinct species. Of the ruminantia several species may be pronounced distinct, particularly a stag superior in size to the elk, common in the marle and peat of England and Ireland, and whose remains have also been found in Italy, France, and Germany; among the elephantine bones of the deer and ox of the caverns, and osseous breccia, which appertain to the same era, we cannot speak so decidedly. It appears, however, pretty clearly that they were not native to the climate; and what is most singular, the bones of the rein-deer, an animal now confined to the inhospitable regions of the north, are located with the remains of the inhabitants of the tropics. We must not, however, omit to notice, that many of the positions from which the bones of ruminantia have been taken, are not sufficiently verified to warrant us in deciding that they were contemporaneous with the larger pachydermata last mentioned. Nay, we are even justified in believing many of them to have been postdiluvian.

In the osseous breccia of the Mediterranean have been found two species of *lagomys*, a genus confined to Siberia; two of rabbit, some campagnols, and rats as small as the water-rat and the mouse; and likewise in the English caverns; also the bones of shrews and lizards.

In the sandy strata of Tuscany the teeth of a porcupine have been found; and in Russia, heads of a species of beaver, larger than any now known, and called *trongotherium*.

The remains of the edentata, above those of all other classes, indicate species of a size far superior to that of their existing congeners, and even of a magnitude altogether gigantic. Such was the *megatherium*, an animal partaking of the generic characters of the tardigrada and the armadillos, and equalling the rhinoceros in size. It has been found only in the sandy strata of North America. The *melagonyx*, found in the caverns of Virginia, and in a small island on the coast of Georgia, very much resembled the *megatherium*, but was not so large. Those two edentata were confined to America. But in Europe one appears to have existed, which, from a fragment remaining, has been considered as not less than four-and-twenty feet in length. This fragment was found in a sand-pit, in the district of Darmstadt, not far from the Rhine, among the bones of elephants, rhinoceroses, and tapirs.

In the osseous breccia are found, but very rarely, the bones of carnivora, which are, as we have seen, far more abundant in the caverns. Those of Germany are principally characterised by the remains of a very large species of bear, much surpassing any existing one in size, the *ursus spelæus*. There are two others, *ursus arctoideus* and *ursus priscus*. There is the fossil hyæna, differing, in some peculiarities of the teeth and head, from the Cape hyæna; two tigers, or panthers, a wolf, fox, glutton, genet, and some other small carnivora.

The bears are not very numerous in the loose strata. The ursus spelæus is said, however, to be found there, in Austria and Hainault. In Tuscany there is a peculiar species, remarkable for its compressed canines, thence termed *U. cultridens*, or *angustidens*. Hyænas are more frequently found in such strata, with the bones of the elephant and rhinoceros.

It appears, then, that during the era of which we are now speaking, the carnivorous order was numerous and powerful. The rodentia, smaller in general, and more feeble, have not so much attracted the attention of collectors of fossil remains. Still, as we have observed, it has presented us with some unknown species in the fossil state.

We have now enumerated the principal animals whose remains are found in the accumulation of earth, sand, and loam, which cover our large plains, and fill many caverns and fissures of rocks, and have been called *diluvium*. They decidedly constitute the population which occupied our part of the world, at the era of the last great catastrophe which destroyed their races, and prepared the soil on which the animals of our own era exist. Whatever resemblances certain of their species may present to those of our days, it cannot be denied that their general character was very different, and that most of their races have been annihilated.

It is, as we before remarked, most remarkable, that in all the strata, and among all the fossil remains now enumerated, no relic has been found of man or monkey. Whether these kindred orders existed at all during the periods in question, or, if they did exist, where they existed, are points which it is yet impossible to decide. But what is quite certain is, that we are now surrounded by a fourth succession of terrestrial animals, and that, after the age of the reptiles, that of the palæotheria, and that of the mammoths, mastodons, and megatheria, the age arrived in which the human species, with the aid of certain domesticated animals, has appropriated and cultivated the earth; and it is only in alluvion, in peat, in recent concretions, in a word, in such formations as have taken place since the last general inundation of the globe, that the bones of man and of existing animals have ever been discovered. To these are referrible the human skeletons, found incrusted in *travertino*, in the island of Guadaloupe. They are accompanied with shells and madrepores of the existing and surrounding seas. The remains of oxen, deer, &c., common in the peat-formations, are in the same predicament; as are likewise the bones of man and domestic animals embedded in alluvia, in ancient burying-grounds and fields of battle. To the era of the last general catastrophe, or to those of any preceding ages, none of these remains are attributable.

The mode in which the fossil bones have been determined, depends upon a principle in comparative anatomy, which regulates the co-existence of organic forms. Every animal may be considered as a whole, all the parts of which are in strict keeping and correspondence with each other. If the animal be carnivorous, it is characterised by a certain system of dentition. The teeth are trenchant, the jaws are powerful, and their condyles peculiarly formed. But such teeth and jaws would be of little service, unless the animal were also provided with claws adapted for seizing and tearing the prey. Claws of this kind necessitate a peculiar construction of the phalanges, a facility of rotation in the fore-arm, and corresponding changes in the humerus. Every animal whose stomach is constituted to digest nothing but flesh, must have every other part of his frame in consonance with this restriction. On the other hand, it is obvious that hoofed animals must be herbivorous, as they possess no means of seizing prey. Accordingly we find that their masticating and digestive organs correspond with this peculiarity. Their teeth are supplied with flat and unequal coronals, to bruise the herbage, &c., on which they feed; and as their system of dentition is generally less complete, so their stomachs are more complicated. This is not the place to enlarge on a subject of this kind; it is sufficient to observe that, by the application of this principle to its utmost extent, assisted by careful observation, the skilful comparative anatomist is enabled, in general, from a single bone, or fragment of bone, to determine with accuracy the form, character, and dimensions of the animal to which it belonged. I do not pretend to say that, in every case of fossil remains, this method is infallible. Fossil fragments may sometimes be too few, too mutilated, or of parts not sufficiently influential to warrant a decided opinion. But in the vast majority of instances, the induction is quite ample enough to justify the conclusion.

It is the study of fossil osteology alone which has led to any precise notions concerning the theory of the earth. Had organic fossils been totally neglected, no one would have imagined that successive eras, and a series of different operations, had taken place in the formation of the globe. By them alone are we certified that the covering of this planet has not always been the same, as it is obvious that, before they were buried in its depths, they must have existed on its surface. We have extended, by analogy to the primitive formations, the conclusion with which the fossils have supplied us for the secondary; and, had the strata of the earth been destitute of organic remains, it would have been impossible to maintain that their production had not been simultaneous.

It is also to the fossils, slight as our acquaintance is with them even yet, that we are indebted for the little that we know concerning the nature of the revolutions of the globe. By them we learn that certain strata have been tranquilly deposited in a fluid mass; that the variations in the strata have corresponded with those of the fluid; that their denudation was occasioned by the translation of this fluid; and that this denudation has taken place more than once. Nothing of all this could have been learned with any certainty, but for the study of the organic remains.

What an immense field for reflection is opened to the mind of the philosopher, by a survey of the discoveries to which fossil osteology has conducted us! We read, in the successive strata, the successive efforts of creative energy, from the sterile masses of primitive formation, up to the fair and fertile superficies of the globe, enriched with animal and vegetable decomposition. We find that there was a time when life did not exist on this planet; we are enabled clearly to draw the line between inanimate and organized matter, and to perceive that the latter is the result of a distinct principle,—of something superadded to, and not inherent in, the former. We also contemplate a progressive system of organic being, graduating towards perfection through innumerable ages. We find the simplest animals in the earliest secondary formations; as we ascend, the living structure grows more complicated—the organic development becomes more and more complete, until it terminates in man, the most perfect animal we behold. And shall we say that this march of creation has yet arrived at the farthest limit of its progress? Are the generative powers of nature exhausted, or can the Creator call no new beings from her fertile womb? We cannot say so. Revolution has succeeded revolution—races have been successively annihilated to give place to others. Other revolutions may yet succeed, and man, the self-styled lord of the creation, be swept from the surface of the earth, to give place to beings as much superior to him as he is to the most elevated of the brutes. The short experience of a few thousand years—a mere drop in the ocean of eternity—is insufficient to warrant a contrary conclusion. Still less will the contemplation of past creations, and the existing constitution of nature, justify the proud assumption that man is the sole end and object of the grand system of animal existence.

In surveying the different species whose remains are found in the fossil state, it will be expedient to deviate from the order of the Animal Kingdom, and to follow that which the Baron has observed in the *Ossemens Fossiles*, as by this means the reader will be better enabled to understand the order of succession, and the respective geological positions of the species described.

Note.—As, in our earlier part of this essay, we have stated, in deference to the opinion of Baron Cuvier, that the existing causes which now modify the earth's surface are insufficient to produce catastrophes on such a scale as that of those we have been surveying, it is but justice to mention that this point has been mooted, with much force of argument, in a very able article on the present subject in the *Quarterly Review* for September, 1826. Some very strong facts are there adduced relative to the action of earthquakes and volcanoes to a very great extent, and also proofs of derangement in comparatively recent strata, which, though partial, was evidently so violent as to prove that the disturbing forces still existed in all their pristine vigour. We must, however, waive any further discussion of this kind, and content ourselves with referring our readers to the article in question, which, for its extent, contains as lucid an epitome of all the latest information on this interesting topic as we have ever had the pleasure to peruse.

THE FOSSIL ELEPHANT,

CALLED MAMMOTH BY THE RUSSIANS.

It would be an endless task, and utterly inconsistent with the plan of our present sketch, to indicate all the places on this globe in which the fossil remains of the elephant have been found. They have been discovered, in fact, in every country, and at every epoch of time. We shall, therefore, content ourselves with a brief geographical view of the principal situations where they have been detected.

We find traces of such discoveries from the time of the ancients. Theophrastus has mentioned the subject in a work now lost, but his testimony has been preserved by Pliny, in the thirty-sixth book of his Natural History.

In consequence of the great resemblance of certain bones in the elephant to those of man, some tolerably good anatomists have been so far imposed on, as to take them for human bones. The pretended discoveries of the trunks of giants, so frequently alluded to by the writers of antiquity, and of the middle ages, are probably to be referred to this source. There are, indeed, bones sometimes spoken of, of the most prodigious magnitude, being eight or ten times the dimensions of those of the elephant; these we might be inclined to refer to the cetacea, if the measurements given by those writers were exactly to be relied on.

It was for some time a prevailing opinion, that the elephants whose osseous remains have been found in certain countries, had been transported thither by man. As long as these discoveries were confined to Italy, and other countries much frequented by the Macedonians, the Carthaginians, and the

Romans, they were sufficiently explained by the prodigious number of elephants possessed by these respective nations. The Macedonians were the first Europeans who possessed any, and Alexander, after the defeat of Porus, brought a sufficient number from India, to enable Aristotle to form very precise notions respecting them. It is certain that this great philosopher was much better acquainted with the elephant's mode of copulation, of sucking, and, in truth, with almost all the other details of its history, than the Count de Buffon. Every thing which he relates concerning these points, has been confirmed by the testimony of recent observers in India.

The princes of the house of Seleucus always maintained a considerable number of these animals. Seleucus himself, surnamed Nicator, received fifty from Sandrocottus, in exchange for an entire district on the banks of the Indus. Plutarch also assures us, that this prince and his allies had four hundred of these animals at the battle of Issus, in which they were victorious over Antigonus, three hundred and one years before Christ.

Antiochus the Great employed two hundred elephants at the battle of Raphia, against Ptolemy Philopater, who had but seventy-three, and fifty-four in that of Magnesia against the Romans, who had but sixteen. This superiority, however, proved but of small utility, inasmuch as he was worsted in both engagements.

Pyrrhus was the first who brought elephants into Italy, and the Romans, who were strangers to these animals, gave them the name of Lucanian oxen, from the circumstance of Pyrrhus having disembarked at Tarentum. Four of these, taken by Curius Dentatus, were the first ever exhibited at Rome.

When Metellus defeated the Carthaginians in Sicily, in 502, A. R., he transported their elephants to Rome on rafts. Authors differ respecting the number of those animals; Orosius makes them one hundred and four, Seneca one hundred and twenty, Eutropius one hundred and thirty, and Pliny one hundred and forty-two. All these, according to Varro, were massacred in the circus, inasmuch as the Romans did not know how to dispose of them.

Elephants soon came to be very generally exhibited in the circus at Rome. They were sometimes opposed in combat to bulls. Germanicus exhibited some, which danced in a clumsy sort of style, and others were shown in the time of Nero which danced on the rope, and performed many other extraordinary feats of address. It is remarkable also that these elephants were born in Rome, which, as well as the fact of their frequent propagation there, we learn from two very remarkable passages in Ælian and Columella.

Down to the time of the emperor Gallienus, the exhibition of elephants was pretty constant at Rome. This prince exhibited the last that were introduced in the Roman games. They were ten in number.

Thus it appears that, at known historical epochs, a considerable number of elephants existed in Italy, and in other countries under the dominion of the Romans. It was therefore not unnatural to attribute the origin of the osseous remains of those animals found in such places, to individuals which existed on the soil in authentic periods of history. There is no doubt that some of them may be referred to this source; but, if we consider the usual circumstances under which such remains are found, we shall be inclined to believe that but comparatively a small number are in this predicament. They are constantly discovered to be intermixed and confused with the bones of the rhinoceros and the hippopotamus, animals which were certainly not transported there by Hannibal or the Roman armies.

We shall now present the reader with a brief exposition of the principal places in Italy where these bones have been discovered.

The largest tusk was found near Rome, in 1789, by the Duke de la Rochefoucauld and M. Desmarest. It was ten feet long and eight inches in diameter, though not entire. About 1664, some remains of this kind were found at the entrance of the Vatican, in digging foundations. Baccius speaks of similar discoveries in this city as long ago as 1582. And it is much more than probable that the body of Pallas, the son of Evander, found, or pretended to have been found, in the reign of the Emperor Henry III., in the middle of the eleventh century, and which was said to have exceeded the walls of the city in height, was nothing but some remains of this description.

In many other parts of Rome elephantine remains have been found, and in whatever direction an observer may proceed from that city, abundance of such bones will be found. They are discovered throughout all the papal domains, and in following the course of the Tiber, the Clanis, and the Arno into Tuscany, they grow still more numerous.

It would be endless to notice all the places where fragments have been discovered of the elephant. We must therefore confine ourselves to the most remarkable facts.

In the Val de Chiano, the Grand Duke Ferdinand II., in 1663, had an entire skeleton disinterred, some of the bones of which are still preserved at Florence, according to Targione. It is proper to notice, that some of those remains have been found in that sort of consolidated sand, called *tufo* by the Italians, which contains marine bodies and foreign wood in a state of petrifaction.

They are so common in the small earth-hills which border the upper part of the Valley of Arno, that the peasants used formerly to employ them, in conjunction with stones, in building those little walls that enclose their farms. But now that they know their value, they preserve them, and sell them to travellers.

Doctor Targione Tozzetti writes to Buffon, in 1764, concerning the elephantine remains found near the castle of Cerreto Guidi, between the lake of Tucecchio and the Arno. He says that they belonged to individuals of very different ages, some of them exceedingly young, and that they were found intermingled with the bones of many other land-animals, such as oxen, deer, and horses.

In consequence of the abundance of these bones, the cabinets of natural history in Tuscany are altogether filled with them. As Hannibal, after the battle of Trebbia, crossed the Apennines, and traversed the whole length of the Vale of Arno, to march against the consul Flaminius at Arezzo, as he halted for a short time near Fiesole, and then must have passed under Arezzo, and pursued his course along the Vale of Chiana, to arrive between Cortona and the Lake of Thrasymene, it was natural enough to consider the first bones of elephants discovered in these districts as the remains of those which were attached to the Carthaginian army. This opinion has been maintained by Stenon, a learned Dane, in his treatise, "De Solido intra Solidum contento." But, according to Eutropius, Hannibal brought only thirty-seven elephants with him into Italy; and we are informed by Polybius, that they all perished from the effects of cold, after the battle of Trebbia, with a single exception. Livy, who is more minute in his accounts, tells us that Hannibal had eight remaining after that engagement, seven of which died very shortly, during the vain attempt of that general to cross the Apennines in winter. But both authors agree, that, in spring, when he descended into the marshes of the Lower Arno, he had but a single elephant, on which he himself was mounted during that severe march, in which he lost his eye from a defluxion.

Now, it is very clear that this single elephant could never have supplied the immense quantity of bones that are scattered through Tuscany. Besides, there are as many bones found of the rhinoceros and the hippopotamus as of the elephant, confusedly intermingled in the same strata; and, consequently, we can never believe that the elephantine remains are derived from animals employed in war.

We may also observe here, that those bones are found usually at the base of those little clay-hills which fill the intervals of the calcareous chains; that in the strata containing them is found petrified and bitumenized wood, which M. Dolomieu supposes to be oak, and that this again is covered by a strata of marine shells, mixed with arundinaceous plants, and by immense banks of argillaceous earth.

The north of Italy does not abound less in elephantine remains. Without, however, descending to particulars, we may remark that the same observations which we have made above, relative to the locale of the others, are equally applicable to these.

There are several relations in authors of the bones of giants disinterred in Sicily, which may, for the most part, be referred to elephantine remains.

From the wretched state of oppression and confusion in which Greece has so long been, we can expect no rational account of her fossil remains. Still, however, that elephantine bones were among the number may be reasonably concluded, from the many accounts we meet with in both ancient and modern writers, concerning the bones of giants discovered both on the continent and in the islands of Greece.

Similar stories are related of Spain. But it is certain that there are some remains of the fossil elephant in the royal cabinet at Madid. In all parts of France these bones have been discovered in the greatest abundance. We know, however, from historical testimony, that comparatively very few of these animals were carried into that country. The only instances of this kind that we read of are, when Hannibal passed through the southern provinces, and when Domitius Ænobarbus marched against the Allobroges and the Arverni. Some of these remains have been found at a depth of eighteen feet in the marly and argillaceous strata near Paris.

The fossil elephants of the Belgic have long been known and described. The absurd notions concerning the bones of giants, were combatted by the learned Van Gorp, as early as the sixteenth century. But he, as was usual in his time, attributed those remains to the expeditions of the Romans. In the environs of Strasburgh many bones have been found, and they are abundantly scattered through the valleys of Switzerland. The history of a giant disinterred near Lucerne in 1577 made almost as much noise as a similar discovery in France, which was considered to be that of the body of Teutobocchus, king of the Cimbri. Felix Plater, a celebrated professor of medicine, made a drawing of a human skeleton, of the supposed height of the being to whom the fossil remains belonged, which was only nineteen feet. The Lucernese have made this pretended giant the supporter of the arms of the city.

These remains are very abundant throughout all Alsace, along the entire courses of the Rhine and the Meuse, and naturally to be expected in the beds of alluvion collected in the mouths of those rivers. We, therefore, find that Holland abounds in those remains.

But Germany is the country beyond all others in which the greatest quantity of fossil elephantine bones have been discovered. This, as the Baron Cuvier remarks, is not perhaps because that country actually contains more of these remains, but because there is scarcely a district in the empire without some man of intelligence and education, capable of scientific researches. As early as 1784, Merk reckoned no less than eighty different places in which those fossils were found. M. de Zach increased the number to more than one hundred, and Blumenbach has doubled it. Were we to enter even into a superficial description of those discoveries in Germany, we might write a volume. Remains have been found in the basins of the Danube, the Elbe, and the Weser; in every part, in short, of this immense empire, in regions where the Roman armies never penetrated, and where, consequently, the presence of such remains cannot be attributed to Roman importation. Indeed, when we consider the immense abundance of those fossils, the situations in which they are generally found, the other animal and vegetable remains with which they are intermingled, we cannot hesitate to renounce an hypothesis of this kind. A skeleton was discovered at Tonna, in the territory of Gotha, in 1696. The physicians of the country consulted by the Duke, declared the bones submitted to their inspection to be mere *lusus naturæ,* and supported this opinion by many profound treatises. Teutzel, however, the librarian of this prince, compared each bone separately with its analogous part in the elephant according to the description of Allen Moulin, and some remarks of Aristotle, Pliny, and Ray, and clearly demonstrated the resemblance. He also proved, from the regularity of the strata under which this skeleton was found, that it could not have been brought there by human means, but by some general catastrophe, such as the Noachian deluge. Yet, so far did the fondness for the contrary hypothesis prevail, that, as long as only isolated discoveries of bones were made in those parts of Germany where the Romans had not been, they were referred to an elephant sent to Charlemagne by the Caliph Haroun-al-Raschid, which arrived as far as Aix-la-Chapelle, and might have been conducted farther!

Let us now pass to our own islands, which, from their situation, could not in ancient times have received many living elephants; and yet we shall find as great a number of fossil remains here, in proportion, as on the continent. It is true, indeed, that Poliænus reports that a single elephant was brought hither by Cæsar; but that will hardly be deemed more sufficient to account for our British fossils, than that of Charlemagne is for the elephantine remains of Germany.

We shall briefly notice the more remarkable detections.

Sir Hans Sloane possessed a tusk discovered in Gray's-Inn-Lane, twelve feet under ground, in the gravel bed. Bones of the elephant, the rhinoceros, the hippopotamus, the deer, and the ox, were discovered near Brentford, mixed with land and fresh-water shells, of which remains a partial description appeared in the "Philosophical Transactions," in 1813. They were found in a bed of gravel, and upon the great stratum of blue argilla which is so considerably extended over England and France.

In 1815 three large tusks and other elephantine bones were found at Newnham, in Warwickshire, and with them two skulls of the rhinoceros, and several stags' horns. All these fossils were found in gravel, thickly mixed with argilla.

In 1803 a large skeleton was found near Harwich, supposed to be about thirty feet in length. But the bones broke immediately on being touched.

To be brief, remains of the elephant have been found in great abundance in various parts of this island, but the most remarkable discovery of them was in the cave of Kirkdale, in Yorkshire, with a variety of other fossils, so ably described by Professor Buckland, in his admirable work "Reliquiæ Diluvianæ." There have been also fossil remains discovered in Ireland. Scandinavia, a country so little adapted to the sustenance of elephants, contains many of their fossil bones. They are found in Sweden, Denmark, and Norway, and even in Iceland, where, according to the report of Torfœus, a cranium and tooth of prodigious size were dug up.

Those immense sandy plains which commence to the east of Germany, give name to Poland, and extend along the breadth of the Russian empire, to the Oural mountains and the Caspian sea, are not less rich in elephantine remains. They are found in the basins of the Oder, the Vistula, and the Dniester, and on the banks of the Hypanis. Through the vast empire of Russia these remains are immensely abundant, especially in those very provinces where we should least expect to find them, in the frozen regions of Siberia. They have been found near St. Petersburgh, near Archangel, in the valley of the Dwina, near Kostynsk, on the banks of the Tanaïs, and in the sandy and ferruginous strata by the Wolga. In this last situation a cranium was unburied, measuring four feet in length. But they are so numerous in all Asiatic Russia, that the inhabitants of Siberia have invented a fable to explain their presence. They have supposed those bones to belong to a subterraneous animal, living like the moles, and unable to endure the light of day. This animal they call *Mammoth*, according to some etymologists, from the word *mamma*, which, in some Tartar idiom, signifies the earth, or, according to others, from the Arabian word *behemoth*, or *mehemoth*, an epithet which the Arabs still attach to the name of the elephant. The Siberians call the fossil tusks the horns of the mammoth, and they are so numerous and so well preserved, especially in the northern parts, that they are employed for the same purposes as fresh ivory, and form so lucrative an article of commerce, that the czars formerly reserved the monopoly of it to themselves.

The Chinese are acquainted with this fable of the subterraneous animal, which they call *tien-schu*, the signification of which word is, *mouse that hides itself*. They describe it as continually remaining in caverns under ground, resembling a mouse in form, but of the size of an ox or buffalo. It is of a dun colour, and has no tail. This is the statement of one writer. Another tells us that its tail is an ell long, the eyes small, and it dies instantly when it sees the rays of the sun or moon. He adds that, during an inundation of the river Tan-schuann-tuy, in 1571, several of these animals were seen in the neighbouring plains. M. Klaproth adds, from a Mantchu manuscript, that these animals are never found but in cold countries towards the northern sea; that the bones resemble ivory, and have no fissures; and that the flesh is of a cold nature, but very wholesome.

Those immense rivers that descend to the Icy sea are continually laying bare the remains of elephants. It was imagined by M. Patrin that they were brought down by those rivers from the neighbouring mountains of India. But there are not fewer remains along those rivers which come from the north, such as the Wolga, the Tanaïs, and the Jaik; nor by the

Lena, the Kolima, the Indigirska, and the Anadir, which descend from the icy mountains of Chinese Tartary. We may add that the Irtish is the only river which approaches near enough to the mountains of Thibet to allow the application of M. Patrin"s hypothesis, and that elephantine remains are found in the peninsula of Kamschatka, whither they could by no possibility have arrived from India.

We are assured by M. Pallas, that in all Asiatic Russia, from the Tanaïs to the promontory of Tchutchis, there is not a single river or stream on whose banks, or within whose bed, the bones of elephants are not found, with those of other animals foreign to the climate. This observation applies to the lower declivities, and to the immense slimy and sandy plains; for the more elevated regions, the primitive and schistous chains, are equally destitute of those remains and of marine petrifactions. Their prodigious abundance would be sufficient to exclude all explication from expeditions conducted by men. But what sets this hypothesis totally aside is, that here, as well as in Germany, Italy, and France, they are found intermingled with the bones of other wild animals: nay, that they are constantly found in strata filled with marine productions.

The most remarkable fact of all is, that in some places the bones of elephants have been discovered with pieces of the flesh still remaining, and other soft parts. The opinion of the Siberians is, that these animals have been actually disinterred with their flesh remaining entire, fresh, and bloody; but this is mere exaggeration, founded, however, on the fact of the flesh having been occasionally discovered in a state of preservation, from the agency of frost.

Isbrand-Ides speaks of a head, the flesh of which was found in a state of putrefaction, and Müller of a tusk, the cavity of which was filled with matter resembling clotted blood. Such facts might be doubted, were they not confirmed by subsequent relations of the most undoubted authenticity. In 1771, near Vilhoui, a rhinoceros was disinterred entire, with the flesh, skin, and hair remaining. Since then an elephant was discovered on the banks of the Alaseia, a river which flows into the Icy sea beyond the Indigirska. It was in an upright position, was almost entire, covered by its skin, to which some long hairs were still attached. Another fact of this kind is so surprising, that it deserves a more detailed account.

In 1799, a Tongoose fisherman observed, on the borders of the Icy sea, near the mouth of the Lena, in the midst of the fragments of ice, a shapeless mass of something, the nature of which he could not conjecture. The next year he observed that this mass was a little more disengaged. Towards the end of the following summer the entire side of the animal and one of the tusks became distinctly visible. In the fifth year, the ice being melted earlier than usual, this enormous mass was cast upon the coast, upon a bank of sand. The fisherman possessed himself of the tusks, which he sold for fifty rubles. Two years after, Mr. Adams, associate of the academy of St. Petersburgh, who was travelling with Count Golovking, on an embassy to China, having heard of this discovery at Yakutsk, repaired immediately to the spot. He found the animal already greatly mutilated. The flesh had partly been cut away by the Yakoots for their dogs, and some of it had been devoured by wild beasts. Still the skeleton was entire, with the exception of a fore-leg. The spine of the back, a shoulder-blade, the pelvis, and the rest of the extremities, were still united by the ligaments and a portion of the skin. The other shoulder-blade was found at some distance. The head was covered with a dry skin. One of the ears, in high preservation, was furnished with a tuft of hair, and the pupil of the eye was still discernible. The brain was found in the cranium, but in a state of desiccation. The under-lip had been torn, and the upper one being utterly destroyed, left the molars visible. *The neck was furnished with a long mane.* The skin was covered with black hairs, and with a reddish sort of wool. The remains were so heavy, that ten persons had much difficulty in removing them. More than thirty pounds weight of hairs and bristles were carried away, which had been sunk into the humid soil by the white bears when devouring the flesh. The animal was a male. The tusks were more than nine feet long; and the head, without the tusks, weighed more than four hundred pounds. Mr. Adams collected, with the utmost care, all the remains of this singular and valuable relic of a former creation. He re-purchased the tusks at Yakutsk, and received for the whole from the Emperor Alexander eight thousand rubles.

These are not the only discoveries of the kind made in the Russian empire; and facts so well authenticated and detailed will not permit us to doubt of anterior testimonies on the same subject, while they also prove a most important point, namely, that these animals must have been arrested by the ice at the moment of their death.

A most extraordinary circumstance, which we read in the account of Belling's voyage, is, that certain islands in the Icy sea abound more in elephantine remains, in proportion to their extent, than any other part of the globe! These islands lie to the north of Siberia, opposite to the shore which separates the mouth of the Lena from that of the Indigirska. That which is nearest to the continent is thirty-six leagues in length. The whole island, with the exception of a few little rocky mountains, is composed of a mixture of sand and ice. Accordingly, when thaw takes place, the bones of the mammoth are found in the most prodigious abundance. Nay, this writer adds, (following the report of an eye-witness,) that the whole island is composed of the bones of this extraordinary animal, of the horns and crania of the buffalo, or some animal which resembles it, and some horns of the rhinoceros. That this is exaggeration there can be little doubt, but it serves to prove the wonderful abundance of those fossil remains.

In a second island, about five leagues farther, these bones and teeth are also found; but in a third, about five-and-twenty leagues to the north, they are no longer discovered.

The south of Asia is very far from having furnished these remains in a similar degree of abundance. The most southern regions of this continent in which the fossil bones of the elephant are said to have been found, are the neighbourhood of the sea of Aral, and the banks of the Jaxartes. Pallas tells us that the Bucharians sometimes fetch ivory from the borders of this river.

But, in fact, every thing relative to those fossil remains in the south of Asia, is mere report and conjecture. The ancients talk of the bodies of giants found in Syria and Asia. Pausanias tells us of the body of Geryon, or Hyllus, found in Upper Lydia, and of another, eleven cubits in length, discovered in the bed of the Orontes. These may be referred to elephants.

It is singular enough that these remains should not be found in climates to which the elephant is native, and has been as far back as historical records extend. Can we suppose that none are buried there, or that the bones have been decomposed by the force of heat? It is very possible that instances of their discovery have been passed over in those countries, as presenting nothing unnatural or extraordinary. It is also by no means improbable that the mammoth was an animal intended to live in the cold climates of the north, from the fact of its having been covered with long hair and a thick wool. Geologists who visit the torrid zone would do well to extend their researches on this subject. It appears, however, that some discoveries of this kind have been made in northern Africa, where elephants do not exist at the present day. They did so, however, formerly, and seem to have been native to Mauritania, according to the report of all the ancients; and we know the Carthaginians had plenty of them. But there are few accounts from this quarter sufficiently authenticated or detailed to identify the species.

We now proceed to the New World, where the elephant has never been found in a living state, and where the species could never have been destroyed in earlier times by the weak and wandering hordes of barbarians that then occupied the country. Still we find the remains of the fossil elephant in very considerable abundance, especially in the northern division of that mighty continent. Buffon imagined that they existed in that part only in consequence of being unable to cross the isthmus of Panama, when the gradual cooling of the earth had impelled them southwards. This hypothesis cannot stand a moment's reflection, for many parts of Mexico are sufficiently warm for the elephant to live in; moreover, the theory does not rest on correct facts, as the bones discovered in Buffon's time did not belong to the elephant, but to the mastodon.

The bones of the true elephant are found in tolerable abundance throughout North America. For this we have the testimony of many recent writers, who clearly distinguish the teeth and bones thus discovered from those of the mastodon. Respecting the peculiar species to which those elephantine remains belonged, all the accounts are not so clear as might be wished. Catesby mentions an instance in which some African negroes recognised the resemblance of some fossil molars, discovered in Carolina, to those of the elephant of their native continent. Mr. Barton talks of teeth and bones dug up in various parts of North America, as resembling those of the Asiatic elephant. But it seems more probable, from the testimony of Mr. Rembrandt Peale and others, that these and the Siberian remains belong to one and the same species. This opinion is corroborated by the remains of elephants sent to the Baron Cuvier, by M. de Humboldt, from Spanish America. These our illustrious author discovered exactly to resemble those of the Siberian species.

A fact worth noticing here is, that the elephantine remains found in Kentucky were far advanced in a state of decomposition: from which Mr. Peale concludes, that the destruction of the elephant in America was considerably anterior to that of the mastodon.

The Spanish accounts of Mexico and Peru are filled with

relations of the bones of giants found in several parts of those countries. But we cannot, with any certainty, apply these stories to the elephant, inasmuch as they are equally applicable to the mastodon, whose teeth have a still closer resemblance to those of man.

We shall close this article with a brief view of the comparative anatomy of the fossil and living elephants.

Without descending to minutiæ, which would be inconsistent with our limits, we shall dwell simply upon the most striking points of difference. The laminæ of the molar teeth of the fossil elephant (comparing teeth of equal length) are narrower and more numerous than those of the Indian species. From this it results, that the number of laminæ employed at a time in trituration must have been more numerous in the fossil elephant.

The lines of enamel which separate the divisions of laminæ are more slender and less *festooned* in the fossil teeth.

The absolute, as well as proportional, magnitude of the fossil teeth is greater.

These three characters are in accordance to the differences of the jaws and crania.

The two first are not invariably constant in all the fossil specimens, but this may be referred to the difference of ages.

We are not aware that similar differences in the magnitude of the tusks, relatively to sex and variety, existed in the fossil elephant as they do in the Indian species. Nor do we know their exact limits as to smallness. But their magnitude was considerably greater than in the living species, more especially than the African. The texture is precisely the same in all, nor is it different in the mastodon.

In a number of fossil specimens the tusks are more curved.

Upon the whole, however, the Baron is of opinion that the tusks do not constitute a character of great importance.

The cranium of the elephant is too cellulous, the osseous plates of which it is composed too slender, to admit easily of perfect preservation in the fossil state. But five crania have been sufficiently preserved, to which we may add that of the Siberian skeleton above-mentioned, to determine the characters.

The first striking difference is the excessive length of the alveoli of the tusks, being found in one instance to be triple that of those in the Indian or African elephant of the same dimensions. Also the triturating face of the molars, instead of meeting the alveolar ridge, must have cut the tube of the alveolus at one-third of its length. This difference, as we shall see, accords with the form of the lower jaw, and must have necessitated a different conformation in the trunk of the fossil elephant: for either the points of attachment for the muscles of the trunk were the same as in the living species, that is, the upper part of the nose, and the lower edge of the alveoli of the tusks, which would make the basis of this organ three times as thick in proportion, or the attachments were totally different, and then the whole structure must have been totally different.

The zygomatic arch was differently figured; the post-orbital apophysis of the os frontis longer, more pointed, and more crooked; and the tubercle of the os lachrymale thicker and more projecting.

These differences were first established upon a drawing of a cranium, by Messerschmidt, and have been since verified on several others.

Another difference not less authenticated, and which accords with those of the lower jaw, is the parallelism of the molars. As to the lower jaws, in the Indian and African species, the lower teeth are converging forward, like those above. The canal, therefore, is hollowed in the middle, and long and narrowed at the anterior point of the jaw.

The teeth in the fossil jaws are nearly parallel. The canal, therefore, is wider, in proportion to the total length of the jaw. This canal is also much shorter in the African and Indian species; for the alveoli of the tusks do not descend below the extremity of the lower jaw, which is consequently advanced between the tusks, and prolonged into a kind of pointed apophysis. But as these alveoli are much longer in the fossil heads, the jaw must have been truncated in front, otherwise it could not close. These observations have been confirmed by an immense number of well-authenticated specimens. In the bones of the spine the differences remarked are not important. But the specimens examined are but few.

The bones of the extremities are generally distinguished by greater massiveness. There are other differences, such as the lower head of the femur being distinguished by a slope between the two condyles, gradually reduced to a narrow line, instead of a wide sinking, as in the two living species. But to follow them minutely would tire our readers, and prove interesting only to the professed anatomist.

From what is known of the soft parts, we find that the skin resembles that of the living elephant, but without the brown points remarkable in the Indian species. Mr. Adams tells us that all of the skin which he had preserved was of a dark grey.

The hair may be said to be of three kinds: the longest from twelve to fifteen inches, brown colour, and of the substance of horse-hair; a shorter kind, from nine to ten inches, more delicate, and of a fawn colour, and the wool, which garnished the roots of the long hairs, four or five inches long, tolerably fine and smooth, though a little frizzled towards the root—it was of a clear fawn colour.

No animal has similar hairs, and therefore there can be no fraud in the case. It cannot, then, be doubted that the fossil elephant, such as it is found in Siberia, was an animal calculated, from the nature of its covering, to endure the temperature of northern climates.

The soles of the feet were found to be rounden, but considerably dilated, as if by the weight of the body, so as to extend beyond the edge of the foot.

We shall now conclude with a recapitulary view of the subject, and a few general reflections.

After what has appeared in our Ninth Number of the Animal Kingdom, it will be unnecessary to refer any more to the living species. Our readers are fully enabled thereby to understand our comparative view.

The *Elephas Primigenius*, fossil elephant, or mammoth, of the Russians, may be thus briefly characterised:

Elongated cranium, concave forehead, very long alveoli of the tusks, lower jaw obtuse, cheek-teeth parallel, broader and marked with narrower stripes than the elephas Indicus. Its bones are only found in the fossil state. It has never been found living, nor have we any reason to suppose that the bones of the two living species have been ever found in the fossil state. The bones are very numerous in many countries, but better preserved in the north. It more resembled the Indian than the African species. It differed from it, however, by the cheek-teeth, the form of the lower jaw, and of many other bones, but especially by the length of the alveoli of the tusks. This last character must have singularly modified the conformation of its trunk, and given the animal a very different physiognomy from the Indian species. The tusks appear to have been generally large, more or less arched into a spiral form, and turned outwards. We have no proof that they differ much according to sex or variety. Its height does not appear to have much exceeded what the Indian species is capable of attaining to; but its proportions were, in general, heavier and more clumsy.

It is manifest, from its osseous remains, that it was a species more different from the Indian than the ass is from the horse, the chacal or the isatis from the wolf and fox.

The size of its ears is not known, nor the precise colour of the skin. The hair we have already described. This was long enough to form a mane on the neck. It would appear, therefore, as we remarked before, that these animals were adapted to sustain a climate, the coldness of which would soon destroy the Indian species.

Their remains are generally found in the loose and superficial strata of the earth, and most frequently in the alluvion which fills the bottom of valleys, or borders the beds of rivers.

They are never found alone, but intermixed with the bones of other quadrupeds of known genera, as the rhinoceros, ox, antelope, horse, and often with the debris of marine animals, a portion of which is frequently attached *upon* them. For this fact, so important, we have the positive testimony of Pallas, Fortis, and many other writers; and the Baron himself has had fragments in his possession, filled with millepores and small oysters. The strata which cover the elephantine remains are of no great thickness, and scarcely ever of a stony nature. These remains are rarely petrified, and but few examples can be quoted of any incrusted in stone, whether coquillaceous or otherwise. They are often found accompanied with our common fresh-water shells.

Every thing, in short, appears to announce that the catastrophe which overwhelmed them was one of the most recent which has contributed to change the surface of our globe. This catastrophe was physical and general. It was also aqueous, as is clearly proved, both by the strata in which they are imbedded, and those which are above them. They have been covered by the waters, and in very many places by waters of the same quality as those of our present sea, as is proved by similarity of marine productions. But it is not by these waters that they were transported to their present situations. They have been found in every country examined by naturalists. An irruption of the sea, therefore, which should have carried them from the habitat of the Indian elephant only, could not have spread them so far, nor dispersed them so equally.

Let us also observe, that the inundation which overwhelmed them was not elevated above the grand mountainous chains of the earth. The strata deposited by it, and which cover these remains, are found only in plains of moderate elevation. It is impossible, then, that the bodies of elephants could have been transported across the Thibetian, the Altaic, and the Oural mountains. Moreover, these bones exhibit no marks of detrition: their ridges, their apophyses, are in perfect preservation. Even the epiphyses of those not yet arrived at perfect growth still adhere, though the slightest effort would be sufficient to detach them. The only alterations that are visible are the result of that decomposition which they have undergone in the bosom of the earth; nor can we say that entire carcasses have been violently removed. On such a supposition, it is true that the bones would have remained perfect, but it is also true that they would have been found together, not scattered. Besides, the marine productions attached to many of them prove that they must have remained some time stripped of their exuviæ and separated at the bottom of the fluid by which they were covered. When this fluid overwhelmed them, they were in the places where we now find them; they were scattered far and wide, as we frequently find the bones of our own animals scattered on their native soil.

It is then much more than probable that the fossil elephants inhabited the countries where their remains are found at present, and that they must have perished by some simultaneous revolution, or a change of climate, which put a stop to their propagation.

Let the cause be what it may, it must have been sudden. The bones and ivory in such fine preservation in the plains of Siberia are only so because they are congealed by the cold, which suddenly arrested the decomposing action of the elements. Had this cold taken place gradually and slowly, the bones, and still more the soft parts, would have had time for decomposition, as has been the case in warm and temperate climates. Above all, it is impossible that an entire carcass could have been found, with the skin and flesh uncorrupted, had it not been instantly enveloped by the ice which continued to preserve it.

Thus we see that every hypothesis, founded on a gradual

refrigeration of the earth, or a slow variation, whether from the inclination or position of the axis of the globe, falls to the ground of itself.

Lastly, had the present elephants of India been the descendants of the fossil refugees in their present climate, at the period of the catastrophe which destroyed the others, how shall we account for the destruction of their species in America? The elevated parts of Mexico offered them a secure escape from such an inundation as we have mentioned, and the climate is hotter than necessary for their temperament. Nor can we suppose that the same cause which extinguished the mastodon, the hippopotamus, and the fossil rhinoceros, would have spared the elephant. That those animals have ceased to exist, can no longer be a subject of controversy.

The Genus Mastodon.

The *Great Mastodon* is one of the most remarkable and apparently the most enormous of all the fossil species. Whatever doubts might have existed concerning the specific difference of the elephas primigenius, and its utter extinction, there could be little controversy of this nature concerning the mastodon, after its remains had been properly examined. Accordingly, we find that this was the first fossil animal by whose remains naturalists were convinced of the possibility of extinct species. The monstrous bulk of the molars, and the formidable tuberosities with which they are provided, could not fail of attracting attention; and it was easy to observe that none of the larger quadrupeds with which we are acquainted possessed teeth of similar conformation or equal volume. Daubenton, indeed, for a time was disposed to believe that some of these teeth might have belonged to the hippopotamus, but he soon renounced this error; and Buffon declared that this ancient species, the first and largest of all terrestrial animals, must be regarded as now extinct. Still he restricted this observation to the enormous hinder teeth, considering the middle and half-worn teeth as belonging to the hippopotamus. The large femur also, dug up in the same place as these teeth, was attributed both by Buffon and Daubenton to the elephant. Dr. Hunter, however, had shown, 1767, that this part, as well as the teeth and the lower jaw, exhibited a different conformation. Still, Dr. Hunter fell into two errors concerning this animal. He confounded it with the mammoth of the Siberians, and supposed, from the structure of its teeth, that it was carnivorous. These errors were sufficiently refuted, the first by Pallas, and the second by Camper. Yet the animal continued to be called *mammoth* both by the Americans and ourselves, and *carnivorous elephant* by some naturalists,—names equally improper. These misnomers occasioned infinite confusion in the accounts of compilers, and determined the Baron to give the animal the name of *mastodon*, compounded of two Greek words, signifying *mammillary teeth*, and expressing its principal character. This was the more necessary, as it is found that the mastodon was not only a distinct species, but a distinct genus, comprehending other species.

It is above one hundred and twenty years since remains of the mastodon were first discovered at Albany, near Hudson river. They are mentioned in a letter from Dr. Mather to Dr. Woodward, in the *Phil. Trans.* 1712. He believed them to be the bones of giants, and a confirmation of the Mosaical accounts of gigantic races of mankind.

Thirty years after, a French officer named Longueil, navigating the Ohio, discovered on the edge of a marsh near this river, some bones, cheek-teeth, and tusks. He brought back to Paris a femur, the extremity of a tusk, and three cheek-teeth, where they are still preserved. These were the first specimens of this animal seen in Europe; and from the place in which they were found, the French called the mastodon *animal of the Ohio*, though its bones have been found in many other parts.

Daubenton declared the tusk and femur to belong to the

elephant, but the teeth to the hippopotamus. But the contrary opinion even then was held by many. Another French officer informed Buffon that the Indians of Canada and Louisiana believed these bones to belong to a peculiar animal which they called *father of oxen*.

The large teeth with eight or ten points, which could not reasonably be confounded with those of the hippopotamus, were already known. One of them was even engraved by Guettard, in the Memoirs of the Academy for 1752.

When our countrymen became masters of Canada, these researches were pursued with fresh activity. Many of these bones were discovered south-east of the Ohio, by Croghan, the geographer, in 1765. The tuberculous teeth and tusks were found mingled together, without any elephantine molar, and the notion of a peculiar animal became more and more confirmed. This last-mentioned gentleman sent several chests of these fragments to London, and from a lower-jaw, which was among them, William Hunter demonstrated that the animal in question differed sensibly from the elephant, and had nothing in common with the hippopotamus.

Buffon first maintained that similar teeth were to be found in the ancient continent. He founded this opinion on a tooth presented to him by the Count de Vergennes, said to have been discovered in Little Tartary. It weighed eleven pounds four ounces. Another, from the cabinet of the Abbé Chappe, was supposed to have come from Siberia. Pallas advanced the same opinion on teeth with six points, found in the Oural mountains. The Baron, however, is inclined to regard these proofs as insufficient, inasmuch as it is by no means certain that the specimens in question did not come from America.

Camper, in 1777, showed again that there was more analogy between the mastodon and elephant, than between it and the hippopotamus. He also deemed it very probable that it had a trunk, A considerable portion of the cranium and some other bones were found by Dr. Brown, in 1715.

Camper, in 1788, retracted his last-mentioned opinion on the mastodon. From some drawings presented him by M. Michaelis, he declared that he had been mistaken; that this animal had a pointed muzzle without tusks; that it did not resemble the elephant, nor could he form any opinion of its true nature. But it afterwards appeared that he took that part of the palate where the teeth approximate, for the anterior part, and, considering the pterygoïd apophyses as intermaxillary bones, found, consequently, no place for the tusks.

A famous discovery, made in the commencement of the present century by Mr. Wilson Peale, founder of the Museum of Natural History at Philadelphia, seems to have set this question at rest for ever, and will require a little more detailed account. In the spring of 1801, he learned that some bones had been dug up the preceding autumn, in the neighbourhood of Newburgh, on the river Hudson. He repaired thither with his sons, and obtained from the farmer who had dug it up a considerable portion of a skeleton, which he sent to Philadelphia. There was a cranium much damaged in the upper part; the lower jaw was broken, and the tusks mutilated. At the close of autumn, after many weeks labour, were found, in the same place, all the cervical vertebræ, many of the dorsal, two shoulder-blades, two humeri, a radius and cubitus, a femur, a tibia, a peroneum, a mutilated pelvis, and some small bones of the feet. These were found between six and seven feet in depth; but many important bones were wanting, such as the lower-jaw, &c. To obtain them, Mr. Peale repaired to another spot, eleven miles distant, where bones had been disinterred about eight years before. He worked for fifteen days, collected many fragments, but not those he wanted. However, on his return he met a farmer, who had found some bones three years before, and who conducted him to the place of his discovery. Here, after much labour, he was fortunate enough to find a complete under jaw, and many other principal bones. With the materials he had thus obtained by three months of laborious research, he formed two skeletons, copying artificially from the bones of one what was wanting in the other, and from the bones of one side what was deficient on the opposite.

Since this discovery, the osteology of this great animal may be considered as completely known, with the exception of the upper part of the cranium. The most complete of these skeletons is placed in Mr. Peale's museum at Philadelphia; the other was brought to our capital by one of his sons, and publicly exhibited.

Besides these materials, there are many in the museum at Paris. The most remarkable constitute the present made to that repository by Mr. Jefferson: these are an enormous tusk, two half lower-jaws, one of which, having belonged to a young subject, is of great interest, as regards the teeth, a tibia, a radius, almost all the bones of the tarsus, metatarsus, carpus and metacarpus, some phalanges, ribs, and vertebræ. The bones of the mastodon, as well as those of the other fossil species which accompany them, are most generally found in marshy places, where there is a sort of brackish water, which attracts wild animals, especially deer, to which places the Americans have, for this reason, given the name of *lick*. The most celebrated of those depôts near the Ohio is called *Big-bone-lick*.

There is a spot in Kentucky, to the south-east of the Ohio, sunk in between small hills, and occupied by a marsh, in which is a small stream of brackish water. The bottom of this consists of a black and stinking mud. Here, and on the borders of the marsh, the remains of the mastodon have been found in the most astonishing profusion. This mud is intermixed with a fine sand, and some ligneous debris are distinguishable in it. On being tried with nitric acid, it exhaled a fetid odour, indicative of some animal principle, and being analysed was found to be composed of about sixty-five parts of argilla, sixteen of sand, and five of sulphate of lime to every hundred. The argilla retained a little carbonate of lime and sulphuret of iron. There was also a little oxide of iron.

Near the river of the Osages myriads of these bones were seen, according to Mr. Smith Barton, and seventeen tusks were collected, some of which were six feet in length and a foot in diameter, most of them far advanced in decomposition.

One of the most remarkable of the depôts of these bones was found at Withe, in Virginia, five feet and a half under ground, on a bank of calcareous stone. One of the teeth weighed seventeen pounds. In the midst of these bones was found a mass of little branches, grass, and leaves, in a half-bruised state. Among these was a species of rose, now common in Virginia, and the whole was enveloped in a kind of sack, which was considered to be the stomach of the animal. The substratum of this soil is a calcareous stone, full of coquillaceous impressions. The caverns there abound in nitre, sulphate of soda, and magnesia.

Not to be tedious as to localities—the bones of the *great mastodon* are found in abundance all over North America, from the forty-third degree of north latitude, north of Lake Erie, as far south as Charlestown, in Carolina, in thirty-three degrees.

As far as we know, the bones of this enormous animal do not exist in any other country of the globe. They are always found at moderate depths, and exhibit few marks of decomposition, and none of detrition; a proof that, like the other fossils, they remained in the places where they are found since the period of the animal's destruction. Those on the river of the Great Osages were found in a vertical position. The ferruginеous substance with which they are tinctured, or penetrated, proves that they have been a long time imbedded in the interior of the earth.

Indications that the sea rested on, or passed over them, are more rare than in the case of elephantine remains. No remains of shells or zoophytes have been found upon their bones, nor, according to all accounts, in the strata from which they have been taken. This is the more singular, as the salt-marshes where they have been most usually found, might be considered as the remains of a more extensive fluid which had overwhelmed them.

Mr. Barton considers that the salt water has been a main cause of their preservation. It would even appear that some soft parts have been occasionally found, which, considering the heat of the climate, is wonderful. Some savages, who saw five skeletons in 1762, informed Mr. Barton that one of the heads had "*a long nose, under which was the mouth;*" and Kalm, speaking of a large skeleton discovered in a marsh, in the country of the Illinois, and which he took for that of an elephant, says that the form of the *beak* was still observable, though half decomposed. These two facts seem to confirm the opinion, that the triturated plants above-mentioned were, in reality, the materials which filled the stomach of the animal.

Many hypotheses have been formed on the origin of these bones, and the causes of the animal's destruction. The Shavanois believe that men of similar proportions existed with those animals, and that the Great Spirit destroyed both with his thunders. The savages of Virginia say, that a troop of these tremendous quadrupeds destroyed for some time the deer, the buffalo, and all the other animals created for the use of the Indians, and spread desolation far and wide. At last "*the Mighty Man above*" seized his thunder and killed them all, with the exception of the largest of the males, who, presenting his head to the thunderbolts, shook them off as they fell, but, being wounded in the side, he betook himself to flight towards the great lakes, where he still resides at the present day.

Such stories sufficiently prove that the Indians have no knowledge of the actual existence of the species in those countries over which they wander.

Lamanon imagined the mastodon to be some unknown species of the cetacea. A.M. de la Coudrenière having found, in some account of Groënland, that the savages of that country pretended to an animal, black and hairy, of the form of a bear, and six fathoms in height, refers not only the mastodon, but the mammoth, to this fabulous monster. This confusion of the two species probably made Mr. Jefferson think that the centre of the frozen zone is the spot where the mammoth arrived at its full perfection, as the countries situated under the equator are the best calculated for sustaining the elephant.

Let us now consider the osteology of the Great Mastodon, beginning with the cheek-teeth, the most important character.

Their form is the most remarkable point about them. The coronal approaches more or less to a rectangular figure. The substance is two-fold, the ivory and the enamel. The latter is very thick, but there is no appearance of that cortical substance, so remarkable in the elephant.

This coronal is divided by very open furrows into a certain number of transversal prominences, each of which is itself divided by a slope into two thick obtuse points, something like quadrangular pyramids, a little rounded. Thus the coronal, as long as it remains unworn, is furnished with thick points, disposed in pairs.

It will be seen that these teeth have no relation to the teeth of the carnivora with one principal and longitudinal edge, divided into notches like a saw. There is only a difference of proportion between these transverse prominences, divided into two points, and the little transverse walls, with edge divided into many tubercles, in the teeth of the elephant. But in the latter, the furrows which separate the prominences are filled with the cortical substance, whereas, in the mastodon, they are filled with nothing. From this it happens, that the coronal of the elephant soon becomes flat from detrition, but nevertheless remains always furrowed transversely, while that of the mastodon is for a long time mammillated. Its protuberances become at first truncated by detrition into a lozenge-form, and at last, when worn flat, its surface is perfectly even, or uniformly concave.

The mastodon must have made the same use of its teeth as

the swine and the hippopotamus, for the structure is similar. It must have fed on tender vegetables, roots, and aquatic plants, and could not have been carnivorous.

The difference in the teeth of the mastodon consist, chiefly, in the number of points, and the relation of length to breadth.

There are three sorts: one almost square, with three pairs of points; another rectangular, with four pair; a third, longer, somewhat contracted behind, with five pair, and an unequal heel. The first is the most worn, the last, on the contrary, very little so. This points out the order of their growth and position. The disposition of the cheek-teeth in the adult is thus, $\frac{4}{4}$, two with six points, and two with eight points above; two with six points, and two with ten points below.

Beside these eight molars in the adult, we find in the young subject, that others preceded them, which fell successively. The succession of growth took place, as in the elephant, from front to rear. When the hinder molar began to pierce the gum, the front one was ready to fall.

The effective number of cheek-teeth which could act together, was eight in the young subject, four in the old. This shows that the mastodon could not have been of the enormous size that Buffon and others imagined, who were led to this conclusion by the error of believing the cheek-teeth equal in number to our own.

These teeth vary in magnitude, and are found in different stages of detrition. But to pursue all their variations would be inconsistent with our plan, and we must, therefore, refer the reader for minuter information to the great work of the Baron.

On examining the lower jaw, we find that, like the elephant and the morse, the mastodon had no incisors or canines below; that the lower jaw terminated in front in a hollowed point of a sort of canal, and that the posterior angle, though obtuse, was yet strongly defined, and not circularly rounded, as in the elephant. The condyle differs little from that of the elephant, but all the ascending part is less elevated in proportion. The coronoïd apophysis is raised to the level of the cóndyle, while it is much lower in the elephant. The longitudinal part is less elevated in proportion to its length, but equally inflated, especially behind. The lower jaw of Mr. Peale's skeleton is two feet ten inches in length, and weighs sixty-three pounds. The principal characters of the cranium are, the divergence of the cheek-teeth in front, while those of the elephant converge, and those of the mammoth are parallel; the osseous palate extends far beyond the last tooth; the pterygoïd apophysis of the palatine bones are thick beyond example. Mr. R. Peale found no trace of orbit in the anterior part of the arch; consequently, the eye must have been much higher than in the elephant. The maxillary bones have much less vertical elevation than in the elephant; the zygomatic arch is less elevated; the occipital condyles, raised in the elephant considerably above the level of the palate, are on that level in the mastodon. The mastodon appears to have had precisely the large cellulæ which give so much thickness to the cranium of the elephant, and are only prolongations of the different sinuses of the nose.

We cannot precisely ascertain the elevation of the top of the cranium; but its weight, that of the cheek-teeth and tusks, cannot permit us to doubt that the occiput was considerably elevated, to furnish adequate attachments for the levator-muscles.

The entire length of the cranium of Mr. Peale's mastodon seems to have been about 1.139.

The tusks are implanted in the incisive bone, like the elephant. They are composed likewise of ivory, the grain of which exhibits curvilinear lozenges.

The tusks of the elephant are often more or less round; those of the mastodon seen by the Baron were elliptical. Their curvature varies as much as in the elephant.

The alveoli in Mr. Peale's mastodon were about eight inches deep. The direction of the tusks, on issuing from the alveoli, is more oblique frontwards than in the elephant. Their position is disputed; Mr. Peale thinks that they were placed with the point downwards, contrary to the position of the elephant's tusks. His reasons are not very cogent, and the Baron comes to a contrary conclusion.

It seems indubitable, from anatomical reasoning, with which we shall not trouble our readers, that the mastodon had a trunk like the elephant. Mr. Peale says, that the spinous apophyses of the three last cervical vertebræ are shorter in the mastodon than in the elephant. The second, third, and fourth dorsal have very long apophyses. They decrease rapidly as far as the twelfth, after which they become very short. In the elephant they are more uniform, which argues more force in the muscles of the spine, and in the cervical ligament.

There are seven cervical vertebræ, nineteen dorsal, and three lumbar. The elephant has one dorsal vertebra, and a pair of ribs additional. The ribs are differently formed from those of the elephant; slender near the cartilage, thick and strong towards the back. The first six pair are very strong, compared with the others, which grow very short in proportion. This character, united to the depression of the pelvis, proves that the belly was less voluminous than that of the elephant.

In general, the long bones of the anterior extremity are much thicker, in proportion, than those of the hinder extremity. The pelvis is much more depressed than in the elephant, in proportion to its width. Its aperture is much more narrow. This conformation must have rendered the abdomen smaller, and the intestines less voluminous. This character, united to the structure of the teeth, would lead us to consider the mastodon as less exclusively herbivorous.

The width of the femur is another distinguishing character of the mastodon. The tibia is also much thicker in proportion to its length, compared with that of the elephant. It also differs in some minuter details.

Very exaggerated ideas were once entertained concerning

the height of the mastodon. The result of the justest measurements, and consequent calculations, seems to be, that it did not exceed ten or twelve feet at most, a stature to which the Indian elephant very commonly attains. The body, however, of the mastodon appears to have been much more elongated, in proportion to its height. The skeleton belonging to Mr. Peale measured fifteen feet, from the end of the muzzle to the posterior edge of the ischium.

The bones of the feet differ, in some minute proportions, from those of the elephant, and we may, in general, remark that they are shorter and thicker. This is especially the case with the metatarsus, and holds good in the phalanges.

To conclude: the *great mastodon*, or *animal of the Ohio*, was very similar to the elephant in the tusks and entire osteology, cheek-teeth excepted. It most probably had a trunk; its height did not exceed that of the elephant, but its body was more elongated, the limbs thicker, and the belly not so voluminous. Notwithstanding these resemblances, the structure of the molars is sufficient to constitute it a different genus. It subsisted, pretty much like the hippopotamus and wild boar, on roots and the stringy parts of vegetables. This kind of food must have attracted it to marshy places, though it evidently was not formed for swimming or living in the water, like the hippopotamus, but was decidedly a terrestrial animal. Its bones are much more common in North America than elsewhere, and most probably are exclusively confined to that country. They are better preserved and fresher than any known fossils, yet there is not the slightest proof that any of the genus are yet in existence.

The Mastodon with narrow Teeth.—A number of teeth have been discovered, from time to time, within the last hundred and fifty years, in various countries, the origin of which was by no means satisfactorily accounted for. These teeth have been found in different parts of France, of Italy, of Germany, and South America. Our author, from a close examination of numbers of these teeth, concludes that they belonged to a mastodon of a different species from the preceding, and which appears to have left its remains in considerable abundance in several parts of the earth.

All these teeth, like those of the great mastodon, are furnished with conic points, more or less numerous, which wear in mastication. The forms of some bones found with these teeth also resemble those of the great mastodon; and there is some reason to believe that they were also accompanied by tusks. From all this, the Baron concludes that they came from an animal of the same genus.

But these teeth are distinguished from those of the great mastodon of the Ohio, by certain specific characters. The principal and most general distinction is, that the cones of their coronals are furrowed more or less deeply, sometimes terminated by many points, and sometimes accompanied by smaller cones on their sides, or in their intervals. Mastication, therefore, produces, first, upon this coronal many small circles, and finally trefoils, or figures with three lobes, but never lozenges.

These trefoils have sometimes occasioned these teeth to be taken for those of the hippopotamus. But, independently of magnitude, the teeth of the hippopotamus have never more than four trefoils, while those of which we are speaking have usually six or ten.

It is more difficult to determine the specific characters of these various teeth, one with another,—they do not altogether resemble. That they have been differently placed in the jaw, may be judged by the number of the points. There are also differences of age, to be determined by the degree of detrition. The Baron examined two teeth, one from the turquoise mines of Simorre, in Gascony, the other from Peru, and he declares that, notwithstanding the distance between the two places, it was impossible not to recognise at once the specific identity of these teeth.

Mastodon with narrow teeth.

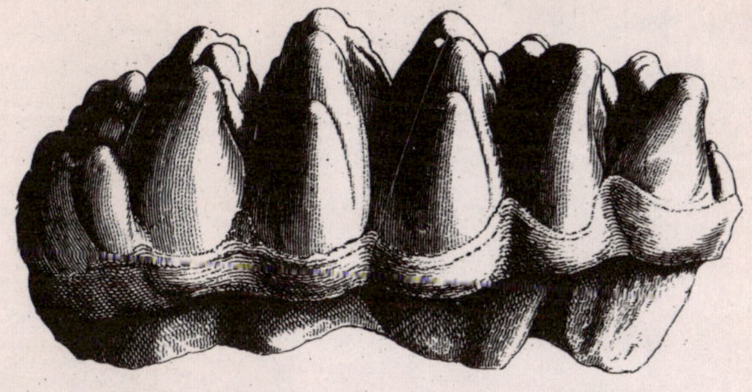

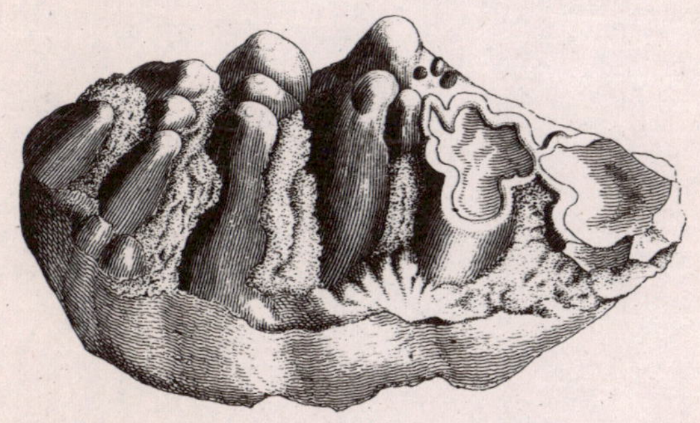

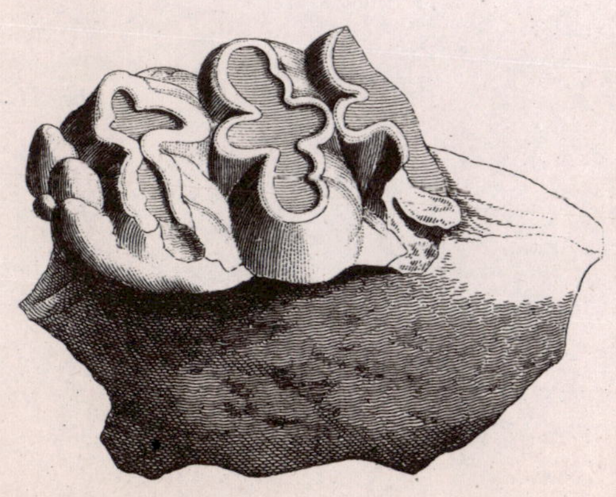

Cheek teeth in different states of detrition.

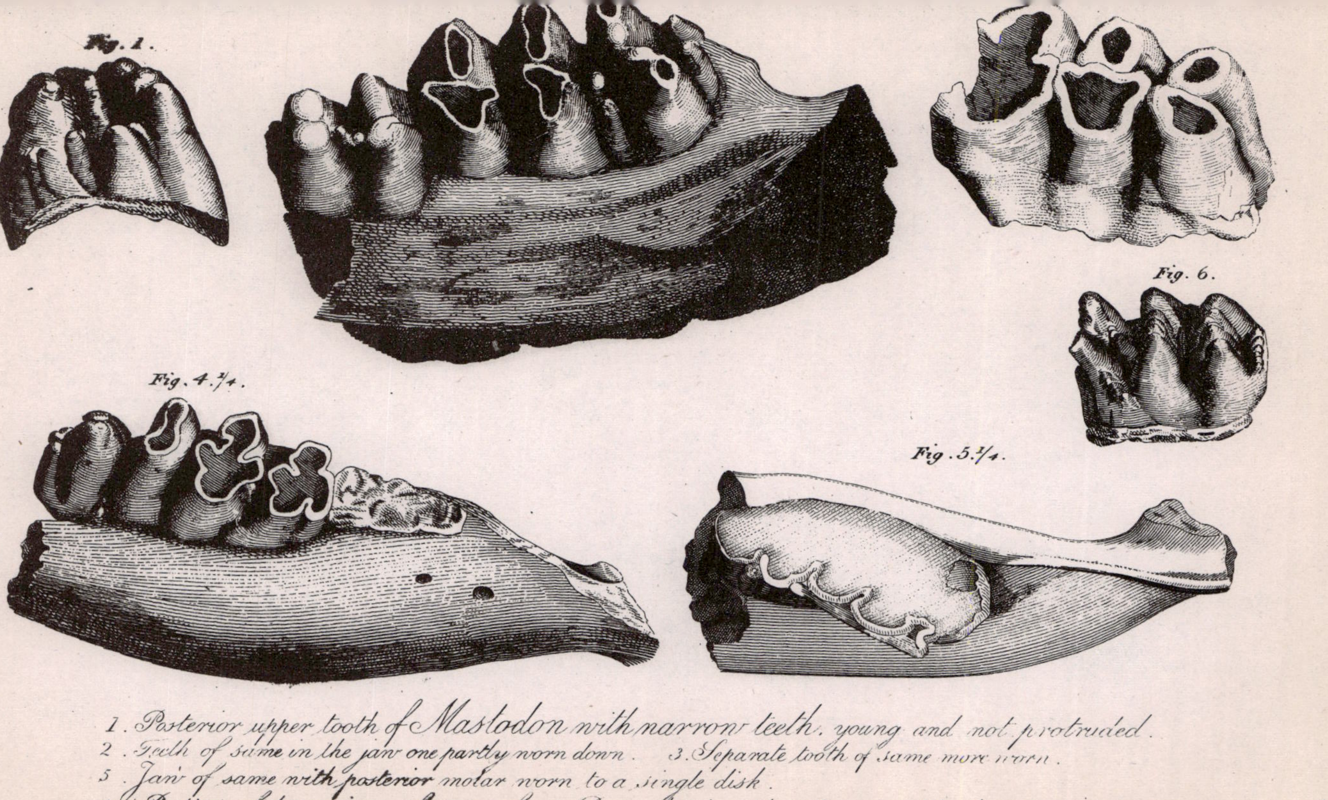

1. Posterior upper tooth of Mastodon with narrow teeth, young and not protruded.
2. Teeth of same in the jaw one partly worn down. 3. Separate tooth of same more worn.
4. Jaw of same with posterior molar worn to a single disk.
5. Portion of lower jaw of same from Perou shewing that the animal had neither incisors or canines below.
6. Tooth of another species of Mastodon.

Among the fragments collected by Dombey in Peru, is a considerable portion of the lower jaw. It terminates in front by a sort of beak, like that of the elephant and mastodon, which proves that, like the two last, this animal had neither incisors nor canines below. There were two teeth in this fragment; one had five pair of points, of which the hinder were the shortest. The two first were joined in quadrilobe figures, the two following were approaching to the same state; the two last and the heel were untouched; the external side was the most worn, the internal most projecting. This must be the case to make the teeth below correspond to those above. Below, the external points form trefoils; above, the internal. This is the result of a general law in the herbivora. When the two sides of the tooth do not resemble, they are placed in contrary directions, in the two jaws. Thus, in the ruminantia, the convexity of the crescents of the upper teeth is inside, and outside in those below.

From a palate, preserved in the British Museum, and which belongs to this species, and not to the great mastodon, we find that the upper molars diverge forward like those of the last mentioned species.

Analogy renders it probable, that the species of which we are speaking, had tusks like that of the Ohio. Daubenton, indeed, says that he could recognise ivory among the fragments sent from the mines of Simorre. Two plates of ivory were found by the Baron among the fragments sent by M. Chouteau from Avary. But we want the direct proof, for no tusk nor alveolus with molar adhering has yet been found.

The lower jaw is certainly that of an animal with long tusks. That from Peru is very like the one of the Ohio. It is, however, less high in proportion. Its lower edge is less rectilinear, and its external surface more inflated. The foramina menti are more advanced. The teeth of this mastodon are much longer, in proportion, than they are broad, which induced the Baron to give it the name he has done.

Compared with that of the elephant, the jaw of the mastodon with narrow teeth has its anterior beak longer, and more narrow in the middle. It is not truncated so vertically. The foramina menti are one behind the other, not one above the other. From another lower jaw, entire behind, we find that this part was more rounded in this species than in that of the Ohio, and more resembled that of the elephant.

A tibia, brought from the Giants-Field, near Santa-fé-de-Bogota, by M. de Humboldt, is the only great bone of the extremities which has been seen. It is much mutilated at the angles, which renders the characters not well determined. A little thicker in proportion than that of the Ohio, it is yet like it in its forms. It is short, by calculation from the proportions of the teeth ; but no definite conclusions can be drawn from a single bone.

It appears that the remains of this mastodon are often found accompanied by marine productions, which is not the case with the others.

The turquoises of Simorre are composed of these and other bones, penetrated and impregnated with some metallic oxide.

The stories of giants in South America seem founded on these remains. The name Giants-Field, above-mentioned, is probably derived from this source, as, according to M. de Humboldt, there is an immense accumulation of these bones there. They are found in that continent, usually, in a very high degree of elevation.

Some teeth have also been found, which seem to indicate other species of the mastodon, differing from the two preceding. M. de Humboldt brought some from South America, the tubercles of which are divided, like those of the mastodon, with narrow teeth, but which have the same canine proportions as those from the Ohio with six points, and might be taken for them, but for the trefoils. They are of two sizes; the largest are equal in magnitude to those of the Ohio. M. de Humboldt found one near the volcano of Imbaburra, in the king-

Fossil Hippopotamus.

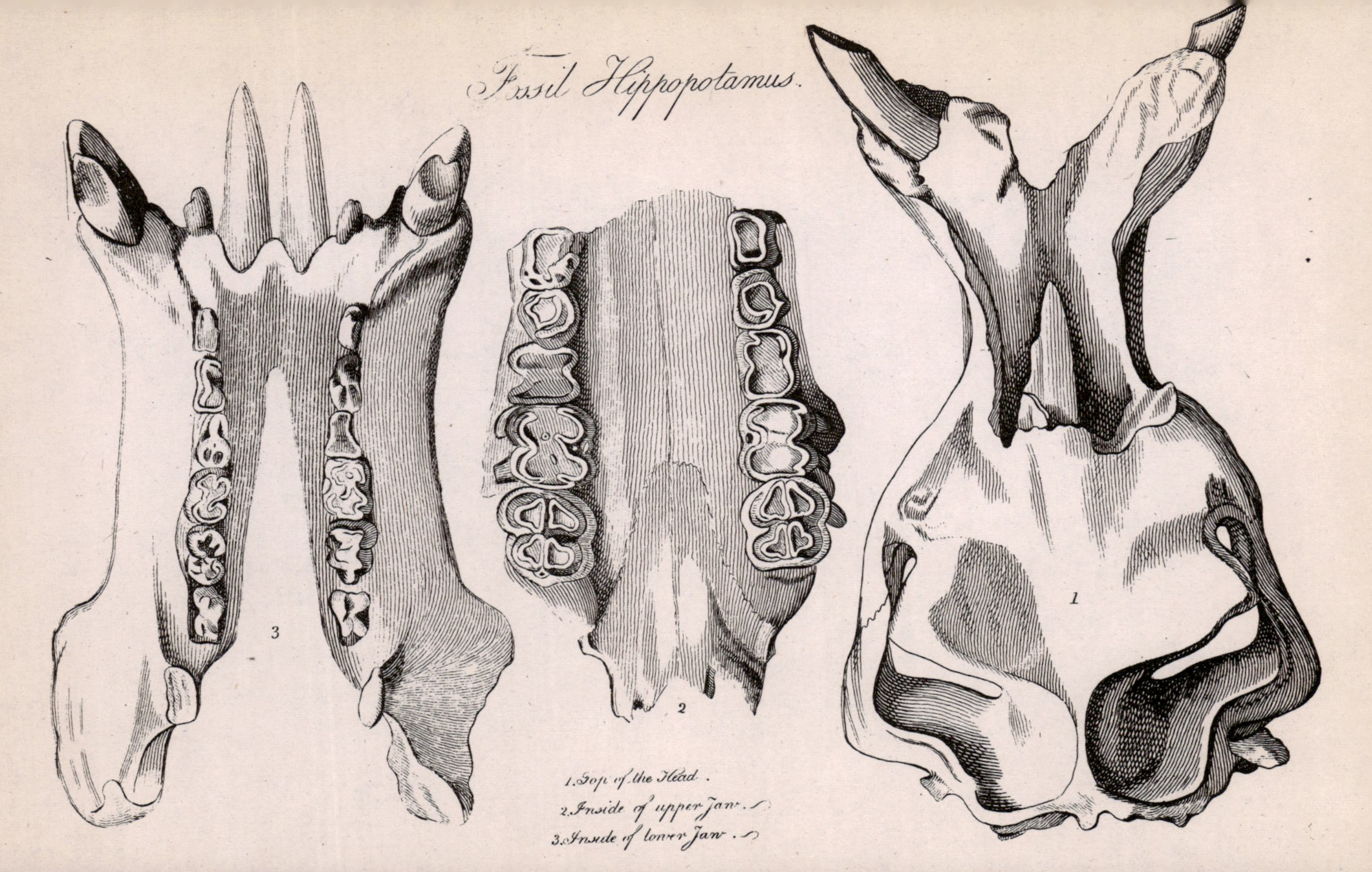

1. Top of the Head.
2. Inside of upper Jaw.
3. Inside of lower Jaw.

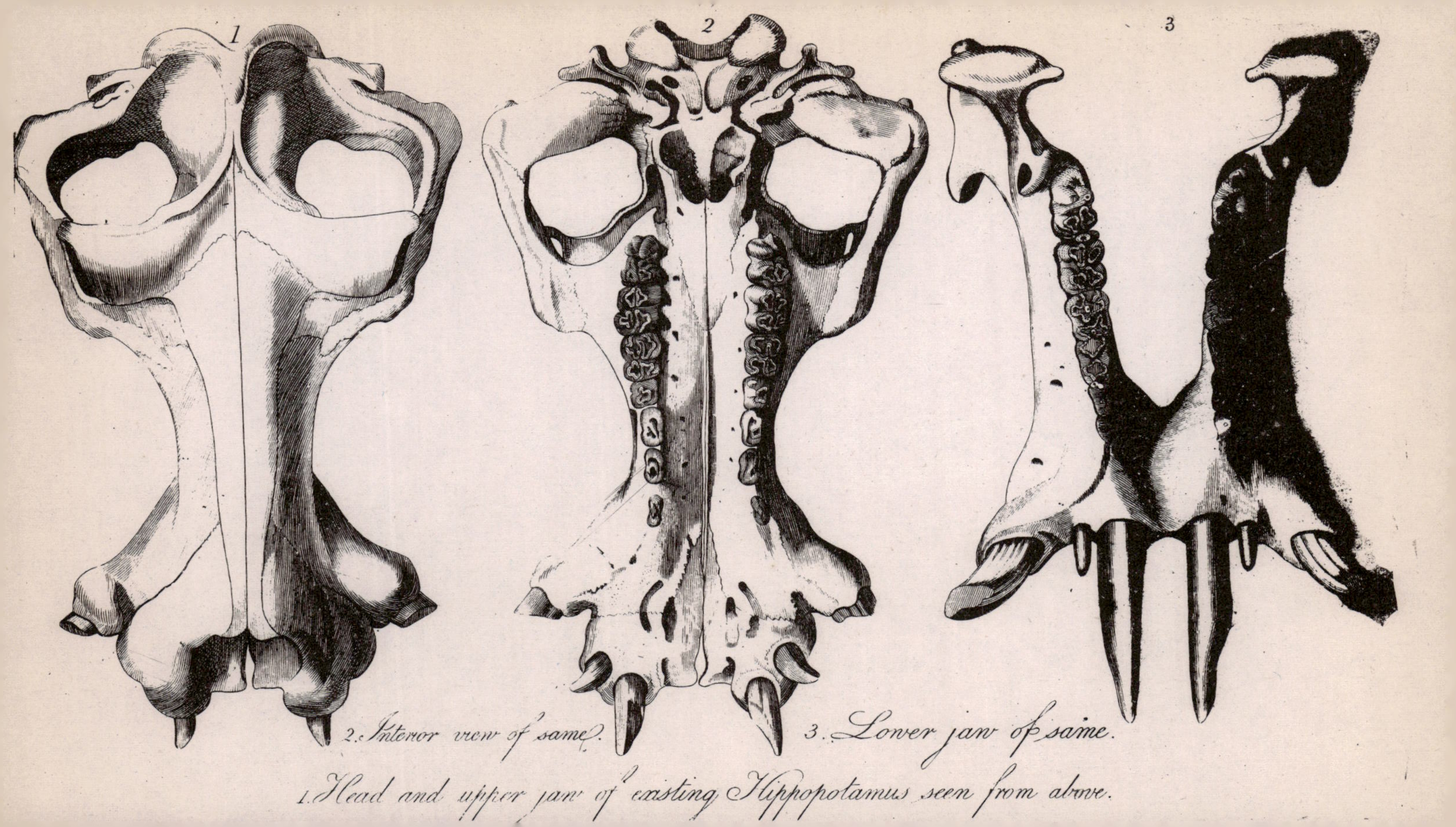

2. Interior view of same.   3. Lower jaw of same.

1. Head and upper jaw of existing Hippopotamus seen from above.

dom of Quito, at an elevation of 1200 fathoms. The other kind, also squared, are about a third less in size; they were also discovered by M. de Humboldt in South America.

In Europe, also, two teeth were found, which appeared too small to the Baron to be classed with any of the preceding species. One was sent from Saxony, but its position was unknown; the other from the neighbourhood of Orleans. The last was found in a quarry of fresh-water limestone, full of shells, &c., and the remains of the palæotherium. Its prominences, simply notched, are not so exactly divided into two points as those of the preceding. This might make us suppose another additional species. These prominences not divided show some relation with the teeth of the great tapir, but still cannot proceed from that genus, whose prominences are more separated, and whose numerous and small indentations bear no resemblance to nipples.

Thus, independently of the great mastodon, and that with narrow teeth, we find indications of four other species. The Baron would call the two from America, when their characters are determined, *the mastodon of the Cordilleras*, and *the mastodon of Humboldt*. To the first, of the European kind, he would give the name of *little mastodon;* to the second, whose hillocks, or prominences, are not completely divided into nipples, that of *tapiroidian mastodon*.

The Fossil Hippopotamus.

The hippopotamus has always been one of those larger quadrupeds whose history has remained in a state of obscurity, and even still is but imperfectly known. Bochart has imagined it to be the behemoth of the book of Job, but its description in that book is too vague to admit of any definite conclusions. The one given by Aristotle of his hippopotamus is so very unlike the animal in question, that it is perfectly inexplicable. He gives it the stature of the ass, the mane and voice of the horse, and the divided hoof of the ox, and says that its astragalus resembles that of the latter tribe; flat muzzle, mouth moderately divided, teeth rather projecting, and tail like that of a wild boar. He adds, that the skin of the back is so thick, that javelins are manufactured out of it. This strange description is almost entirely borrowed from Herodotus, who makes an additional error in asserting that the tail of the hippopotamus is like that of the horse.

It would, however, be inconsistent with our plan and limits to enter into the various descriptions given of the living hippopotamus, or into its osteology. We must be contented with such references to both, as may be necessary to the complete elucidation of our remarks on the fossil species.

There is but one existing species of the hippopotamus hitherto known; but Cuvier has discovered two, and he thinks even four, in the fossil state. The first so nearly resembles the living species, that this great naturalist found some difficulty at first to distinguish them. The second is about the size of the wild boar; the third would seem to be intermediate between the other two, and the fourth, of which some traces have been discovered by the Baron, might have been about the size of a guinea-pig.

For the knowledge of the smaller species we are entirely indebted to the illustrious writer just mentioned, and even for the full and complete authentication of the largest of the fossil hippopotami. One of his most immediate predecessors in this very walk of science has declared, that, in the course of his researches, he could find no proof of the discovery of fossil remains of the hippopotamus, previously to the period in which he wrote. There was not, however, this very absolute dearth of information on the subject; but still the most accomplished naturalists have fallen into a very gross error, in attributing to the hippopotamus certain fossil remains which had no sort of connection with that animal. We have already adverted to the mistake of Daubenton, in referring the molar teeth of the mastodon to the hippopotamus. Peter Camper appears to have

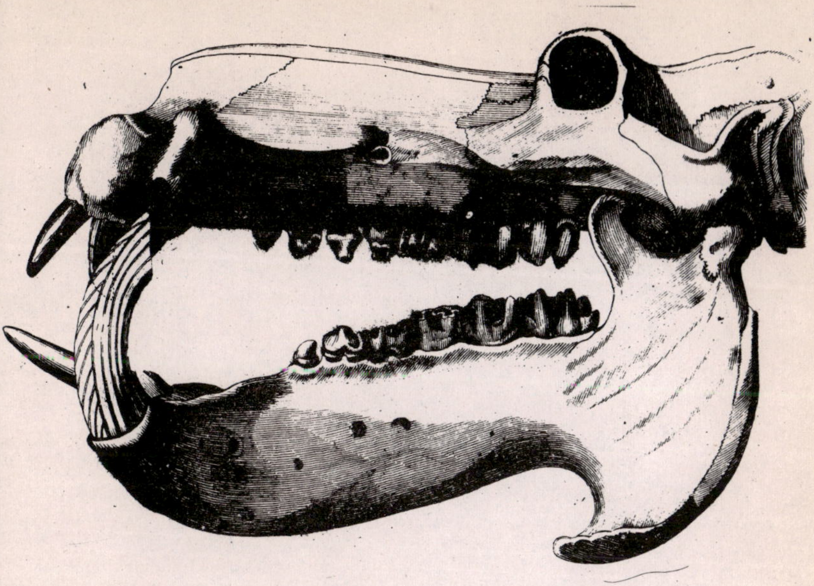

2

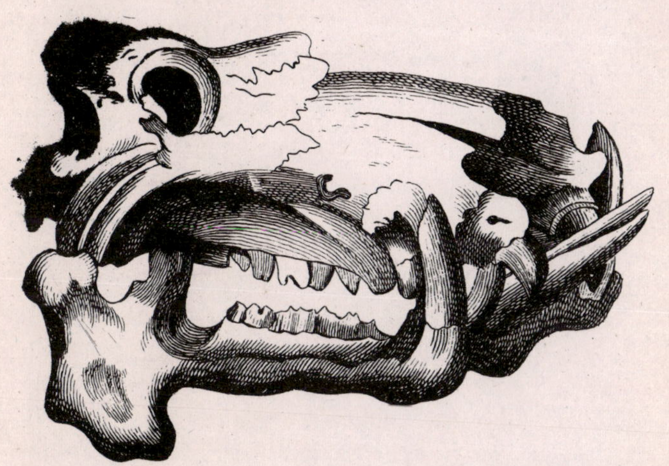

1. Head of existing Hippopotamus.
2. Head of Fossil Hippopotamus found in Kirkdale Cave.

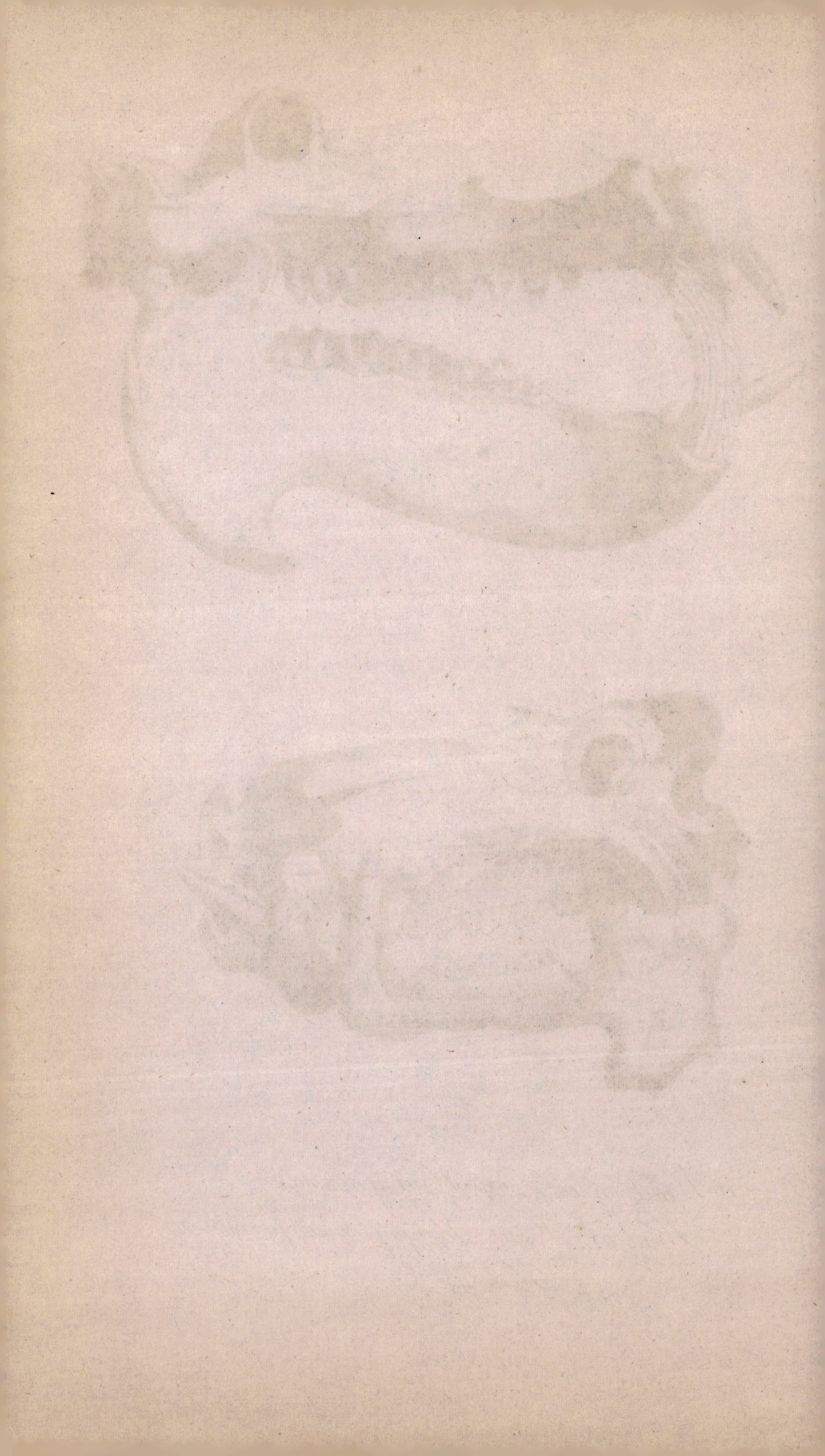

fallen into a similar error in a description, addressed to Pallas, of a tooth in the British Museum. He could not have confounded this tooth with that of the great mastodon, because he has proved himself elsewhere to have been well acquainted with this latter animal. But as he was ignorant of the distinction between the mastodon with narrow teeth, and that of the Ohio, he might have been deceived respecting a single tooth. The teeth of that mastodon, as we have observed in the proper place, at a certain stage of detrition, exhibit some resemblance on a larger scale to those of the hippopotamus, being marked with trefoils in a similar way. But at all events, the tooth in question could not, by any possibility, have belonged to the living hippopotamus, nor to that one usually found in the fossil state, since it is four times as large as theirs.

Many more errors of this description have been committed. M. Merck describes a molar found in the environs of Francfort-on-the-Maine, as belonging to the hippopotamus, but which turned out to be an intermediate tooth of the mastodon of the Ohio, the summits of which were a little worn. M. de Luc speaks of an hippopotamus's tooth, found among volcanic productions in the same neighbourhood, but this M. Merck assures us belonged to the rhinoceros. Charles Nicholas Laing, in a work published in 1708, *Historia lapidum figuratorum Helvetiæ*, has made a still greater mistake, in attributing to the hippopotamus the teeth of the horse. It is singular enough, by the way, that lithologists should have been repeatedly deceived concerning the teeth of this latter animal, notwithstanding that the species is so very common.

Davila, a very modern writer, in the catalogue of his cabinet, describes the jaw of an hippopotamus, with five molar teeth, found in the plaster-quarries in the neighbourhood of Paris; but the Baron, whose authority respecting this *locale* is indisputable, assures us that no remains of the hippopotamus were ever found there. He is persuaded that this same jaw was a fragment of the great *palæotherium*.

Mistakes have also been made on the opposite side : genuine teeth of the hippopotamus have been, by some authors attributed to other animals. Aldrovandus, for instance, attributed to the elephant some of the teeth of this same animal ; and Besler has characterized one of its petrified molars by the simple name of *dens maxillaris lapideus.*

The first specimens by which the Baron was convinced of the presence of the hippopotamus among the fossil remains, were in the Royal Museum, and had been pointed out by Daubenton. These specimens are two in number : the first, a portion of the lower jaw on the right side, containing the penultimate and antepenultimate molar ; the second, a penultimate molar of the upper jaw, in a moderate stage of detrition. A third fragment came from the cabinet of M. Joubert, treasurer of the states of Languedoc. It is a fragment of the upper jaw, containing the last molar and the last but one, in the left side, precisely in that state of detrition in which the figures of the trefoils and other lineaments of the crown render them most easy to be distinguished.

After this the Baron found an astragalus among various fossils collected by M. Miot in the vale of Arno. He observed that its form could not sanction its attribution to the elephant or rhinoceros, nor its magnitude to any smaller animal. And as its figure resembled that of the astragalus of the swine, which of all animals approximates most to the hippopotamus in organization, no doubt was left on his mind as to the propriety of its allocation. However, for still further confirmation, he compared it with the astragalus of a fœtus of the living species, then in his possession, and found no difference except in magnitude.

From M. Fabbroné he subsequently received the drawings of those teeth, which had evidently belonged to the hippopotamus ; one of these was a fragment of a tusk, or lower canine. It may not be uninteresting to mark the difference of the tusks in those animals which possess them. No other animal has tusks formed like those of the hippopotamus. Those of the elephant are larger, but neither angular nor striated; those of the morse, which are also larger, are very much striated towards the root, but not angular; the tusk of the narwhal is straight, but twisted spirally by the striæ of the surface. The texture of the osseous substance is likewise very different. In the elephant we find brownish traits, which cross into curvilinear lozenges, of a very regular form; in the morse there are brown grains, moulded, as it were, into a whiter substance; in the narwhal all seems homogeneous; in the hippopotamus there are five striæ, concentrical to the contour of the tooth.

M. Fabbroni adds, that the diameter of the tusk in question bore a nearer proportion to its length than that of the African hippopotamus, and its spiral curve was also much more marked. These teeth were found in the upper vale of Arno.

The Baron travelled into Tuscany in 1809 and 1810, and found in the museum of Florence, and in that of the academy of the vale of Arno, such an abundance of fossil remains of the hippopotamus, that there could be no difficulty in recomposing a skeleton. He brought back a considerable quantity to Paris, which he had bought from the Tuscan peasants. A skeleton nearly entire, and fragments of eleven individuals, have been in the cabinet of the Grand Duke since 1816. In fact the bones of the hippopotamus are nearly as numerous as those of the elephant in the upper vale of Arno, and more numerous than those of the rhinoceros. They are found confused with both in the small sand-hills, which form the final links of that mountainous chain by which this beautiful valley is engirded.

The bones of the hippopotamus have been found at Rome and in other parts of Italy, also in France and England, but more abundantly in the vale of Arno than elsewhere.

The distinguishing characters of the great fossil hippopotamus are not quite so strongly marked as those of the elephant and rhinoceros, but still quite sufficient, on a comparison of all the bones, to prove a decided difference from the living species. In the fossil *head* the occipital crest is more narrow, the zygomatic arches less separated behind, and that portion of the cranium bounded by these arches on the sides is larger in proportion; the junction of the cheek-bone to the muzzle is made by an oblique line, and not by a sudden slope; the occiput is more suddenly raised, so that the fall of the sagittal crest towards the interval of the orbits is more rapid, and, of course, the vertical height of the occiput is greater. In the lower jaw the interval between the two branches is more narrow.

The Baron possessed five *vertebræ*, none of which were completely similar to their correspondents in the living species.

The articular face of the *shoulder-blade* was more rounded, and had a coracoïd tubercle more blunt and curved inward. This fragment must have belonged to an individual fifteen feet long.

It is unnecessary to pursue this comparison any further, as it could not prove interesting to the general reader. It is sufficient to observe that, notwithstanding the general resemblance of all the bones to those of the living species, there are differences in every one of them, sufficiently marked to warrant a distinction of species.

The little fossil hippopotamus was discovered by M. Cuvier, in a block which was for a long time in one of the magazines of the Museum of Natural History, and the origin of which was unknown. It resembled considerably the osseous *breccia* of Gibraltar, Dalmatia, and Ceuta, except that the paste, instead of being calcareous and stalactitic, was a homogeneous sand-stone, filled with fragments of bones and teeth, which formed an incomparably greater portion of the mass than they do in the breccia.

With infinite labour and care, were at length disengaged from this block the remains of an animal, the existence of which had never been previously suspected.

From a block of the same description, submitted to the disposal of the Baron by M. Journu-Aubert, additional information was derived concerning this fossil species. But no remains were found in the place from which the block was taken.

From these two blocks M. Cuvier obtained nearly all the teeth, which were found in all respects similar to those of the hippopotamus, except that they were one-half smaller in all their dimensions. Certain fragments of the jaw showed indications of that crotchet so characteristic in the lower jaw of the hippopotamus. An astragalus, a scaphoïd bone, a portion of the humerus, another of the femur, a part of the pelvis, also exhibit analogous conformations, but of smaller dimensions, proportional to those of the teeth. The state of dentition and ossification sufficiently proved that this animal was adult, and consequently belonged to a species distinct from that which inhabits the South African rivers.

Some remains of an animal, which the Baron calls the *middle fossil* hippopotamus, were found in a department of the Maine and Loire, in a calcareous tufa, apparently the production of fresh water.

These fragments were not much larger than their analogous parts in the little hippopotamus; but as, in other respects, they had no more resemblance to them than to those of the great hippopotamus, there is no doubt of their having belonged to a different species; at the same time their relation to the hippopotamus is sufficiently marked to refer them to that genus. Some other teeth were found in France, which seem to indicate a species bordering on the hippopotamus, but smaller than the swine; but the Baron did not pass any definitive judgment concerning them, for want of other bones.

The Fossil Rhinoceros.

Singular as is the genus of the rhinoceros, it is yet less isolated in the animal kingdom than that of the elephant. It is nearly allied in its osteology to the daman, the tapir, and the horse; and among the fossils there are many genera to which it exhibits a partial approximation.

The fossil remains of the rhinoceros, though not quite so numerous as those of the elephant, are yet extremely abundant. Both are most generally found together, but the teeth of the rhinoceros, being less voluminous, have been less generally remarked. These animals have not the enormous ivory tusks of the other, which could not fail, in all cases, of attracting attention; and most probably, in many instances, these constituted the motive to form collections of elephantine remains.

Having remarked that the bones of the rhinoceros are generally found in the same strata, and very often in the very same places as those of the elephant, we may observe, that there exist two, or perhaps three, large species, without reckoning one or two much smaller than the others. But, says the Baron, as this distinction is recent, it would be difficult to introduce an historical detail of the particular places where the specific bones were found. It will be sufficient to observe, that the major part of those which have been found in middle and northern Europe, as well as in Asia, belong to the species most anciently discovered, in which the nostrils are separated by an osseous partition; that it is only in Italy that fragments incontestably belonging to the other species, in which the nostrils are not separated by bone, have hitherto been discovered; and finally, that we know nothing of the third large species, and of the small ones, but by some pieces belonging to each, found in a single spot.

The first specimens of the fossil rhinoceros mentioned by writers were found in England, near Canterbury. They are described in the Philosophical Transactions, 1701. In the number were two teeth of the rhinoceros, which the author of the article believed to belong to the hippopotamus.

South of the Hartz, on the side of Hanover, were discovered in 1751 a number of remarkably large bones. They were first believed to belong to the elephant; but *Hollman* showed, by comparison with the descriptions of that animal then published, that this could not be the case. The description of the skull of the hippopotamus, given, in 1724, by Antoine de Jussieu, excluded this animal; and finally, Mickel having compared one of the teeth with those of a living rhinoceros which he saw at Paris, immediately recognized the resemblance, and thus the genus was determined.

Pallas, in 1768, found in the fossil remains accumulated in the cabinet of St. Petersburgh four craniums and five horns of the rhinoceros. The most perfect of the four craniums was without the teeth.

Fifteen years after he published a relation of the astonishing discovery which he had made in Siberia of an *entire rhinoceros* with the skin found buried in the sand on the banks of the Wiluji, in 64° north latitude. He adds the description of a much more perfect cranium than any of the former, found beyond the Lake Baïkal. In many other parts of his travels he speaks of fossil remains of the same species.

There are not less of these bones found in Europe than in Siberia. Besides those already mentioned, Zuckert has published an account of some discovered at Quedlimbourg in 1728, in the same place where the pretended unicorn was discovered in 1663, of which Leibnitz speaks in his Protogea. This same unicorn, by the way, was spoken of before Leibnitz by Otto de Guerike, the celebrated inventor of the pneumatic machine. It was found in a calcareous and gypseous hill at one league distance to the south-east of Quedlimbourg. The bones had been in a great measure broken, until the time in which the remnants were collected and deposited in the abbatial palace. A sketch was then made of the animal, such as it was pretended to have been found entire in the quarry; but a glance is sufficient to show that this sketch was done by very ignorant hands, and taken after parts most incongruously joined together. The bones of the horse seemed to have formed a principal portion of the composition. The bones described by Zuckert, being a considerable portion of the muzzle, a portion of the humerus, a lower tooth, and an unguical phalanx, belong, doubtless, to the rhinoceros with bony partition of the nostrils.

In the country of Darmstadt, on the banks of the Rhine, were found a cranium, and several other bones, accompanied by many bones of the elephant and ox; another in the department of Worms; and a third by Prince Schwartsburg-Rudolstadt, at Cumbach; of all which Merck makes mention in his letters.

It would be quite inconsistent with our plan, and not very interesting to the reader, to particularize every place in Germany where such remains have been found. It appears that, as early as 1786, there had been fragments of at least twenty-two individuals found in that country; and since that period vast numbers of others have been found.

France has not furnished nearly as great a quantity of the remains of the rhinoceros as Germany, or perhaps those which have been discovered have not been so universally described. It will be sufficient, therefore, to observe, though considerable numbers have been discovered, that for the most part, teeth excepted, they were rather in a fragmentary state.

Italy, so abundant in fossils of all kinds, possesses the remains of the rhinoceros in immense quantities. They are found in the Vale of Arno, and most of them appertain to a second species, distinct from that whose remains are most commonly to be found in Germany and Siberia. They principally abound in the Vale of the Upper Arno. They are found in the same strata as the bones of the elephant and hippopotamus, namely, in those clay and sand-hills which constitute, as it were, the first step of the mountains. They have also been found on this side the Appenine chain; but the most considerable and interesting discovery of these bones was made in 1805, by M. Cortesi of Plaçenza, on a hill parallel with Monte Pulgnasco, where he had also discovered an elephant.

The skeleton of the rhinoceros was about a mile from that of the elephant, and the gangue was the same, though at a much greater depth. There was over it at least two hundred feet of sand. An entire head was found there, ten vertebræ, fourteen ribs, two shoulder-blades entire, and two fore-legs, all authenticated to have belonged to a species different from the Siberian.

M. Cortesi afterwards discovered two petrified humeri at some distance, and a lower jaw, in a state of great preservation, on the Monte Pulgnasco itself.

These bones, as well as the elephant's, are in strata filled with marine shells. The two last-mentioned humeri are replete with oysters, and, singular to relate, close to the lower jaw is found the radius of a whale. This would lead us to believe that a part of this stratum had been overturned; for the cetacea discovered by M. Cortesi were in other beds, and much deeper than the last mentioned.

In our own country, the remains of the rhinoceros have been found in vast abundance: at Chatham, near Brentford, in the neighbourhood of Harwich, at Newham, near Rugby in Warwickshire, recently at Lawton in the same county, at Oreston, near Plymouth, in a cavern, (as we mentioned before) &c., all found in the same sort of loose strata, in the blue argilla or gravel; all which Professor Buckland designates as the diluvial detritus.

We may, in fact, conclude that the bones of the rhinoceros are found in almost every country where the bones of the elephant are found; that these two kinds of bones accompany each other, and are found with the bones of other large species; that almost always they are detected under the same circumstances; that their degree of preservation is similar; and that they have been placed by the same geological causes in the position in which they are found.

Without entering very minutely into osteological details, we shall simply mention that the fossil species are four in number; the first was originally called the rhinoceros of Pallas, from the circumstance of its having been first described by that celebrated writer. It is the fossil rhinoceros of Siberia, called by the Baron Cuvier *rhinoceros tichorinus.* We have already noticed the fact of this animal having been found in the sand on the banks of the Wiluji, nearly entire; the probability that these animals inhabited the places where their remains are found; that they were destroyed by the effect of some sudden revolution, or by some change of climate which prevented their further propagation. The long and thick fur with which they were clothed seems to prove that they might have inhabited a cold climate, though not one of the present temperature of Siberia.

This rhinoceros was of a considerably greater size than the two-horned species of Africa. Its head, extremely elongated, seems to have sustained two very long horns, the anterior of which was situated on a vast vault formed by the nasal bones, and consolidated by an osseous, vertical middle partition, which is wanting in the living species. There were no incisors in the jaws; the hair which covered the body was of a brown colour, and particularly abundant on the limbs; while the Indian and Cape rhinoceroses are totally deficient of hair in these parts. It would also appear that the head is not only absolutely much larger, but particularly so in proportion to the height of the limbs, and that the general form of the animal was much more low and compact than that of any living species.

So, then, it was owing to some peasants of Siberia that we happen to know this species of the ancient world, as exactly as we do those of our own times! With a little more precaution, the entire body might have been preserved, as well as the head and feet. Fortunate, however, it is, that the most essential parts of this ancient and singular monument have been saved from destruction.

The second species is the Baron's *rhinoceros leptorhinus*, called by other naturalists, in honour of its illustrious describer, *rhinoceros Cuvierii*. Its remains abound most in those parts of Italy above-mentioned. It had two horns on the nose, no incisors, nor were the nostrils partitioned, as in the last species. All these characters are peculiar to the bicorned rhinoceros of Africa. But the nostrils, in proportion, were much more

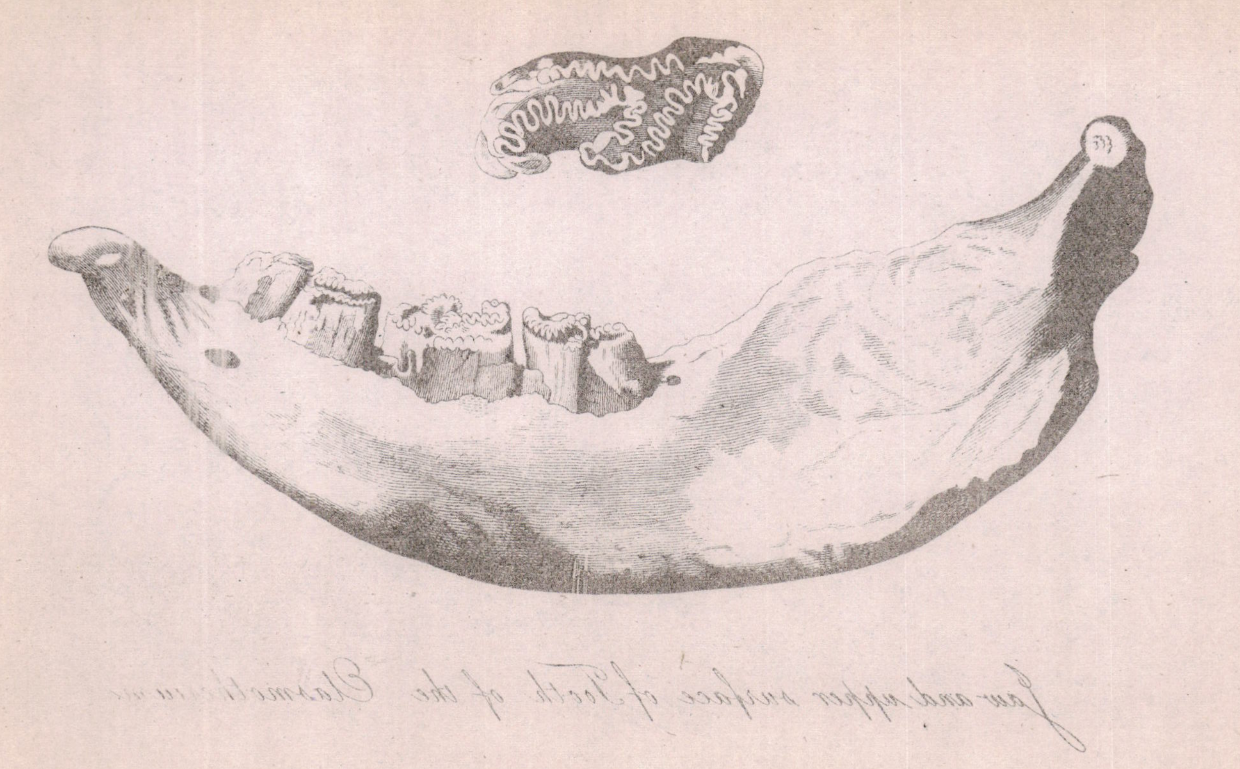

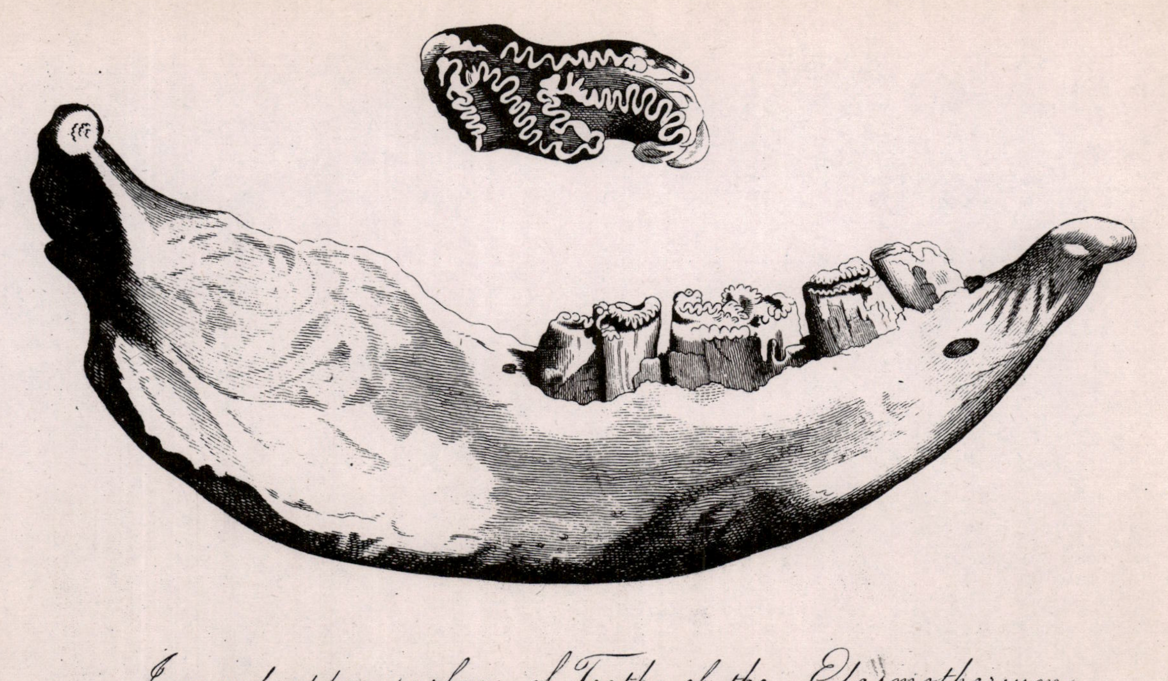

Jaw and upper surface of Tooth of the Elasmotherium.

slender, and the bones of the nose analogously more minute. It appears to have been, generally, lighter, more elevated and less massive in the limbs than the species with partitioned nostrils, and to have had the head less elongated.

The third species is the rhinoceros with incisive teeth *(rhinoceros incisivus)*. This rhinoceros is a species founded solely on the discovery of some incisive teeth in Germany by Camper. As neither the fossil rhinoceros with partitioned nostrils nor the rhinoceros leptorhinus could have such teeth, as their jaws afforded no lodgment for them, the Baron has assigned them to a third species; and though he has found no other bones which he can positively refer to it, he does not hesitate to inscribe it in the list of fossil animals.

The last fossil species of rhinoceros is the *minutus*, also provided with incisive teeth ; but its size could not much have exceeded that of the hog, or one-third of that of the common rhinoceros. This species rests on some teeth and divers bones belonging to adult, and even old individuals, found in 1821 at St. Laurent, near the town of Moissac, in the department of Tarn and Garonne, nearly seventy feet deep, after having successively penetrated the vegetable earth, a strong and compact marl, a bed of gravel, a bed of sandstone, and many others of sand and gravel. The stratum containing these bones had the appearance of our common river-gravel, and also contained the bones of crocodiles and tortoises, and many debris of adult rhinoceroses, some of the ordinary size, and others two-thirds or one-half less.

The state of these bones has led the Baron to conclude that they may appertain to many different species, differing not only in size, but also in many other minute characters, which our present limits will not permit us to detail.

THE ELASMOTHERIUM.

The Elasmotherium is an extinct genus of the pachydermata, known unfortunately but by a single piece. M. Fischer, aulic counsellor to the Emperor of Russia, observed among the presents made to the University of Moscow, (of which he was professor,) by the Princess Daschkaw, a portion of a jaw resembling that of the fossil rhinoceros, but presenting very peculiar characters. He soon perceived that it belonged to a different animal, and published an account of it at Moscow in 1808, and another in 1809.

The general disposition of this jaw is nearly the same as that of the fossil rhinoceros, and it has, likewise, in front, a prominent part without teeth, but not quite so long as in the rhinoceros. The branches where the teeth lay seemed more convex. The lower edge forms an almost uniformly elliptical curve, and does not form underneath a right line, and then an angle, on which the ascending branch rises almost perpendicularly, as in the rhinoceros. The coronoïd apophyses seemed also less elevated, and the ascending branch goes more obliquely backward. The articular facet of the condyle is transverse, rather cylindrical, and a little wider at the external edge, nearly like the rhinoceros.

In this jaw were found four molar teeth, increasing in size from the first to the fourth, and the alveolus of a fifth was just visible. These teeth are prismatic, like those of the horse in his prime, and the bottom of their shaft not yet divided into roots. The length of their coronal is double its width.

What distinguishes the elasmotherium from all known animals is, that the laminæ of these teeth form a very elevated shaft, which grows like that of the horse, preserving a long time its prismatic form, and that they descend vertically through the entire extent of this shaft, not dividing into roots until after a considerable time, while, in other animals, they unite promptly into a single osseous body, which is itself speedily divided into roots ; and also that the plates of enamel are channelled over their entire elevation, and that their section has its edges festooned like those of the transversal bands of the molars of the Indian elephant.

These characters, whatever the age of the animal to whom this jaw belonged might have been, or whatever the complete number of teeth might have been in the animal's full developement, clearly indicate a distinct genus, and also prove that its regimen was more exclusively graminivorous than that of the rhinoceros, and more resembled that of the horse and elephant. It is also extremely probable that it bore, in other respects, very strong relations to the rhinoceros and to the horse, and formed, perhaps, an intermediate link between these two genera. It appears to have been about the size of the former.

The enamel of the teeth is of a beautiful white and very hard. It strikes fire with steel. The osseous substance is yellow at the coronal, brown below. It effervesces with acids.

It is not known in what part of Siberia this precious relic of a former world was discovered.

Horses.

On the fossil horses it would be useless to dilate. Their remains occur in a vast variety of places in the same strata with the bones we have been reviewing. This proves that they were contemporaneous with those extinct pachydermata; but nothing has as yet been observed respecting them, to indicate a distinct species from the existing. The Baron merely remarks that the bones are not as large as the bones of our large horses, but more approaching the size of those of the zebra, &c.

No well-authenticated remains of the *genus* of the swine have been found, except in the peat and other very recent strata, and of course not different from our present races.

The Gigantic Tapirs.

These are the first of a series of fossil animals which exhibited a strong affinity with the existing tapir in the transverse hillocks of a part of the molar teeth, and also in their general structure. We shall not pursue the Baron in his extensive and luminous survey of the osteology of the daman and living tapirs, which, for his purpose, was a necessary introduction to the examination of the rest of the ancient fossil pachydermata, but would suit neither our plan nor limits. We shall proceed at once to these extinct animals, and endeavour to compress our comparisons within as brief a space as possible.

Remains of these gigantic tapirs have been found in a variety of places in the South of France. These remains, with the exception of a radius, consist entirely in molar teeth. These teeth exhibit the closest affinity to those of the existing tapirs in all their forms, but more especially in their transverse hillocks; but they were all of them of dimensions far surpassing those of the teeth of our living tapirs, and clearly indicating species of gigantic size. Still it would be necessary to find the incisors and canines to establish the complete resemblance of the dentition of these animals to the tapir. For, in truth, the tapir is not the only animal possessing the kind of teeth above described. The lamantin and kangaroo are in the same predicament. In the kangaroo, there are two hillocks, and even a line descending obliquely to the external edge, exactly resembling a germ of one of the tapiroïdian teeth above mentioned. The kangaroo also has six molars in youth, and the first compressed and triangular. The lamantin has nine molars, the first of which only is triangular, the others are square, with two crenulated hillocks, just like the tapirs in question, and two heels, one before and one behind.

But the radius which I mentioned before, and which was found with some teeth at Carlat-le-Comte, a small town near the Pyrenees, was sufficient to determine the Baron to decide that the animal to which it belonged appertained to neither of the above genera. Its short and rounded form corresponded to no animal but the tapir, and its size bears a proportion to that of the teeth, nearly analogous to what is observed in living tapirs. Besides, the lamantin has this bone much more triangular; and the bone in question, by its thickness, would indicate a lamantin of a size much too large to have the corresponding teeth; for, in the cetacea, all the parts of the superior extremity are greatly contracted. To bring the radius of the kangaroo into the question would be absurd, the slender forms of which are so totally different.

Every observation, then, concurs to approximate the animals to whom these remains belonged to tapirs; and until it is proved that the incisive and canine teeth do not correspond with those of this genus, they must be referred to it.

The Baron uses the name, *gigantic tapirs*, in the plural number, inasmuch as he observed sufficient discrepancy in size in some of the teeth to warrant him in supposing that there might have been more than one species. That species to which certain of the larger teeth belonged, judging by comparison with living tapirs, especially with that of India, he thinks might have been eighteen feet long and eleven feet high, equal in size to the largest elephants and to the great mastodon of America. The other individuals, whose teeth were discovered at Carlat and Cheville, might have been somewhat less, but most certainly were very formidable animals.

It appears that these gigantic tapirs belonged to the same era as the fossil elephants and mastodons; that they lived with them, and were destroyed by the same catastrophe: their bones are found in the same strata, and sometimes mingled with the others. M. Lockhart observes, that at Avary these bones were found beyond the valley of the Loire, not inclosed in the regular rocky strata; they were in a bed of sand, immediately supported by a calcareous fresh-water formation. This bed is formed of a very variegated sand, composed of small calcareous fragments and rolled quartz, of different sizes and colours. This bed is surmounted by the stratum of vegetable earth.

THE LOPHIODON.

We now arrive at those numerous ancient pachydermata, whose remains are concealed in the bosom of the earth, and which differ more or less from all genera existing at the present day. Accordingly, we find ourselves approaching the deeper strata, those more completely covered by marine formations, and which seem to appertain to eras more remote than those in which the animals we have hitherto been surveying existed.

The species which constitute the genus lophiodon are not, however, so widely removed from the tapirs, though, for the sake of precision, the Baron has separated them. Like the tapirs, they have six incisors and two canines in each jaw, and the greater number of their molars exhibit the same transverse hillocks. But in the first upper molars there are not two of these, but one. In all, the hillocks are more oblique, and the base of the teeth, especially of the last ones, is less rectangular; the molars have three hillocks, instead of two. The anterior have hillocks much more unequal; finally, in some species, these more oblique and arched hillocks approach the crescented form peculiar to the daman and rhinoceros, and thus conduct us by degrees to that most extraordinary genus, the palæotherium.

The remains of the lophiodon have been found in such numbers and variety, and in so many different places, that our readers must pardon us if we do not enter very minutely into their geographical localities, and other circumstances under which they were discovered. They have been found abundantly at Issel, a village in France, along the declivities of the Black Mountain, in the department of the Aude; near Argenton, in the department of the Indre; near Buchsweiler, department of the Lower Rhine; along the eastern declivities of the Vosges mountains; at Soissons, Orleans, the Laonnois, and the Vale of Arno.

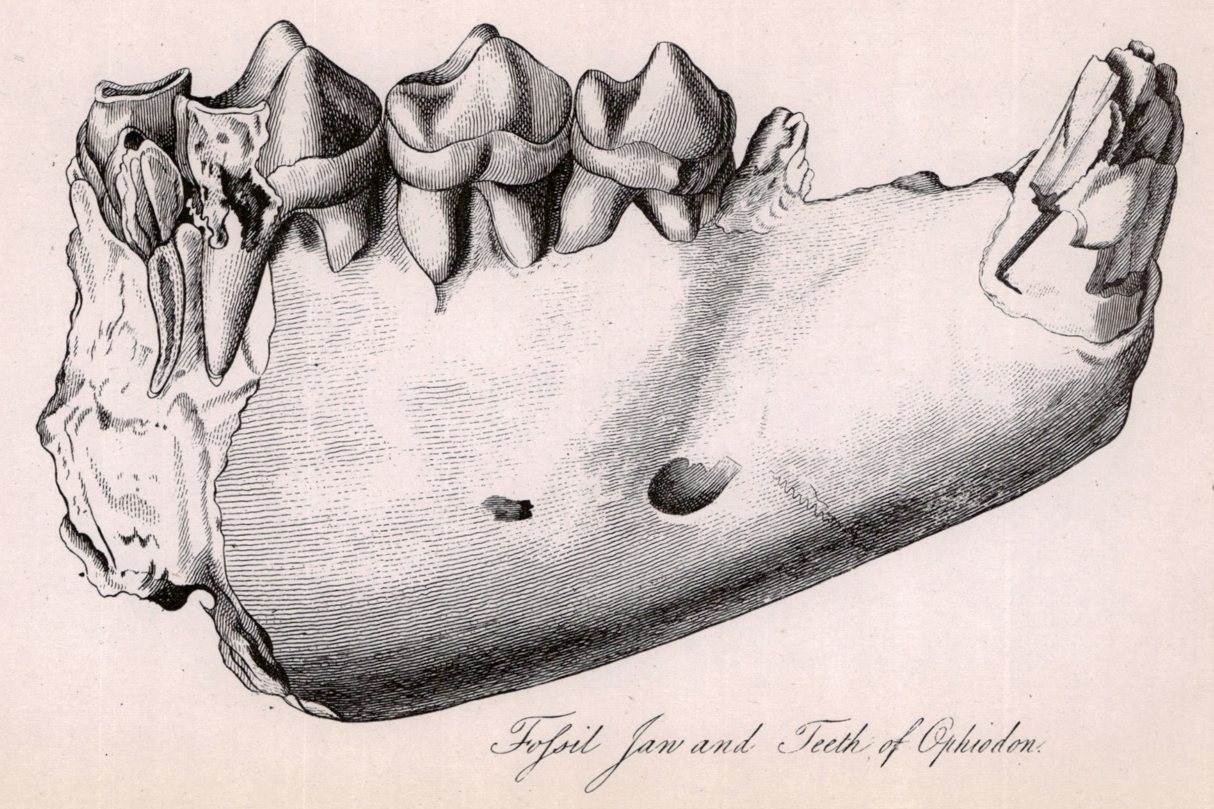

Fossil Jaw and Teeth of Ophiodon.

The generic characters of the lophiodons are as follow :—

Six incisors and two canines in each jaw; seven molars in each side in the upper jaw, and six in the lower, with an empty space between the canine and the first molar, points in which they resemble the tapirs.

A third hillock to the last lower molar. This is wanting in the tapirs.

The anterior lower molars are not provided with transverse hillocks, as in the tapirs, but exhibit either a longitudinal series of tubercles, or one isolated conical tubercle.

The upper molars have their transverse hillocks more oblique, and in this respect resemble those of the rhinoceros, from which, however, they differ in the absence of crotchets on these same hillocks.

The rest of the osteology of the lophiodon, as far as it is known, exhibits close relations with the tapirs, the rhinoceros, and sometimes with the hippopotamus. But it must be remarked, that many essential points of this osteology are yet unknown, particularly the number of toes on each foot, and the form of the nasal bones.

There are twelve species of the lophiodon pretty well determined: three found at Issel, the largest of which was again found at Argenton; three other species at Argenton, altogether distinct from the former; two species at Buchsweiler; one of Montpellier; two of Montabusard near Orleans, the largest of which was of gigantic size, being calculated by the Baron to have been nine feet (French) in length; and one in the Laonnois. Besides which, there was a humerus found in the last-mentioned place, and a pelvis in the Vale of Arno, which M. Cuvier as yet regards as doubtful.

The most important point relative to these animals is connected with the theory of the earth. Their debris, wherever it has been possible to authenticate their localities, are enveloped in rocks or earths, exclusively filled with shells of the fresh water, and which, consequently, have been deposited in fresh water. The animals whose debris are found with theirs, are either terrestrial animals, and, like them, unknown; or crocodiles, trionyces, and emydes, the genera of which inhabit the fresh water at the present day in the torrid zone. Finally, in many situations well determined, these strata are again covered by strata of an origin indubitably marine. Consequently, we find this genus of the lophiodon closely united to those of the palæotheria and anaplotheria, and other unknown genera yet to be described, and which demonstrate the certainty of an anterior state of animal creation, which occupied the surface of our present continents, and which an irruption of the sea overwhelmed, and covered again their debris with rocks of a new origin.

Previously to entering on any account of those more ancient races, for whose discovery and description we are exclusively indebted to M. Cuvier, it may be as well to take a retrospective glance at what we have already done, and present the reader with a short recapitulation of this portion of the subject.

We find, then, that, in the loose or ancient alluvial strata, by Dr. Buckland called *diluvial,* those depositions which fill the bottom of our valleys, and cover the superficies of our continents, there have been discovered, in the order of the pachydermata alone, the remains of seventeen or eighteen species: one elephant, six mastodons, three, or perhaps four, hippopotami, as many rhinoceroses, an elasmotherium, one species perhaps of the horse, and, at all events, one gigantic tapir.

Of all these animals, the horse is the only species that has not been clearly proved to be absolutely foreign to the climates where these remains are found.

The mastodons constitute a genus by themselves now unknown, but very much approximating to the elephant. The elasmotherium is also most assuredly a separate and extinct genus. All the others belong to genera now existing in the torrid zone exclusively. Four of these genera, the elephant, hippopotamus, rhinoceros, and horse, are found only in the

Old World. The fifth, that of the tapirs, is the only one existing in the two continents at the present day. It is only in the old continent that the bones of the tapir, the rhinoceros, and hippopotamus have been found. In the New have been discovered some bones of the elephant; the mastodon, in one species, is common to both.

The species belonging to known genera are distinct from the existing species, and cannot be considered as simple varieties. This point is beyond the shadow of a doubt, as regards the little hippopotamus, the little rhinoceros, the rhinoceros with partitioned nostrils, and the gigantic tapir. It is not quite so obvious in the case of the elephant and the other rhinoceros, but there are more than sufficient reasons to convince the experienced anatomist of the fact. The same conclusion must be extended to the great fossil hippopotamus, and, by analogy, to the fossil horses, although their remains have not yet furnished the clearest evidence of specific distinction.

Such may be considered the osteological result of the researches that have been made on the fossil pachydermata; the geological consequences are these:—

The different remains of which we have been treating, are all embedded in depositions nearly similar. They are often found in company with the bones of other animals pretty similar to those which exist at present. These depositions are, for the most part, loose and incompact, whether composed of sandy or marly substances, and always more or less approximating to the surface of the soil.

It seems probable from this, that these bones were enveloped by the last, or one of the latest, revolutions of this globe.

In a very great number of situations they are accompanied with the accumulated debris of marine animals; in others these debris are wanting, and, in some places, the sand or marl enveloping the bones contains only fresh-water shells.

It does not appear, from any account worthy of the slightest credence, that they are ever covered by the regular rocky strata, which are filled with marine shells. The consequence deduced from this is, that the sea did not rest upon them for any length of time, and that the catastrophe which overwhelmed them was a grand, but transitory, marine inundation.

It is the opinion of the Baron, as we noticed before in treating of the elephant, that this inundation did not rise above the higher mountain ranges. Strata analogous to those covering the fossil remains in question, he declares are not to be found in such situations, nor yet the fossil remains, not even in the more elevated valleys, except in some parts of South America. It is proper to remark, that this opinion has been controverted by Dr. Buckland, to whose work we must content ourselves with referring the reader, without presuming to decide upon the question.

The bones are generally neither rolled, nor are the skeletons found perfect; they are most usually scattered in disorder, and partly fractured. This proves that they were brought from no great distance by the inundation; that they were found by it in the places where they are disinterred; and that the animals to which they belonged must have existed in those places. In a few instances, as we have seen, the skeletons were discovered perfect, and even invested with the softer parts, which proves the suddenness with which, in those instances, they were overwhelmed.

Previously to the catastrophe in question these animals, then, lived in the climates where their bones are disinterred. This catastrophe covered with new strata the bones which it found scattered on the surface of the soil. It destroyed and buried the individuals which it found existing; and as the same species are no longer any where observed, it follows, as a necessary consequence, that their races were totally annihilated.

Species of genera peculiar to the torrid zone existed once in the northern climates; but it by no means follows, that the species now existing in the torrid zone descended from these ancient animals, gradually or suddenly transported towards the equator. They are not all the same, and no well authenticated fact authorises us to believe that such transformations ever take place, more especially in wild animals. Neither is there any decided proof that the temperature of the northern climates has changed very materially, since the era in which those animals existed. The fossil species do not differ less from the living, than certain existing animals of the north differ from their congeners of the south. There is even proof that some of them were destined to inhabit a cold climate, from their having had, like all northern animals, two sorts of hair, and a wool next the skin.

All these results, which hold good respecting the animal remains discovered in the ancient alluvial strata, will not apply to the lophiodons. Most part of their remains, perhaps all, appertain to rocky strata of more ancient date. They form, too, a connecting link between the tapirs and those more extraordinary and ancient pachydermata, which it now becomes our business to describe.

Cuvierian Fossil Pachydermata.

Under this general title we shall describe the remaining extinct genera of this order, characterizing them by the name of their illustrious and indefatigable describer, to whom all our knowledge of them is exclusively due. In doing so, we only mean to offer a small tribute of respect to one, to whose laborious researches, and invaluable services to the cause of science, no language of ours can render adequate justice.

The Baron declares, that when his attention was originally drawn to the subject of fossil osteology, by the accidental view of the remains of the elephant and bear, and the idea suggested itself of applying the general rules of comparative anatomy to the reconstruction and determination of the fossil species, he had little notion that he was then treading on a soil replete with relics of a far more extraordinary description than any he had hitherto witnessed. He little imagined that his labours were destined to bring to light so many entire genera of animals, all utterly extinct and buried for myriads of ages in the bosom of the earth. He had even paid no attention to the partial accounts given of these remains, in the environs of Paris, by certain naturalists, who did not even pretend to determine the species, or seem to have suspected that there was any thing singular about them. Guettard had simply announced their existence: Pralon described the strata of Montmartre in a summary way, and spoke generally of the bones therein contained; Lamanon gave a partial account of a few bones, as did likewise Pazumot. This constituted the amount of all that had been done concerning them, previously to the commencement of our author's researches.

His attention was first directed to them by M. Vuarin, who presented him with a few specimens, which not a little excited his astonishment. He immediately obtained access to the collections of several gentlemen, and every relic he met with there of these bones, excited his curiosity more and more, and determined him to proceed in his inquiries. He finally set about forming a collection himself from the plaster quarries of Paris, and by his liberality to the workmen, and indefatigable zeal, he soon succeeded in accumulating an immense quantity of materials on which to commence his operations.

It was more easy, however, to collect the materials than to arrange them—more easy to accumulate the bones, than reconstruct the skeletons, which was yet the only means by which a just idea could be formed of the species. From the first he perceived that the species whose remains were found in the gypsum were considerably numerous. Soon after he discovered that they appertained to different genera, and that the species of the different genera were often of the same size, so that the relative magnitude would prove rather a source of embarrassment than of assistance. He had in his possession the mutilated and incomplete remains of some hundreds of skeletons, all mixed and confused together, and it was absolutely necessary that each bone should be placed with those to which it naturally corresponded, before any satisfactory result could be obtained —a task the stupendous difficulty of which the reader can easily appreciate. It was necessary, (to use the eloquent language of the Baron himself) that a sort of resurrection in miniature should take place; and he had not at his disposal the all-powerful trumpet, at whose sound the scattered fragments should re-unite, and each resume its proper place. But stupendous as was this task, it was yet accomplished. On the immutable laws prescribed by nature to living beings, he reconstructed those ancient animals, and the voice of comparative anatomy was the trumpet of this scientific resurrection. He has no language, he says, to depict the pleasure he experienced, as he observed, on the discovery of each peculiar character, the consequences which he had predicted from it develope themselves in gradual succession. Thus, for example, the feet corresponded with the peculiarities announced by the teeth, and the teeth with those indicated by the feet. The bones of the legs, thighs, &c., all proved conformable to the judgment he had formed beforehand, from the consideration of other parts. Each species, in fact, seemed, as it were, to be reproduced from a single one of its component elements.

The Baron enters into a very minute and detailed account of the steps which he was obliged to pursue, in the restoration of these monuments of a former age. This, indubitably, is the only plan for enabling the reader thoroughly to appreciate the difficulty, extent, and value of his labours. By this means, too, we are put in possession of the strongest demonstrations of the truth and justice of those principles which conducted him to his conclusions. This part of his work contains a multiplication of examples of the precision with which nature in all cases observes the laws of co-existence, and is of inestimable value to the natural historian. Such are the researches which have raised zoology to the rank of a rational science—which have banished from it those absurd and arbitrary combinations, dignified with the title of systems, and based it on the natural and necessary relations which link together the various parts of all organized bodies.

These researches, also, led to geological conclusions of vast importance. They proved that the sea had covered, for a long period, all that country in the neighbourhood of the French metropolis, and tranquilly deposited there a variety of different kinds of strata: it then abandoned it to the fresh water, which then expanded over the soil in vast lakes. In these lakes were found the gypsum and the marly strata alternating with it, or immediately covering it. The peculiar animals, with whose bones this gypsum is replete, lived on the banks, or in the islands of these lakes, swam in their waters, and fell in there when they died. At a more recent era the sea re-occupied its ancient domain, and deposited sand and marl, filled with shells. Finally, after its last retreat, the surface of the soil, as well in its elevations as its valleys, was again, for a long period, covered with ponds or marshes, which have left thick strata of stone, abounding with shells of the fresh water.

This pecular stone of fresh-water formation, neglected or unknown by geologists, is one of the most remarkable results of the Baron's labours. Its existence has since been ascertained in almost every part of France, but its alternation with the marine strata is no where so evident as in the neighbourhood of Paris.

When animals approximating to those of the Parisian environs are found elsewhere, they are invariably in a stratum of fresh-water formation, but not always in the gypsum. The calcareous depositions of Orleans and Buchsweiler, which contain such remains, also contain abundance of lacustral shells; and those of Buchsweiler are covered, like the gypsum of Paris, with marine coquillaceous strata. This parity of phenomena proves, the vast extent of the catastrophes which produced them.

It may be as well, before we enter on any specific notice of these ancient animals, to say a word on the peculiar state in which their bones were found in the gypsum-quarries.

They were either entire or broken, according to the degree of resistance which they opposed to the pressure of the strata resting on them. The bones of the carpus and tarsus, whose interior is solid, were generally found entire, except in cases where they must have been mutilated previously to incrustation. The femora, the tibiæ, and the other long and hollow bones, more especially those belonging to the larger species, were seldom entire, except in the extremities, which are solid. The skulls were generally broken or crushed, or but one half of them frequently found. As for the skeletons, those of the very small animals were almost always entire, having the ribs and frequently the bones of both their sides. In the animals of middle size, the ribs of one side alone were to be found, and the skeletons of the very large species were almost always disunited. The reason of this appears to be that a longer time was necessary to incrust them with a coat of plaster of sufficient thickness, while a small animal might be incrusted before the tendons were rotten and the bones detached. When the animal was a little large, and resting on one side, the uppermost ribs had time to detach themselves from the skeleton, while the under ones were in a process of incrustation. These bones are scarcely ever worn or rolled, which sufficiently proves that they were carried from no great distance. They were occasionally fractured, and sometimes evidently gnawed, previously to being incrusted, which proves that carnivorous animals existed contemporaneously with the herbivora in question. Nor are they found in a state of petrifaction, but simply fossil, and still preserving, after so many ages, a portion of their animal substance. On analysis they were found to contain sixty-five parts phosphate of lime, eighteen sulphate of lime, and seven carbonate of lime, and they still had a portion of gelatine, as they were blackened by the action of fire.

It is astonishing that, in a country of such extent as that

occupied by the plaster-quarries, more than twenty leagues from east to west, that scarcely any bones but those belonging to a single family should be found. The remains of the small number of species different from this family discovered there, are rare in the extreme. It is not to be doubted that the number of bones of each species found in the fossil state, bears a relative proportion to the number of animals which once existed on the soil, for it is impossible to imagine any destructive agency, that could have overwhelmed or incrusted in the gypsum, the bones of any one species in preference to those of another.

In the present state of the globe we find animals of almost all families inhabiting the countries which compose all our large continents, according to the degree of latitude, and the nature of the soil. But this is not at all the case with the larger islands of the earth. The actual state of New Holland, in particular, may serve to illustrate the probable state of that part of France, which was once inhabited by these ancient pachydermata. Five-sixths of the quadrupeds of that island belong to one and the same family, namely, the marsupialia. There are six genera of these approximating very closely to each other, and having nothing analogous in the animal world, except the didelphes of the warmer regions of America. To these we may add the ornithorhyncus and echidna, which exhibit close relations to the pouched animals. The number of species in these genera are more than forty, and there are but eight or ten species of other mammalia to oppose to them in the whole country.

Here, then, we find in a considerable, but isolated, region, a proportion in the number of its quadrupeds very similar to what appears to have obtained formerly in the country inhabited by those ancient animals. Among a dozen or fifteen pachydermata, we find but two or three carnivora. This resemblance has led the Baron to conjecture, that, at the era in which those animals lived, the country which they occupied was environed on all sides by the sea, and that, like all the greater islands, it possessed its own peculiar population. This even holds good with respect to its vegetation: for the debris only of plants now unknown in Europe are found in the strata of which we are speaking.

The Baron in his grand work on the "Ossemens Fossiles," first gives a geological survey of the environs of Paris, and then a very detailed account of the steps which he pursued in restoring the skeletons of the ancient animals. It would be utterly out of the question for us to attempt to follow him in this narrative of his amazing labours. We must limit ourselves to giving the results in as brief a space as possible.

Of the fossil quadrupeds, whose remains were found in the plaster-quarries, M. Cuvier has formed two distinct genera under the names of PALÆOTHERIUM and ANOPLOTHERIUM. The first approximates to the tapirs, in the number and disposition of the teeth, and more particularly in the conformation of the nasal bones. The second is peculiarly remarkable in not having the canine teeth projecting, and in all teeth forming a continuous series, in the same manner as in the human species.

Two species only of the palæotherium furnished a sufficient quantity of bones, to enable the Baron pretty nearly to reconstruct their skeletons entire, and, consequently, to draw conclusions respecting the forms of these skeletons, respecting those of the soft parts, and to advance some probable conjectures respecting the mode of living of these animals. The greatest number of the other species have only been recognised by portions, more or less considerable, of heads furnished with teeth, or by certain bones of the extremities. The reality, however, of these species is rendered incontestable, by the differences which their debris exhibit, when compared with those of the two restored species.

The bones of the first species of the palæotherium were found, as we said, in the Parisian calcareous gypsum. Since then, other species of the same genus have been found in several parts of France, as well as in the different strata in the neighbourhood of the metropolis.

The characters of the genus palæotherium may be thus described: Six incisors in each jaw, ranged in one and the same line, angular, and tolerably strong; four canines, one in each side in each jaw, conical, and so distant as to cross each other when the mouth was closed; seven molars on the right and left in each jaw, the upper of a square form, with four roots and three crests on the external side, leaving between them two channels — they have a furrow on the internal side; their coronal pretty analogous to that of the upper molars of the rhinoceros and daman, presents, on its external edge, a sort of projecting figure, in the form of a W, to which are united internally two oblique hillocks, proceeding to the two extremities of the W, leaving between them a valley, also oblique, and the entire base of the tooth is surrounded by a cincture. The lower molars show their enamelled outlines in the form of a double crescent, *i. e.*, two crescents, one at the end of the other, more or less oblique. The general form of the head is like that of the tapirs; the nasal bones are very short and slender, jutting out only on the lower part of the nasal aperture, and very probably having formed a point of attachment for the muscles of a small and mobile proboscis. The orbital and temporal foramina were separated above by a well-marked projection, and the first was very small, and less elevated than the second, proving that the eye must have been small, and situated low. The zygomatic arches rather projected. The cranium was very narrow at the elevation of the temporal foramina, which are enormously large. The glenoïd cavity is level, as in the tapirs. The meatus auditorius very small, and not elevated, whence M. Cuvier concludes that the ear was attached very low down. The occipital facet was very small, and the crests of the occiput strongly projecting. Ribs (in one species, ***Pal. Minus)*** true and false, fifteen pair; extremities slightly elevated; cubitus distinct from the radius; peroneum distinct from the tibia; three toes

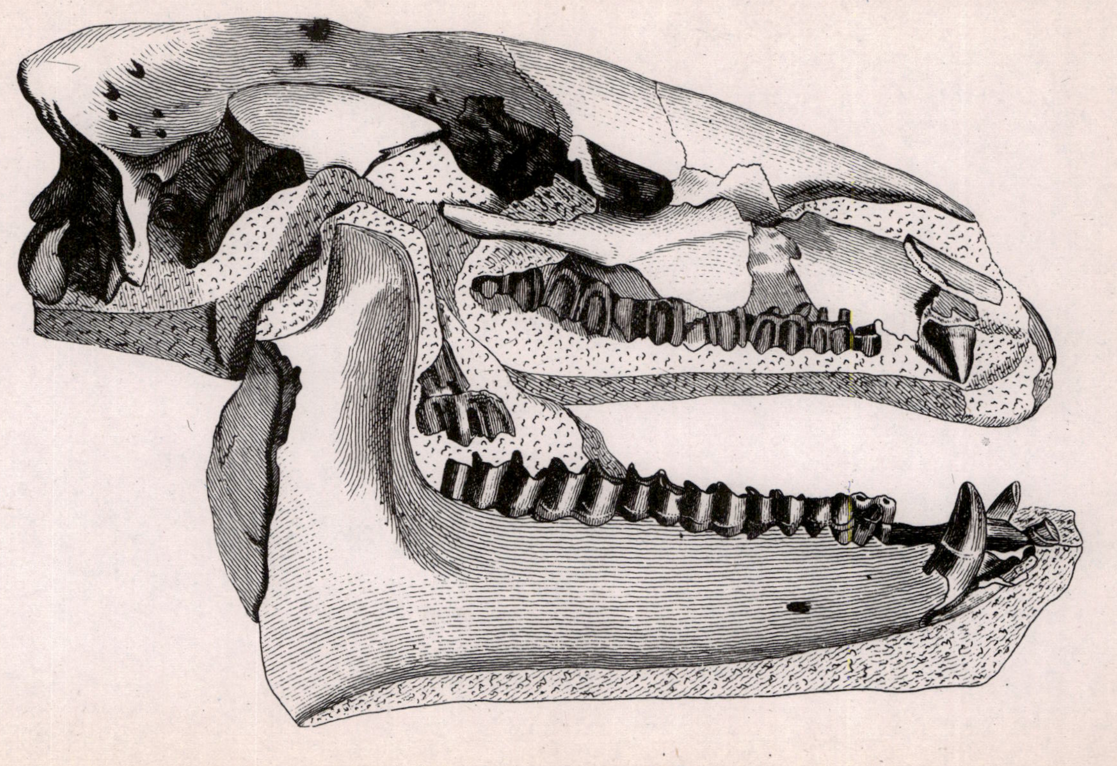

Head of the middle sized Palæotherium.

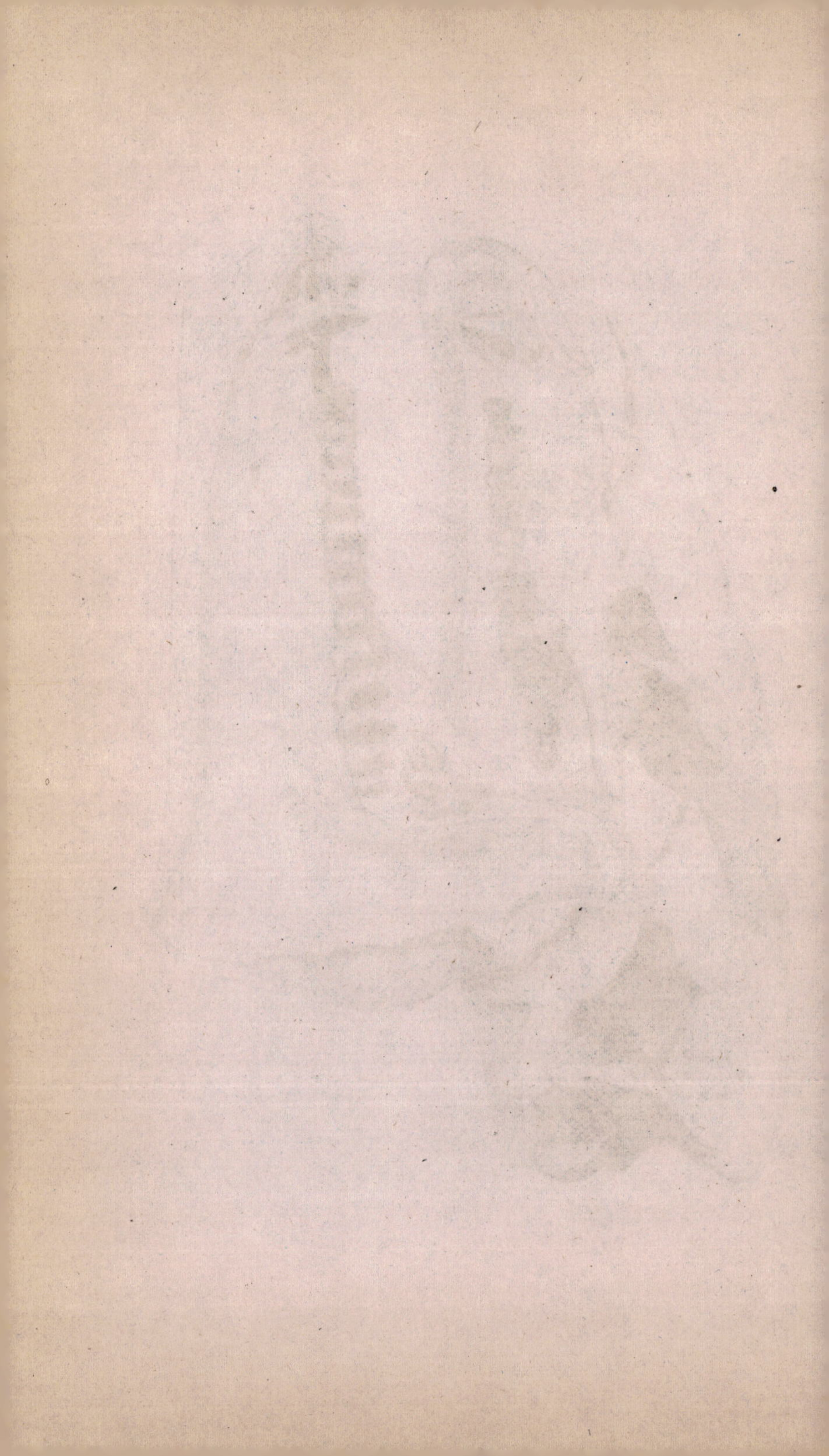

on each foot, the middle one the largest, the two others nearly equal; tail of a moderate length.

The palæotheria in the environs of Paris, says the Baron, do not differ either in teeth or number of toes. It is almost impossible to characterize them otherwise than by size. But among those which have been found elsewhere there are general characters of conformation sufficient for distinction.

The Parisian species are seven in number.

The Great Palæotherium (Pal. Magnum.)—This animal was of the size of the horse. The head and feet have been restored, but the trunk is in a great measure wanting. Of this species, of which the Baron has given a figure with the external forms he attributes to it, it is easy to form a conception. We need only imagine a tapir of the size of a horse, with some differences in the teeth, and a toe less on the fore-foot. Arguing from analogy, the hair should have been close and smooth, or, perhaps, in no greater quantity than it is on the elephant or tapir. It was more than four feet and a half in height to the wither, just the height of the rhinoceros of Java. It was more squat in its proportions and general figure than the horse. The head was more massive, and the extremities thicker and shorter.

The middle-sized Palæotherium (Pal. Medium), was almost the size of a hog. The feet were rather long and slender; the bones of the nose were shorter, from which the Baron conjectures that the proboscis was more long and mobile than in the following species. It resembled a tapir with slender limbs, and might have been, in its own genus, something analogous to what the babyroussa is among the swine. The height to the wither might have been thirty-one or thirty-two inches. In addition to the remains of the head were found the cubitus, the radius, the fore-foot, the tibia, and the hind-foot.

The thick-footed Palæotherium (Pal. crassum.)—This species, of the same size with the last, had the feet shorter and thicker in proportion. It might have been about thirty inches in height, and of all the fossil animals of the Parisian gypsum most resembled the tapir in its conformation, though inferior in size. Of this species there is a head in very great preservation, and the superior and inferior extremities.

The broad-footed Palæotherium (Pal. Latum.)—The fore-arm and the feet alone were found of this animal, which in general conformation seemed exactly opposed to *Pal. Medium.* From the shortness and breadth of its extremities, the Baron judges that it must have been singularly slow and clumsy in its movements. It appeared to hold a similar place in this family with the phascolome among the marsupialia. It was probably not more than from four-and-twenty to six-and-twenty inches in height, but its proportions were as large, and its members as thick, as those of the preceding species.

The short Palæotherium (Pal. Curtum.)—M. Cuvier collected of this species only the head and some portions of the feet, by which he judges that it very much resembled the *palæotherium latum,* but was considerably smaller, not being larger than a sheep.

The small Palæotherium (Pal. Minus) was found almost complete at Pantin, and many lower jaws and feet referable to it were found elsewhere. The pelvis, the sacrum, and the tail remained incomplete, and also the top of the head. But the form of the last may be well presumed from the heads of the other species. " Could this animal," says M. Cuvier, " be as easily re-animated as its bones have been collected, we should behold a tapir smaller than the roe-buck, with slight and slender limbs, for such, to a certainty, was the figure of the animal."

The very small Palæotherium (Pal. Minimum) was only about the size of a hare, and had very small and slight feet. Nothing has been found of it but some bones of the extremities.

A fragment of the lower-jaw of the palæotherium, furnished with teeth, was found at Puy, in Velay, in a gypsous stratum, by M. Bertrand-Roux. But this single fragment was not

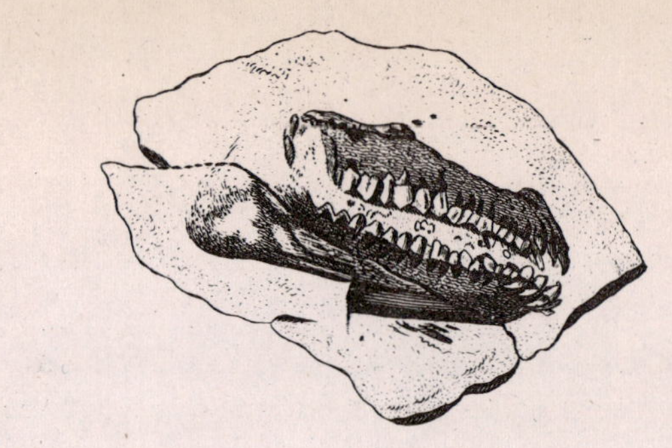

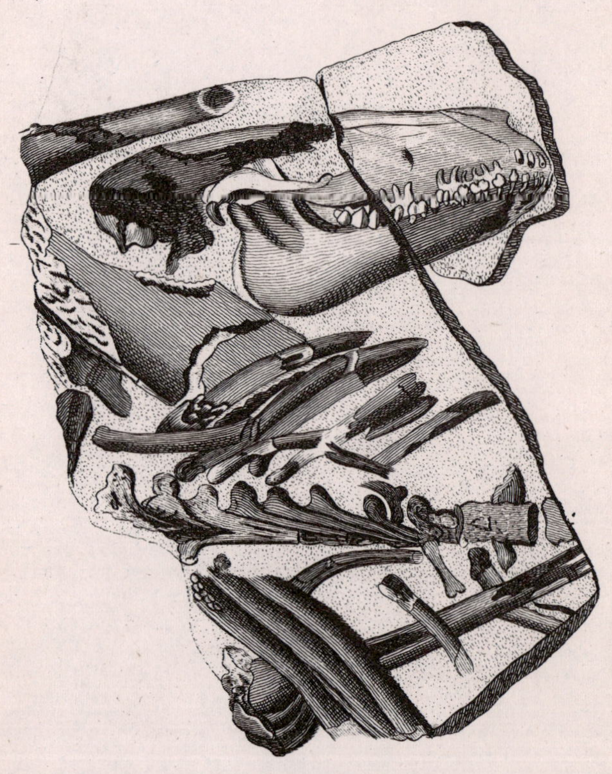

1. Head of the Anoplotherium found at Montmartre.

2. Part of Skeleton of Anoplotherium found at Antony.

sufficient to convince the Baron of the identity of this palæotherium with any of those of the Parisian environs. He rather inclines to believe it distinct. In the environs of Montabusard, near Orleans, where, as we have mentioned, the bones of lophiodons were found, were also found two species of the palæotherium, different from those of Paris. One must have been rather smaller than the *palæotherium crassum.* Some other debris, apparently of this species, were found near St. Geniez, three leagues from Montpellier. A fragment of a left lower-jaw, containing the four last molars, was found at more than thirty feet deep, in a coquillaceous, hard, and compact stone, which M. Cuvier supposes to be of fresh-water formation. Lastly, in the declivities of the Black Mountain, near Issel, were also contained the bones of a palæotherium, extremely similar to that of Orleans, and it is not improbable that some debris found in this last place are referable to the species of Issel.

These species have been called by the Baron, *Pal. Velaunum*, *Pal. Aurelianense*, and *Pal. Isselanum.* The two last differ principally from the others by the lower molars having their intermediate re-entering angle divided in two at its summit.

The ANOPLOTHERIA, up to the time in which the Baron's last edition was published, were found only in the plaster-quarries of Paris. We have noticed, in a preceding part of this article, some subsequent detections of them. They have two characters not observable in any other animal: feet with two toes, in which the bones of the metacarpus and metatarsus remain distinct, and are not soldered together as in the ruminantia, and teeth in a continued series, without any intervening gap. Man alone possesses teeth of this description, whose contiguity is uninterrupted by any vacant interval. The anoplotheria have six incisors in each jaw, one canine and seven molars, on each side, as well above as below. The canines are short, and similar to the outer incisors. The three first molars are compressed, the four others in the upper-jaw are squared, with transverse crests, and a small cone between them. In the lower they are formed into a double crescent, but without collar, or neck, at the base. The last has three crescents. The head is of an oblong form, and does not indicate the existence of a proboscis. The composition of the tarsus is the same as in the camel.

This extraordinary genus, to which there is nothing analogous in existing nature, is subdivided by the Baron into three subgenera: the ANOPLOTHERIA, properly so called, in which the anterior molars are tolerably thick, and the hinder ones in the lower-jaw have their crescents with a simple crest; the XIPHODONS, in which the anterior molars are slender and trenchant, and the hinder ones below have, opposite the concavity of each of their crescents, a point which, in the course of wear, also takes the form of a crescent, so that then the crescents are double, as in the ruminants; and the DICHOBUNES, whose exterior crescents are also pointed in the commencement, and which have, on their back-molars in the lower-jaw, points arranged in pairs.

The *Anoplotherium commune*, so called from its remains being the most usually found, was an animal about the height of a wild boar, but much more elongated in form, and bearing a very long and thick tail. Its proportions much resembled those of the otter, but on a larger scale. It seems probable that it swam well, and frequented the lakes, in the bottom of which its bones have been incrusted by the gypsum there deposited. There is another, a little smaller, but in other respects similar to the last, and called by the Baron, *An. Secundarium*.

As yet but one xiphodon is known. This the Baron formerly called *An. Medium*, but has finally given it the epithet of *An. Gracile*, from the peculiar elegance of its proportions. It was a remarkable animal, of the size and form of the gazelle. The lightness of its form causes M. Cuvier to conjecture that this species lived after the manner of the deer and antelopes;

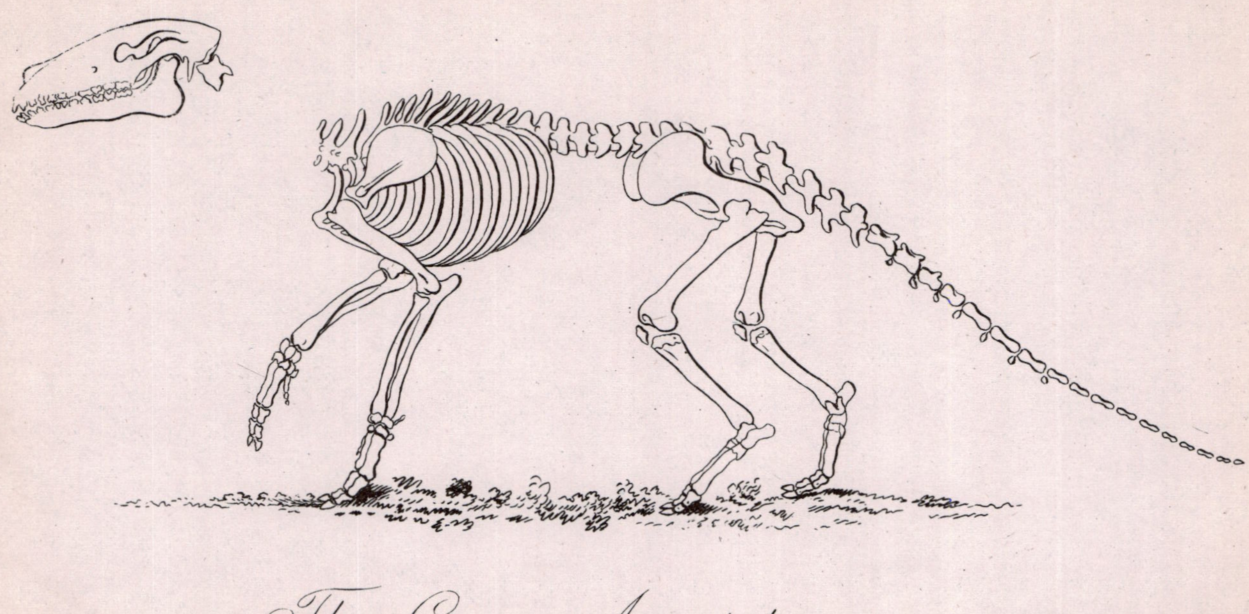

The Common Anoplotherium restored.

Several of the cervical vertebræ have not been found.

that its senses were analogous to theirs, and that it was covered with hairs somewhat similar to those ruminants.

There is one dichobune pretty nearly the size of the hare, and named *An. Leporinum.* Besides its subgeneric characters, it differs from the anoplotheria and xiphodons in having two small and slender toes on each foot, at the sides of the two great toes. It is not ascertained whether these two lateral toes exist or not in the two other dichobunes. They must have been very small species, and scarcely larger than the *aperea.*

Another genus of the pachydermata, found in the gypsum of Paris, is the CHÆROPOTAMUS. This is known only by the teeth and some parts of the head. The incisors, if there were any, are lost. The lower canine is pointed, and tolerably large. Between it and the first molar is an empty space. This molar is conical, pointed, and slightly compressed, but by no means trenchant, and has two thick roots, which separate as they sink into the alveolus. The second is rather more compressed, and has also two roots; and behind its point, which is blunt, are other points, much lower, and scarcely projecting, which form a second lobe. Then come two teeth, which are tuberculous. There are four principal tubercles on the coronal, which is nearly rectangular. In the middle of these tubercles are two smaller ones, and there are some other inequalities about their bases. They resemble the third and fourth molars of the babyroussa, and these teeth, in general, seem to indicate an animal of the swine family. But no known swine has the first molar of this conical form, and the pecari alone has a canine so small as the chæropotamus, and is, besides, a smaller animal than the individual to whom the teeth we have been describing belonged.

From these and some other fragments, the Baron concludes, that the plaster-quarries inclose the remains of an animal more approximating to the genus porcus, than the anoplotherium or palæotherium, and which did not yet resemble precisely the living swine. He suspects that the dichobunes, whose feet so much resembled those of the swine, approached very near this new genus, and, perhaps, formed the link between it and the anoplotheria proper.

The ADAPIS is another of these extraordinary and numerous genera. It is only known by some debris of the head. Its general form appears to have been something like that of the hedgehog, but a third larger. Four incisors were discovered, trenchant, and rather oblique, like those of the anaplotherium; then, above and below, a conical canine, thicker and rather more projecting than the other teeth; the upper one a straight cone, the lower oblique, and couched forwards; the alveolus of the upper was very deep. The molars appear to have been seven in number in each. Six were discovered in the upper jaw, the first trenchant, the second surrounded by a crest, the third apparently so; the fourth and two last were like the hinder molars of the anoplotherium. In the lower jaw, the two first molars are pointed and trenchant, the third similar, but longer and wider; the three following were wanting in the lower jaw discovered. The last is oblong, and seems to have had the tubercles in the form of unequal transverse hillocks. The animal might have been about the size of a rabbit, and approximated to the anoplotherium.

The last of these extinct pachydermata is the ANTRACOTHERIUM. Of this genus two species were discovered in the lignites of Liguria, at Cadibona, near Savone, and a third in the fresh-water formation of the environs of Agen. It is impossible for us to follow the Baron through his account of these discoveries and of their osteology. The jaw-teeth exhibited considerable analogies with those of the chæropotamus and the dichobunes. But besides that these molars presented of themselves specific distinctions, the large and projecting canines with which they were accompanied, left no doubt of the existence of a new and distinct genus. The first species approached to the size of the rhinoceros, the second was considerably smaller, and the third rests upon the fragment of a jaw found in the

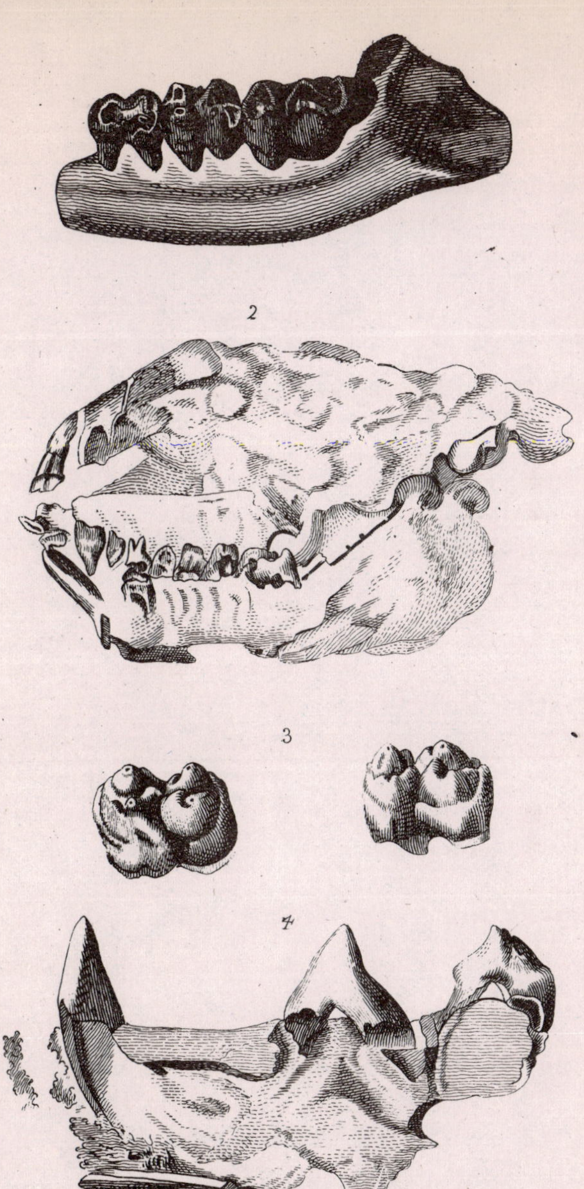

1. Jaw and Teeth of *Antracotherium*. 2 Head of *Adapis*.

3. and 4 Teeth and Jaw of *Chæropotamus*.

department of Lot-et-Garonne which exhibited certain peculiarities.

The genus antracotherium, according to the Baron, held an intermediate place between the palæotheria, anaplotheria, and swine.

This is the place to observe on a very striking fact, which the study of the fossils has served to explain. It was an old observation of the Baron's, that the order of the pachydermata, less abounding in genera than other orders, and in which the genera are less naturally connected together, must have suffered losses to which those intervals that interrupt its series are owing. Accordingly, we find the fossil remains of this order in the most immense abundance, developing new and singular forms to our observation. The living species bear no sort of proportion to the lost. Those shades which approximate genera to each other, those intermediate forms, those steps from one genus to another, so common in the other families of the animal kingdom, are wanting here. It was reserved for the science of fossil osteology to recover them from the entrails of the earth, among the races which completed the grand system of animated nature, and whose destruction has produced such wide and striking intervals. Thus we see that, without the study of the fossils, zoology itself must have remained an imperfect science, and the laws of inter-approximation, on which natural methods are founded, must have still wanted the most complete and satisfactory evidence of their truth.

Fossil Ruminantia.

As the species of ruminantia which occur in the fossil state have been already so accurately described by Major Smith, in that department of the "Animal Kingdom," the reader must excuse me for confining myself altogether to general observations on this part of my subject.

The remains of the ruminantia are excessively abundant

among the fossils, but their study involves a world of difficulty, both in an osteological and a geological point of view. Among the living ruminants, the species are by no means easily to be distinguished; for in this family, though so strongly separated from all others, the inter-resemblance of its members is so great, that naturalists have been forced to adopt parts of comparatively little importance as generic characters: the horns, for instance, an external character, variable in the same species, according to sex, age, and climate, in form and in size, and even under many of these circumstances totally wanting.

It is easy, then, to conceive how difficult it is to pronounce whether any isolated specimen belongs to an existing species or not. If horns, their nuclei, or the frontal bone be wanting, our judgments are always liable to doubt and uncertainty.

The fossil ruminantia are found in the depositions of many different eras. The Baron, indeed, states that he discovered no remains of this kind in the gypsum of Paris, with the ancient pachydermata there incrusted. But they are coeval with the lophiodons, in the calcareous fresh-water formation of Orleans, which also incloses the debris of palæotheria. They become exceedingly numerous in the extensive ancient alluvial strata where the elephant and rhinoceros are found, even as numerous as the bones of horses. The caverns which are filled with the osseous remains of carnivora also contain, at times, the debris of ruminants. Finally, where they most particularly abound is in those apertures which traverse certain mountains in the south of Europe, and which are filled with what we have already described as the osseous breccia.

The ruminants, then, were clearly coeval with the other mammifera of the ancient world; and they existed in a numerical proportion sufficiently great to produce an abundance of their bones in various depositions. But this, which holds true of the order, does not apply to all the genera which compose it. The bones of many species of the deer and ox are found abundantly among the fossils. But as for the bones of sheep, goats, antelopes, cameleopards, camels, &c., the Baron declares, that in his researches he has been able to meet with none. He does not say that, among such abundance of fragments, there may not be some isolated piece belonging to one of these genera, because from such pieces a genus cannot be determined; but he asserts that, in the course of twenty years of constant research, both by himself and others, no frontal bone, no nucleus of horn, no front part of the jaw, no occiput, in short, no bone characteristic of one of these genera has been discovered. This could never have happened had they been but a tenth part as numerous among the fossils as the deer or oxen.

Pallas, indeed, mentions the horn of the antelope among the Siberian fossils in the academy of St. Petersburgh; and at the British Museum, Camper took a drawing of a fragment of the lower-jaw, which he judged to belong to the camel. But there is no authentic testimony that these were genuine fossils, and most probably they got into these respective collections by some mistake.

There is nothing in the actual state of the globe which can explain the absence of these genera among the fossils. Climate will not do it: for the antelopes, like the elephant and rhinoceros, are the natives of warm countries; and the mouflon, the chamois, and the wild goat, like the ox and deer, are inhabitants of the north. Nor is it difference of size, for there are antelopes superior in stature to the stag, and the wild goat and the mouflon are larger than the fossil roebuck; to say nothing of the diminutive rodentia and insectivora, whose littleness did not prevent their detection.

Amongst all these singularities, there is a fact perhaps still more singular than the rest. The fossil ruminants appertain precisely to the genera and sub-genera at present most common in the northern climates; to the aurochs, the musk-ox, the elk, and the rein-deer: while the fossil pachydermata, the elephant, the rhinoceros, the hippopotamus, and the tapir, are limited at present to the torrid zone.

In the loose strata the remains of six species of deer are found, one of which at least, namely, that with gigantic antlers, has totally disappeared from the face of the earth. In the osseous breccia are four others, three of which exist no longer, at least in our climates, and bear no analogy to any of the tribe at present in existence, except the deer of countries far remote from ours.

The regular rocky strata, which inclose the most ancient pachydermata, have also furnished us with one species of deer, that of Orleans. It is entirely unknown at the present day, and may be said even to exhibit characters of almost generic distinction.

We also find the distribution of this genus in the different strata regulated by the same laws as that of the pachydermata. There is a most important observation of the Baron's connected with this subject; and that is, that if the fossil rein-deer was of the same species as the existing, or had the same habits, its coexistence with the rhinoceros in the cavern of Breuges, and with the mastodon near Etampes, renders more and more probable the opinion that these large pachydermata inhabited the countries where their remains have been found. It also proves that these countries can have undergone no very great change of temperature.

A similar observation is applicable to the remains of the ox, which also accompany those of the elephant. It is clearly proved that this genus existed coevally with the pachydermata of the ancient alluvion. It appears that there were at least two species, one with slender limbs, resembling the aurochs, another with more massive members, like the domestic ox, or the buffalo. Craniums have been found of the former, and other bones, in various localities, which the Baron is inclined to refer to the same source; but as the bones and craniums were not found together, it is impossible to speak with absolute certainty on this correspondence. It is also to be observed, that the distinction between these said crania and those of the European aurochs, or the American buffalo or bison, is not yet very clearly made out.

The crania, similar to those of the domestic ox, have not been found, authentically verified, except in the turbaries and other very superficial formations. It is therefore far from improbable that they are of more modern origin than the bones of the elephant and rhinoceros; nay, it is far from unlikely that they might have appertained to the original wild stock from which our native oxen are descended.

As yet no relic has been found among the fossils which resembles any variety of the Indian or Cape buffalo. Consequently, if the fossils are derived from existing species, it is from the species peculiar to cold, and not to hot, climates.

The crania resembling that of the American musk-ox, having been seen but three times, and on the coast of Siberia, there are doubts respecting not only their identity of species, but also regarding the question, whether they are truly fossil, or might have been transported from America, during the thaws, by currents, on floating ice.

Bones belonging to this genus of ruminants have, as we before hinted, been also found in certain caverns, with other osseous remains.

From all the researches which have been made respecting the ruminantia, it appears that some species existed in tolerable numbers, contemporaneously with the elephant, rhinoceros, &c., of the fossil species; but it still is extremely doubtful, that, a few excepted, they can be with confidence referred to species no longer in existence.

The Fossil Carnivora.

Bones belonging to this order are not found so abundantly in the alluvial strata; but, as we have seen before, they exist in immense quantities in the caverns, and are also found in the osseous breccia. The Baron has entered very deeply into osteological details respecting both the living species and the fossil remains, into which, as our limits must prevent our following him, we are forced to content ourselves with giving a brief view of the discoveries which have been made, and shall consider them under the head of each genus successively. We shall begin with the

FOSSIL BEARS.

The vast abundance of the bones of this animal in the caverns of Germany, had long ago attracted the attention of the curious, and many authors on the materia medica have spoken of them under the title of the *fossil unicorn*. The first truly osteological notice of them was given by J. Paterson Hayn, in the "Ephemerides of curious Matters in Nature," in 1672. He describes many of these bones, which he has figured respectably enough, under the whimsical title of the bones of *dragons*. Amongst his figures the humeri of two species are distinguishable, half a pelvis, a portion of cranium, one-half of the lower jaw, an axis, two other vertebræ, and some bones of the metacarpus. These bones were found in the first cavern of the Crapach mountains, not far from a convent of the Chartreux, near the river Dunajek. The same author mentions, in another place, some more bones found in a cavern of Liptov, near the Rag river.

In the same collection there is another notice of these bones by H. Vollgnad, who also terms them the bones of dragons, and even goes so far as to pretend that true dragons were then to be found living and flying in Transylvania. There is, however, accompanying this notice, a good figure of an entire head of our large species of bear, with the convex front. Vollgnad also gives two figures of unguical phalanges, but they belong not to ursus, but felis.

For near a century after we find nothing precise respecting these remains, in an osteological view; nothing, in fact, but an occasional notice of their existence, by some mineralogist or describer of caverns. Some taken from the cave of Schartzfels are mentioned by Mylius; Leibnitz, in his Protogæa, gives three fragments from the same place. Bruckmann, in his description of the caverns of Hungary, declares that their remains do not differ from those of the caverns of the Hartz; and he appears to have been the first who compared them to the bones of the bear.

As a proof of the low state of comparative anatomy in those days, we find Kundmann mistaking two teeth taken from Baumann's Höhle, one for that of the horse, the other for that of a calf, whereas the first belonged to the bear, and the other to the hyæna; and we find Walch attaching to his figures, of half a lower jaw and two canines of ursus, the pleasant observation that they bore a certain resemblance to those of the hippopotamus!

Esper's description of the caverns of Franconia contains a great number of exact figures of portions of the head; and though there is no complete head, yet the fragments are sufficient to distinguish the species from which they come, and which may amount to three or four. This writer, however, from his superficial knowledge of comparative anatomy, multiplied them far too considerably, making them nine in number. Some that belong to ursus he sometimes refers to hyæna, sometimes to phoca. There are, however, fragments belonging to other genera than the bear; some, for instance, to that of the lion or tiger, one of the wolf, and some of the hyæna.

M. Esper says, in a subsequent publication, that, having procured the head of a polar bear, he recognised its decided identity with those of the caverns; and M. Fuch, governor of the pages of the king of Prussia, declares that, having had occasion to see craniums of the fossil and polar bear together, he found the strongest resemblance between them. These assertions only prove how easily the most remarkable forms of skulls may be mistaken; for of all the bears the polar is precisely the one that has the least resemblance to the fossil.

Accordingly we find that celebrated anatomist Camper, in a very early stage of his researches, putting a most decided negative on this pretended identity. His principal reason is, the want of that little tooth, which the common bears, and the polar among the rest, invariably have behind the canine. But as there were many other reasons for this negative, and many still more convincing, it became a matter of interest that some one should employ himself in collecting them. This was done by M. Rosenmüller, an anatomist of Leipsic, first in a Latin description, published in 1794, and afterwards in a little German book, called "Materials for the History and Knowledge of the Fossil Bones," in 1795. He gives a figure of a complete head of the large fossil bear, with convex forehead, the lower jaw of which appertained to an individual of larger size. This cranium came from Gaylenreuth. M. Rosenmüller enters into a careful comparison of this cranium with that of the brown bear, and with Pallas's description of that of the polar bear. The result of this proves the three animals to have been totally different. But the author makes no mention here of the other bones of this bear, nor of the other species of ursus, with whose bones its remains are intermingled. In 1804, however, he published, in French and German, a much more detailed description, with very numerous figures, of the fossil osteology of the bear.

Peter Camper seems to have been the first who recognised any distinction between the fossil species among themselves. His researches were followed up by his son Adrien. M. Blumenbach expressly distinguishes two; that most anciently known he calls *ursus spelæus*, and a second, *ursus arctoidevs*, because he found in it much more resemblance than in the first to the brown, or rather black, bear of Europe. These two last, as is known, were confounded by Linnæus, under the name of *ursus arctos*.

Such was the state of ursine fossil osteology up to the first publication of Baron Cuvier's work. Though remote from the actual localities of these bones, he was fortunate enough, by his access to valuable collections, and the assistance of his friends, to be soon enabled to treat the subject in a manner infinitely more complete than any of his predecessors had done.

It is not only in the caverns that the bones of bears are found. Similar remains, though comparatively few in number, are found in the loose strata. Many specimens, for example, have been found in the Vale of Arno; but the Baron declares them different in species from the cavern bears.

Dr. Buckland found, in 1820, in the collection of the convent of Krems-Munster, in Upper Austria, certain crania and bones, which he judges to have belonged to the large fossil species with convex chaffron. They had been disinterred from a gravel-quarry which had been consolidated into a pudding-stone, employed in that part of the country for the purposes of building. The bones of bears have also been found in the cavern of Oreston, near Plymouth, which we mentioned in an earlier part of this essay.

As to the osteology of the fossil bears, we shall only trouble our readers with a few remarks on the teeth and crania, these constituting the most material specific characters.

The teeth of which the caverns of Germany have furnished so many myriads, have been clearly proved to have all the generic characters of the bear. The first point which indicates difference of species is their magnitude. The largest living teeth are either less than, or, at the most, but equal to, the smallest fossil; and, in general, one-fourth smaller than the largest. This constant superiority was a sufficient indication of a difference and a superior size of species, which the other parts have since confirmed. Those other parts have also proved what the teeth alone could never do, at least not very clearly, namely, that the remains of more than one species of bear existed in these caverns.

These teeth are, in general, less worn, and have preserved their enamel and their eminences better than those of the living bears, which proves that the species from which they came were more exclusively carnivorous.

Among the fossil crania, we find the cheek teeth worn only in the oldest and largest subjects.

But a more marked difference between the fossil and living crania, is in relation to the little molar situated immediately behind the canine, both above and below, and to the first of the molars in the series of the upper jaw. The little molar aforesaid is never wanting in the living bears at any age, and has never yet been found in the fossil of the large species, young or old, in the upper jaw. It exists, however, in the cranium of the inferior species lately described by M. Goldfuss; nor is it always wanting in the lower jaw of the other. The other difference regards the second little molar in the upper jaw, which in the living bears is immediately situated in front of the antepenultimate, and forms, with it, a continued series. Its alveolus has been found by the Baron but twice, in fragments of craniums from Gaylenreuth and Sundwich; but there appears no vestige of it in any other fragment, nor in any of the entire craniums seen by the Baron, or any body else. From this he concludes, that these bears had usually but three cheek-teeth above, in a continued series, and but thirty teeth in all; the living bears have generally thirty-six, and sometimes forty.

Still, however, the circumstance of these little teeth being occasionally found at all, seems to prove that the specific character of these bears was, that they lost these teeth early, not that they wanted them altogether.

The most common cranium found in the caverns having decidedly all the generic characters, and being in perfect correspondence with the structure of the teeth, all that remained was the determination of the species to which these craniums belonged. These species were at least two in number.

The first is characterized by a prominent and convex chaffron, a strong elevation of the forehead above the root of the nose, and two convex prominences on this same forehead, while no bear has so flat a forehead as the polar, to which, as we have seen, the fossil has been absurdly compared.

These crania are also remarkable for the great projection

and prompt approximation of the temporal crests, and also for the length and elevation of the sagittal, both indications of vast force in the crotaphite muscles. Now, in the polar bear, these parts are the least strongly developed. In these points the black bears, both of Europe and America, more approach the fossil; but they are as remote from it as the others, by the flatness of the front of the head.

In the peculiar serpentine line of the profile, the brown bear approaches to the fossil, and so does that other bear, which Shaw placed among the sloths; but, in other points, there is no sort of comparison.

The fossil head measures one-fourth more, from the spine of the occiput to the incisives, than the largest of the living heads; one-third more than that of the polar bear.

These peculiarities might be supposed to attach to the age of the individual, but they have been found to hold good in subjects that other characters evidently proved to have been young.

Other crania have been found equally large, but less convex, than the last; and a third one, smaller, and partaking more of the characters of our brown and black bears.

From all this the Baron concludes, that these caverns have furnished three distinct forms of the adult head. Those with convex forehead; those equally large, but flattened; and a smaller one, resembling the brown bear.

It is sufficient to observe, without entering into further details, that the entire osteology of the fossil bear justify the following conclusions:—

The most common bones in the caverns belong to this genus.

The largest of the crania, and some of the other bones, present such striking differences, that they must be regarded as belonging to species distinct from those of the present day.

Among these large crania, some are less convex, and, most probably, belonged to a different species from those which are more so.

Among the other bones, there are decidedly found those of at least two species.

Some bones of one of these more resembled those of our living bears than some of the other. A humerus, &c., in particular, were scarcely distinguishable. This was also the case with some other small bones belonging to both species.

The bones of bears are also found in the loose strata. Those observed in Tuscany differ from the cave bears, and approach more to the brown.

To the large species with convex front, the Baron gives the name of *Ursus Spelæus*. The large species with flattened front, he hypothetically terms *U. Arctoideus*. The one with small cranium he calls *U. Priscus*, and the one of Tuscany he first named *U. Etruscus;* but afterwards, on account of its compressed canines, he changed the name to *Cultridens*.

The Fossil Hyæna.

So much is already known and written concerning this animal that we must be brief. One hyæna has most assuredly been very abundant among those ancient animals which we have been describing. Its bones have been found not only in the caverns where we have seen the remains of the bear so abundant, but also in the alluvial strata containing the debris of elephants. There are sufficient proofs in print of the very ancient existence of some one species of this animal in three different places in Germany, in the cave of Gaylenreuth, in the sand-hills near Eichstadt, and in Baumann's Höhle. For this we have the testimony of figures given by Esper, Collini, and Kundmann, though all three mistook the animal to which the bones they figured had belonged.

Hyænas' bones have also been discovered in other parts of Germany, and in France; but by far the most abundant depôt of them was found in the cave of Kirkdale, which we have already noticed.

The Baron is satisfied, from an examination of vast numbers

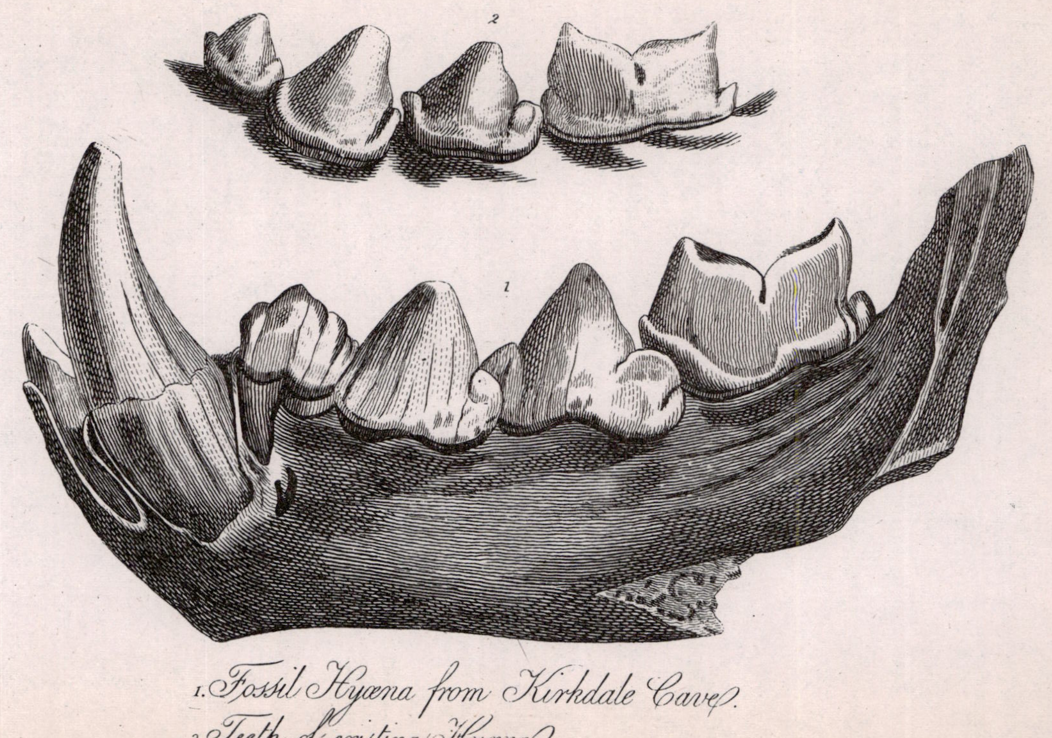

1. Fossil Hyæna from Kirkdale Cave.
2. Teeth of existing Hyæna.

of these bones, that they belonged to a species different from those which now exist.

The habits, however, of this ancient hyæna exactly resembled those of the modern hyænas. A passage of Busbequius, descriptive of their habits, has been quoted by Dr. Buckland: "Sepulchra suffodit, extrahitque cadavera portatque ad suam speluncam, juxta quam videre est ingentem cumulum ossium humanorum, veterinariorum et reliquorum omne genus animalium."

If the Kirkdale hyænas did not accumulate human bones with those of the herbivora abounding there, it is very clear that the human species did not exist in the time and place in question.

The bones of hyænas have been found in other caverns of England, such as Oreston and Rugby, with the large bones already described, proving the coeval existence of this carnivorous animal with the great pachydermata in our island, as well as on the continent.

It is enough to observe, in a few words, that the fossil hyæna was nearly one-third larger than the largest of the existing species. In the structure of the teeth it more nearly resembled the Cape than the Abyssinian hyæna. The muzzle was also shorter and stronger than in either of the last, and the bite must consequently have been more powerful.

We shall conclude by extracting from Dr. Buckland the various localities on the continent in which hyænas' bones were found.

In the caves of Muggendorf, in Franconia, with bears and tigers.

In the Hartz forest, with similar bones, in Scharzfield, and Baumann's Höhle.

At Sundwich, in Westphalia, with the bones of carnivora, and some remains of deer and rhinoceros.

In France near Fouvent, in the department of Doubes, with remains of elephant, rhinoceros, and horse.

At Kostritz, in the valley of the Elster, in Saxony, with carnivorous and herbivorous bones.

At Candstadt, in the valley of the Necker, in Wirtemberg, in the same company.

In Bavaria, on the west base of the Hartz Forest, and in the Val' d'Arno.

In the four last-mentioned places they were embedded in the ancient alluvial strata.

Fossil Felinæ.

One very large animal, and another less in size, of the genus Felis, have left their remains in the caverns and in the loose strata. Proofs of this, as far as respects the caves of Hungary, we find as long ago as the memoir of Vollgnad, mentioned in our last article. Also from the cavern of Schartzfels we have a portion of cranium, represented by Leibnitz in his Protogæa. According to M. Sœmmering, this cranium entirely resembled that of a lion of middle size, and differed in no less than thirty-six points from that of the cavern bears. But most of these have as much relation to the genus felis in general, as to the lion species in particular.

In Esper's figures from Gaylenreuth, there is one-half of the upper jaw and many teeth easy to be recognised as belonging to felis, and the resemblance of which this author himself had recognised. M. Rosenmüller, in his treatise of the bear, mentioned before, announced that he should soon publish a work containing a description of the bones of an unknown animal of the lion family, and adds, that those bones appeared not exactly to resemble those of the existing lion. M. Goldfuss, in his description of the environs of Muggendorf, has given a figure of a complete head, evidently of the feline genus, but of an unknown species. He says that in Gaylenreuth the isolated bones and teeth of felis are not more rare than those of hyæna. In Kirkdale, Dr. Buckland only found two teeth, between which and those of the existing lion the Baron can discover no difference.

The Baron, as we have said, has decided on there having been two extinct species of felis. The first and largest he calls

Felis Spelæa; the second and smaller, *Felis Antiqua.* It is totally unnecessary, and it would be quite uninteresting, for us to enter into the osteological distinctions of those from the living felinæ, which differences, though sufficient for the naturalist, would not, perhaps, be deemed very striking by the general reader. It is sufficient to remark, that M. Cuvier has discovered in these remains a closer analogy to the jaguar than to any other living felis.

The rest of the fossil carnivora may be soon despatched. Bones of the wolf, the fox, the glutton, the weasel, the genet, and other small carnivora, have been found in the same situations as the animals we have been describing. It must be observed, however, that in no case have these bones been clearly established as belonging to species distinct from those now existing. Neither can we say that, in every instance, identity with the existing species has been proved. It is certain that these bones, be they what they may, are in the same state with those of the bears, the hyænas, and the felinæ; they have the same colour, the same consistence, and are similarly embedded. Every thing indicates that they must all be referred to the same era, and that they were all overwhelmed by the same catastrophe. The differences between them and their living congeners principally consist in relative magnitude. Two teeth were found at Avary, in France, with the bones of the mastodon, rhinoceros, and tapir, which seemed to prove the ancient existence of canis; but one of gigantic size. The Baron calculates, that it could not be less than eight feet from the extremity of the muzzle to the root of the tail, and at least five feet in height to the shoulder. But, in general, it may be said of these animals that the resemblance to the living species is much stronger than the differences. The same observation is applicable to the same genera of carnivora whose remains were found near Paris, with few exceptions. One large animal, however, of this order was there found approximating to the racoons and coatis, but certainly not referrible even to any known genus.

But the most remarkable of this order discovered in those rich depositaries of the earlier works of animated nature, the gypsum-quarries of Paris, was a small species of SARIGUE, an animal the family of which is now confined to the tropical regions of the New World, and to that newer world, Australasia. The reflections of the Baron on this head are so just and striking, that we cannot deny ourselves the pleasure of a short extract.

"The rich collection of the bones and skeletons of an ancient world, which nature appears to have assembled round our city, and reserved for the instruction of the present age, is doubtless a most striking phenomenon. Some new relic is discovered every day; every successive day presents new materials for astonishment, and additional demonstration that nothing which once constituted a part of the past population of our soil constitutes a part of the present. There is little doubt, too, that proofs of this description will multiply in proportion as more interest is taken in, and more attention given to, their production. There is scarcely a single block of gypsum in certain strata that does not contain bones. How many millions of these bones have been destroyed since those quarries first began to be worked, and the gypsum to be used as materials for building! How many, even at this present moment, may not be destroyed by negligence, and how many, by their minuteness, may escape the eye of the most attentive collectors! The fragment in question is a proof of this. The lineaments imprinted in the gypsum are so slight, that it requires the very closest inspection to be enabled to trace them; and yet how valuable are such lineaments! They bear the impression of an animal, of which we find no relic elsewhere, in the same district; of an animal which, buried perhaps for myriads of ages, now re-appears for the first time to the eye of the naturalist."

The impression of the skeleton was found nearly complete on two stones, one covering the other, and thus dividing it, as it were, between them. The animal is there, or its outlines rather, nearly in the natural position. That it belongs to the genus of the sarigue, now exclusively appropriated to America, the Baron has proved, by a long and scientific comparison of it with the different genera of this numerous and extraordinary family in both America and Australasia. But whether it is an extinct or a living species he has not, from our imperfect knowledge concerning this family in general, been able completely to determine. He is, however, decidedly of opinion, that it belongs to no species of that family of which we know sufficient to furnish data for a satisfactory comparison.

Fossil Rodentia.

It does not appear that this order bore a less relative proportion to the other animals of former worlds, than it does to the population of the present. The majority, however, of the species were small, as are those of the present day, and it is only under peculiar circumstances that their remains have been remarked and collected. In fact, they have been generally found incrusted in stones, or in such concretions as have preserved them from decay. The genus of the castor alone seems to have, from its magnitude, escaped destruction under other circumstances, and some remains of it have been discovered in the loose strata.

We shall simply notice, without dwelling on them, that two species of the RABBIT have been found in the osseous breccia of Gibraltar, Cette, and Pisa; remains of the LAGOMYS in that of Corsica and Sardinia; and of the CAMPAGNOL in that of Sardinia, Corsica, and Cette: besides two species of the LOIR, or dormouse, in the plaster-quarries of Paris. Without entering into any specific account of these, we shall at once proceed to the fossil rodentia of other localities. The small bones of the caverns have been generally too much neglected. This, however, has not been the case with the cave of Kirkdale. Dr. Buckland, in his account of that, particularizes the bones of rabbits, of campagnols, and of mice found in this cavern.

Those of one species of campagnol (the *hypudæus*), about the size of a water-rat, are there in the most immense abundance. In these bones the generic characters of the campagnol, more especially of that particular subdivision to which the water-rat belongs, are easily to be recognised. Still, with the exception of the jaws and teeth, all the other bones are somewhat smaller, which causes the Baron Cuvier to consider the species as not the same. This campagnol of Kirkdale is found, on comparison, considerably smaller than those of the osseous breccia of Sardinia, Corsica, and Ceuta.

There are also in this cavern campagnols of another species, not exceeding in size the *mus arvalis.* There are also teeth found there which indubitably appertain to the genus of the rat, properly so called. There are, besides, some bones which Dr. Buckland gives as those of the rabbit, but M. Cuvier is more inclined to think (if not of an unknown species) to belong to the hare.

In the turbaries the bones of castors have been found, but in such formations scarcely any but the bones of indigenous animals are preserved. This animal formerly inhabited all the great rivers of Europe, as it does many of them still. It is not, therefore, surprising that its bones should be found in peat formations, and preserved by the same causes by which the aquatic mosses are preserved there. This proves nothing with respect to the antediluvian existence of this genus.

In the loose strata, however, near the sea of Azof, in the neighbourhood of Taganrok, was found the head of a castor, apparently of a lost species. M. Fischer, indeed, attributed it to a lost genus, which he named *Trongotherium.*

The teeth and all the forms of this head possess the characters of the castor, but the head is about a fifth larger than our European castors, which exceed the American in size. In the whole order rodentia no animal has a larger head, except the cabia. Another head was found near the lake Rostoff, inferior in length, incontestably belonging to a castor, and fully agreeing in all the details with the existing castors. The stratum in which it was found is unknown.

It is singular enough that among the innumerable fish that, in various situations, fill the laminæ of the calcareous and marly schists, are found, though very rarely, some viviparous quadrupeds belonging to the order rodentia.

The most numerous and considerable have been taken from the celebrated quarries of Œningen, where it was for a long time imagined that no animals were incrusted but those indigenous to the country.

Three species of rodentia have been drawn from thence: one is the *domestic mouse*, of which M. Karg assures us he has found several individuals. Another is the *muscardin*, of which there is one individual in the cabinet of Mersbourg: it is five inches long, but its limbs are wanting, and is so bent and compressed, that its determination is next to impossible.

There is a third preserved in the collection of M. Ziegler, at Winterthur, which is thought to be an *aperea*. In short, we we may with safety say that, concerning these rodentia of the *fissile* strata of Œningen, nothing can be learned with the least certainty. Let us pass on to the

Fossil Edentata.

As yet but one genus, and, at most, but two species, belonging to this order have been discovered in the fossil state. But this genus has the closest and strongest analogy with the living genera of the edentata. The animals of which we are about to treat belonged to the family of the Tardigrada, constituting a distinct genus, to which M. Cuvier has given the name Megatherium. It comprehends two species, the *megatherium*, properly so called, and the *megalonyx*.

The skeleton of the first of these animals is known almost entirely, and its examination proves that it had more analogy to the sloths than to any other living beings, especially in regard to the system of dentition, to the form of the head, and the composition of the extremities.

As for the megalonyx, but one tooth and a few bones belonging to the limbs have as yet been collected; but those relics are sufficient to prove its approximation to the megatherium, though it must be considered specifically different.

Both were at least as large as the ox. Their limbs were robust, and terminated by five thick toes, of which only some were provided with an enormous claw, arched and crooked like the claws of some tatous, ant-eaters, and bradypi. The megatherium, of which a clearer idea may be formed than of the megalonyx, had a small head, short muzzle, terminated, perhaps, by a short proboscis, the mouth furnished only with molars, whose coronals were marked with transverse hillocks. The neck was moderately short; the body voluminous and heavy; the limbs extremely robust, and the anterior ones provided with powerful clavicles. Recent observations seem to prove, that if it had analogies with the bradypi, in the forms of the head and dentary system, and with the ant-eaters in the conformation of the extremities, it also resembles the tatous in the nature of its teguments. Its skin, thick and, as it were, ossified, was divided into a number of polygonous scales, approximating one to the other, like the pieces which enter into the composition of mosaic work.

The form of the molars, and the size of these animals, seem to indicate that they fed on vegetables and roots. The conformation of their limbs shows that their walk must have been slow and equal. Their debris have been found only in America.

The megatherium, sometimes called the *animal of Paraguay*, was discovered towards the end of the last century. The skeleton, almost entire, was found nearly at one hundred feet of depth, in excavations made in the midst of an ancient alluvial stratum, on the banks of the river of Luxan, a league south-east of the town of the same name, which is three leagues west-south-west of Buenos Ayres. It was sent to the museum

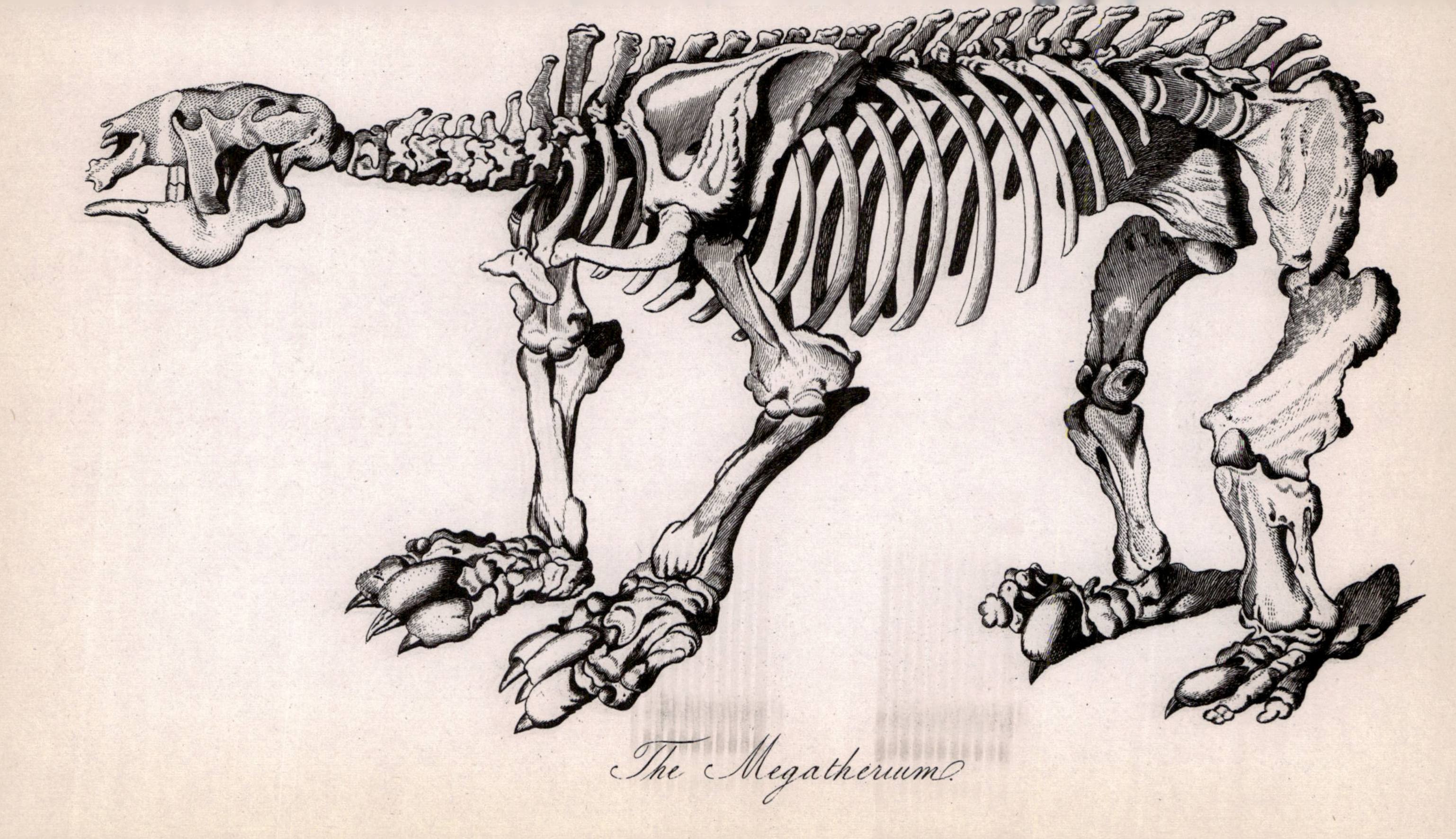

The Megatherium.

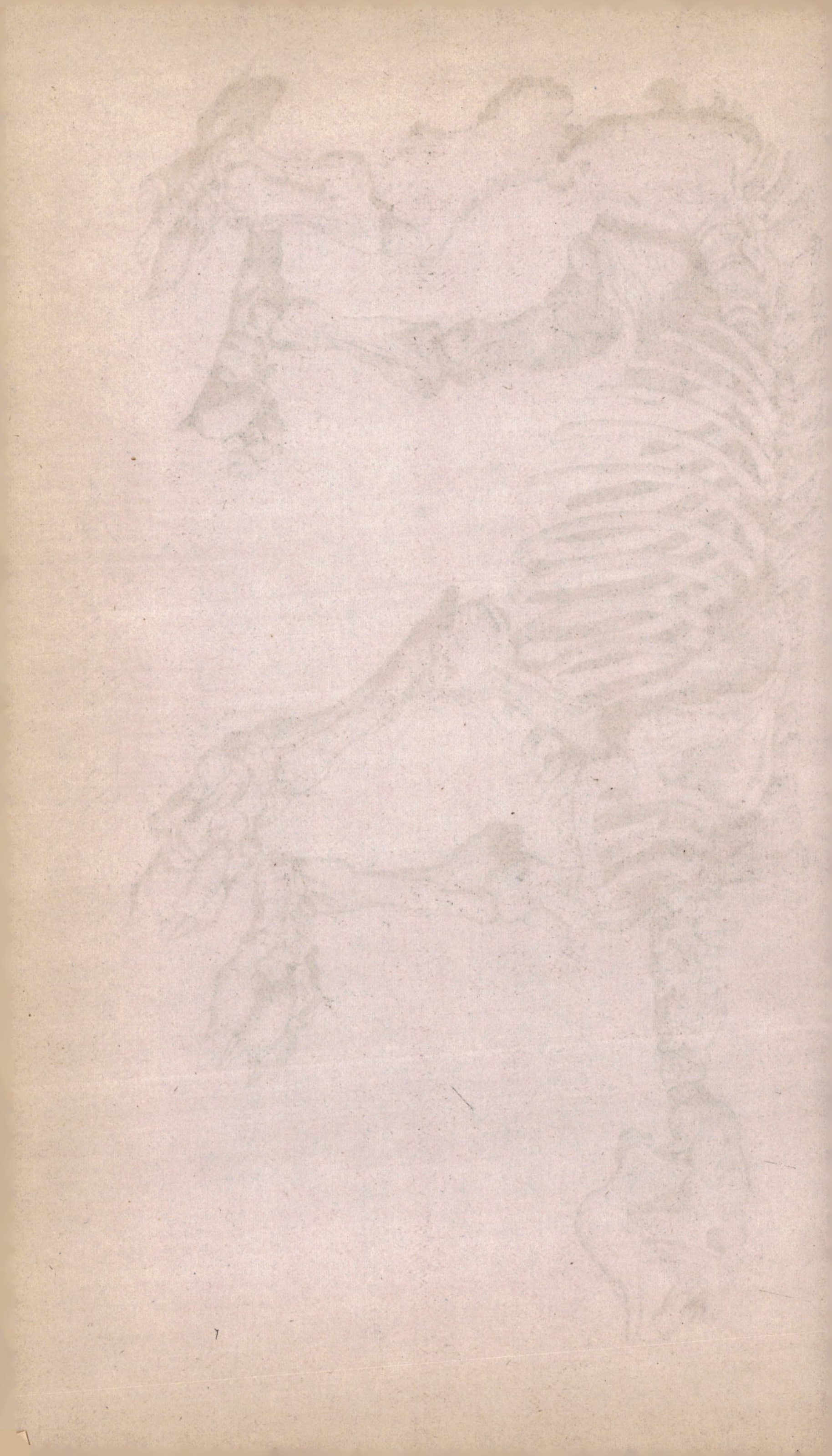

of Madrid in 1789. A second skeleton, less complete, forming part of the same collection, was sent there from Lima, in 1795. A third has been found in Paraguay. Bru put together at Madrid the skeleton of Buenos Ayres, and had some good figures of it engraved. M. Cuvier, on the examination of these figures, unfolded the affinity of this animal to the sloths and other edentata. Afterwards, Garriga, who translated Cuvier's article into Spanish, added to it the original and very extended description given by Bru. The Baron has left nothing wanting on the subject in his "Ossemens Fossiles."

Many other writers have treated concerning this animal. Abildgaard, who knew nothing of the researches of Cuvier, did yet, like him, refer the megatherium to the edentata, or the bruta of Linnæus. Shaw more tardily adopted this opinion, which Lichtenstein and Faujas combatted without success. After that, for a long space of time, nothing was added to our knowledge of this fossil animal, until Don Damasio de Laranhaia acquainted the Philomathic Society with the discovery of certain parts of the head analogous to that of the tatou, and which appeared to have belonged to the megatherium.

The general forms of the head of the megatherium resemble considerably that of the bradypi. But the most striking trait of resemblance consists in a long descending apophysis, flattened, and situated at the basis of the zygomatic arch. This arch is entire, while in the bradypi it is interrupted behind. The under part of the lower jaw has on each side a very remarkable projection, to which there is nothing analogous but in the lower jaw of the elephant, though there it is much less perceptible. The symphysis is considerably elongated, which renders the muzzle much more projecting than that of the aï or unau. The bones peculiar to the nose being very short, it is suspected that the animal might have had a trunk like those of the elephant or tapir. But, if so, this trunk must have been short, as is indicated by the length of the neck. There are neither the usual incisors, nor tusks, nor canines.

The molars, four in number on each side of the jaws, approximate to each other, are prismatic and squared, and the coronal exhibits two transverse hillocks, separated by a furrow. The bradypi have the molars separated, and preceded by a canine, in the form of a pyramid with three faces.

The cervical vertebræ appear to have been seven in number, like the unau, and not nine, as in the aï; sixteen dorsal vertebræ have been collected, and, consequently, sixteen pairs of ribs. There are three lumbar vertebræ, and the caccygian, the knowledge of which is owing to Don Damasio, seem tolerably numerous. The bones of the ilia form a semi-pelvis, rather wide, which indicates that the belly was large. The pubis and the ischion are wanting in the skeleton of Madrid.

The anterior extremities are longer, but more slender than the posterior, but have not the immeasurable length of those of the aï, nor even of the unau. The femur is thicker than that of any known animal, and its length is only double its greatest thickness; the tibia and the peroneum, also very thick and short, are cemented at their two extremities; the shoulder-blade has the same proportions as those of the bradypi; the existence of the clavicle, and the length of the phalanges of the unguiculated toes, proves that the anterior extremities might have been used for grasping, and even for climbing. The humerus is very wide in its lower part, in consequence of the great development of the crests to which the motores muscles of the toes are attached. The radius, distinct from the cubitus, had the power of rotation upon it. On the olecranon was a projection considerably marked. The hand which leant altogether on the ground, had the metacarpus very short, and composed of separate bones. The three middle toes, very thick and long, are terminated by an enormous unguical phalanx, the extremity of which is composed of an arched and conical axis, which sustains a claw, and of a deep sheath, which incloses the base of this claw, and strengthens it. The two lateral toes are shorter, appear to have had no claw, and were doubtless rudimentary. The hind feet are smaller than the fore, and are articulated with the tibia by a broad astragalus, in a manner much more oblique than in the bradypi. In the Madrid skeleton, but one of their toes is provided with a claw at all comparable to those of the fore-feet. Next this toe are two external rudimentary ones, and there is none visible on the interior side. M. Cuvier suspects that these feet are not completely reconstructed; for it is a rule without any exception, that all unguiculated animals have five toes, either visible or rudimentary. There is reason, then, to believe that the two internal toes are wanting, and it is possible that all were provided with claws. According to the measurements reported of the different parts of the megatherium, it must have been pretty nearly about the magnitude of the rhinoceros.

The *megalonyx* was thus named by Mr. Jefferson, the celebrated President of America, who was the first describer of some of its bones, in the thirtieth number of the "Transactions of the Philosophical Society of Philadelphia." It has also formed an object of the Baron's researches.

Its debris were found, for the first time, at a depth of two or three feet, in one of the caverns of the calcareous mountains of Green-Briar, in Western Virginia. They consist of bones of the extremities, particularly a fore-foot, the forms of which are almost absolutely identical with the analogous parts of the megatherium; but these bones are one-third smaller, though they bear evident marks of having belonged to an adult subject. A tooth, reported to be American by M. Palisot de Beauvois, has been recognised by Cuvier to be precisely and rigorously the tooth of a bradypus. It was a simple cylinder of osseous substance, enveloped in a case of enamel. Its coronal was hollow in the middle, with projecting edges; as to the form of this tooth, the megalonyx differed remarkably from the megatherium, in which the coronal of the molars is marked with transverse hillocks.

In his memoir on the megalonyx, M. Cuvier has given the

most minute details on the forms and relative localities of these different debris. He has taken particular pains to prove the resemblance which they exhibit to the analogous parts of the ant-eaters and bradypi, and has discussed and refuted the opinion of Mr. Jefferson and M. Faujas, who considered the megalonyx as a large carnivorous animal with acerated claws, and probably appertaining to the genus felis. He has particularly compared the unguical phalanges of the lion with those of the megalonyx, and shown that their difference is enormous, while between the latter and those of the edentata there is the strongest analogy.

Some time since, Mr. Clinton, of New York, attempted to prove that the debris of the megalonyx belong to the living species of the large gray bear of America. But he does not support this opinion by an exact and detailed comparison of these debris with their correspondent parts, the only process from which any just result can be expected. He confines himself to remarking that the bones of the megalonyx are not really fossil, because they have been discovered at a little depth in the loose earth, in some caverns of the United States. That the size of the megalonyx was nearly the same as that of the gray bear ((the magnitude of the ox !); and that the last-mentioned animal must have the unguical phalanges extremely robust, to support the enormous claws with which it is provided.

If the gray bear differed no more from other bears in the conformation of the bones of the extremities, than they differ from each other, which is extremely probable, this opinion of Mr. Clinton's would be totally overturned by a simple comparison of these bones with the extremities of the megalonyx.

To sum up : M. Cuvier approximates the megalonyx to the megatherium, and considers that these two animals must have constituted an intermediate genus between the bradypi and ant-eaters. He considers them both as herbivorous, and the megalonyx especially as herbivorous after the manner of the sloths, since its teeth were conformed precisely like theirs. From the resemblance of their feet he concludes that their gait was similar, their movements alike, with the differences that so considerable a volume of body in the one might have occasioned. Thus, he observes, the megalonyx could but seldom have climbed up trees, because it must rarely have found any sufficient to support its weight. This difference of habit from the bradypi, he considers no more surprising than what is found to exist in the habits of animals of the genus felis, the small species of which, such as the wild-cat and the lynx, climb trees with facility, while the larger ones, such as the lion and the tiger, rarely, if ever, do so.

Before we altogether dismiss the fossil edentata, it is necessary to notice an unguical phalanx, which appeared to have belonged to some unknown animal of this order, probably of the genus of the pangolin, but like the megatherium, of gigantic dimensions.

Those who understand the laws of comparative anatomy, and have thoroughly studied the researches of the Baron, must be satisfied that this single fragment is sufficient to prove that animals, unknown at the present day, existed in more ancient eras, and that some catastrophe has caused them to disappear from the countries which they inhabited, and in all probability, annihilated them throughout the entire globe.

The knowledge of this fragment, the Baron says, he was but recently indebted for to M. Schleyermacher, librarian and private secretary to the Grand Duke of Hesse, who sent him a model of it in plaster. It was found, with many bones of rhinoceros, mastodon, hippopotamus, and tapir, near Eppelsheim, a canton of Alzey, in that part of the ancient palatinate which at present belongs to the Grand Duke of Hesse, in a pit of sand and gravel, supposed to have been accumulated by the alluvions of the Rhine.

On the first view, this fragment exhibits two very distinguishing characters of the order edentata. Its hinder facet for the articulation with the last phalanx but one, is after the fashion of a double pulley, hollow on each side, with a projecting crest in the middle, which constitutes a serrated gynglymus, as in the edentata. The concave arch formed by this pulley goes more in the rear of its upper part, which prevents the phalanx from straightening again as in the felinæ, but forces it to bend underneath, as in the edentata. These characters clearly prove it to be an unguical phalanx of this order.

There are two others, which as clearly determine the genus.

1. On the unguical phalanges of the ant-eaters is a furrow, indicating a disposition to bifurcation ; but the pangolins alone have this bifurcation decidedly marked, and deepening vertically through the entire elevation of the bone, as far as the middle of its length. Now, the fossil bone has this bifurcation still more strongly marked. Though one of the branches of the fork is broken towards the root, yet the entire bottom of the fissure which separates them is visible, and it must have occupied more than half the length of the bone.

2. The unguical phalanges of the pangolins want those osseous sheaths which, in the sloths, and partly in the ant-eaters and the tatous, rise on the sides of the base, and envelope the root of the claw. The largest of these unguical phalanges is marked by nothing but a slight enlargement under the base, which forms on each side a small longitudinal edge. The fossil bone is precisely similar ; no osseous sheath is visible upon it.

Thus we find that this bone has nothing analogous in nature, excepting the correspondent parts of the pangolins ; and from all the laws of co-existence, no other conclusion can be drawn but that the animal which possessed it belonged to the same genus with these quadrupeds. Nor was this the largest of the unguical phalanges, for it has not the slight edges of the large phalanges of the pangolins—the under part of the base is only a little rugous and inflated. The holes through which the larger vessels pass are not pierced underneath, but towards the bottom and hinder part of each lateral facet.

It is impossible to establish an exact comparison of the size of this animal with that of the pangolin, without knowing to what foot and what toe this unguical phalanx appertained. But supposing it to have belonged to the second or fourth toe of the hind foot, the fossil animal must have been more than eight times the size of the adult pangolin; and supposing the general proportions of both to have been nearly analogous, the former must have been four-and-twenty feet in length.

It is impossible to avoid remarking here, nor can it be too often impressed on the mind of the reader, how scientific a character fossil osteology has received under the hands of Cuvier. We find, from the instance just now mentioned, that a single fragment, certainly of a characteristic part, is sufficient to determine the order and genus of an animal with a precision amounting almost to mathematical certainty. We arrive, too, by the same means, at least to a strong probability regarding the dimensions of the skeleton;—a probability sufficient to warrant the inference of a specific distinction: for, in wild animals not subject in, by any means, the same proportion to those varieties which domestication induces, a considerable discrepancy of size furnishes a sufficient basis for a distinction of species.

Fossil Marine Mammalia.

We now arrive at the last order of this class, in considering the fossil remains of which we will, after the example of the Baron, though contrary to the order observed in the Animal Kingdom, make some preliminary remarks on the

Fossil Phocæ.

It was natural enough, at a time when all the kinds of strata were confounded together, and all considered as the productions of the sea, to attribute to marine animals the osseous remains so abundantly found in certain formations. Accordingly we find the describers of fossil bones continually referring them to the genus of the seals. But at present, when it is so amply demonstrated that the mammiferous remains inclosed in such a variety of different strata belonged to land animals, we must expect to find among them very few debris indeed of the marine genera.

In truth, nothing is more rare than the bones of seals and lamantins among the fossils. No remains of the morse, clearly established, have been found; and if there are rather more of the larger cetacea, they are always found, like those of the lamantins and phocæ, in strata evidently of marine origin, with bones and teeth of the usual fishes, and shells, or simply in alluvions very recently abandoned by the sea.

The only well-authenticated remains of phocæ have been found in the neighbourhood of Angers. They consist in the upper part of a humerus, and in the lower part of another of smaller size. They bear all the characters of the genus, and the first seems to have belonged to a phoca, about twice and a half as large as *vitulina*, and the second to a smaller one.

Fossil Lamantins.

Some debris of fossil bones have been found, which the Baron Cuvier has recognised to belong to this genus. The principal portions have been discovered by M. Renou, in a very coarse shelly limestone, of which a part of the hillocks bordering the river of Layon, in the department of the Maine and Loire, is composed. These debris consist of fragments of the head, anterior extremities, and ribs, and they were accompanied with other fragments which appeared to have belonged to the bones of phocæ and cetacea. All these bones were changed into a ferruginous reddish limestone, which was found to contain some fluate of lime.

These debris of lamantins differ sufficiently from the analogous parts of the known species, to justify the inference of specific distinction. It would also appear that they belonged to a lost species, like all the mammalia found in similar marine depositions. This species seems to have been remarkable for its size, and the peculiar forms of the head.

Other debris of lamantins, not so well characterised, consisting of remains of ribs, have been found by M. Dargelas, at Capian, about fifteen leagues from Bourdeaux. They were also in a marine coarse limestone, and had suffered the same change as the former.

Some more fragments of ribs of lamantins were found at Marly, enclosed in the plastic argilla which is generally found above the chalk formation, wherever it exists in the environs of Paris.

They have also been found in a few other localities, which it would be superfluous to enumerate.

It is quite certain, however, that an animal of this genus, now proper to the torrid zone, inhabited the ancient sea which covered Europe with its shells, at a period subsequent to the chalk formation, but anterior to that of the gypsum in which the ancient pachydermata were found.

Fossil Dolphins.

A dolphin, approximating to the grampus and globiceps, was discovered by Cortesi, on the acclivities of the Apeninnes, to the south of Fiorenzuola, in 1793. The skeleton was found almost entire, in a bluish argilla, filled with marine shells. The head was nearly complete, and also one of the branches of the lower jaw. Even the bones of the ear were in their proper places. There remained thirty-three vertebræ; twenty ribs, thirteen of one side and seven of the other; three quadrangular bones, supposed to belong to the sternum; and some small bones, more or less mutilated, which M. Cortesi supposed to appertain to an anterior extremity.

There were twenty-eight teeth in each jaw, fourteen of a side, in all fifty-six. It does not appear that any grampus or globiceps possesses an equal number. The largest of these teeth were two inches long.

In other points this fossil dolphin differed essentially from the congeners we have named. The head was much more narrow in proportion to the length, the muzzle much longer in proportion to the cranium, the orbit smaller, and the indention before the nostrils more narrow and more hollowed.

In the dimensions of the other parts similar differences existed, and M. Cortesi calculates, with great probability, that the animal was nearly thirty feet long.

On the whole, the Baron concludes that it belonged to a species different from those of any of the genus now existing.

Another dolphin, with a peculiarly long symphysis of the lower jaw, was discovered in the falun of the department of Landes, at Sort, a village two leagues from Dax. This falun is very abundant in shells and other marine productions.

A tolerably complete jaw of this dolphin is in a collection made by M. de Borda d'Ovo, and attached to the town of Dax, along with another fragment containing some teeth.

The peculiar character is the length of the symphysis, to which no living dolphin approaches, except the *frontatus*, and the species of the Ganges. Both these, however, are much smaller, and differ somewhat in the form of this part and of the teeth, and in the number of the latter. The conclusion from these and other considerations is, that the species to which this dolphin belonged is unknown, at least as far as our very imperfect knowledge of the cetacea will bear us out in such a conclusion.

In the formation above mentioned was also found a portion of lower jaw, indicating a species very near the common dolphin, but with some slight differences.

A portion of upper jaw, found in the *calcaire grossier* of the Orne, seemed also to belong to a dolphin different from any known species. The characteristic distinction is, that the pyramidal and descending projection of the back nostrils begins to be visible opposite to the last molars. This is not the case with any living dolphin with which we are acquainted.

Fossil Narwhals and Cetacea,

Approaching the Hyperoodontes and Cachalots.

The osteology of the cetacea was too little known, even a comparatively short time back, to distinguish the *Narwhal* by any character but the long tusk. This, however, is one not easily mistaken, if met with. Still the examples of it are very rare, and, what is more, by no means well authenticated.

Mr. Parkinson speaks of two fragments of it in the Leverian museum, and suspects that they were found on the coast of Essex.

Georgi, in his description of the Russian empire, speaks of a fossil tooth from Siberia, in the museum of St. Petersburgh; of another from the banks of the Indigirska; and a third found in a marsh near the Anadir.

The geological position of these, and one or more other fragments, is totally unascertained: it is, therefore, superfluous to speak of them further than as affording a motive to ulterior researches.

The petrified head of some unknown genus of cetacea, approximating to the hyperoodontes and cachalots, was discovered in 1804 on the coast of Provence. To this genus the Baron gives the name of Ziphius. The head differs from that of the hyperoodon, in the maxillary bones not forming vertical partitions on the sides of the muzzle, and in the partition behind the nostrils not only rising vertically, but also curving, so as to form a kind of half cupola over these cavities. The species to which this head belonged, M. Cuvier calls *Ziphius cavirostris*.

These portions of petrified heads, found in excavating the basins of Anvers, presented generic characters, like those of the last, with sufficient indications of a distinct species, to which our author gives the name of *Ziphius planirostris*.

A petrified fragment, preserved a long time in the Paris

museum, indicates a species approaching to the preceding, but with much more elongated muzzle. This is Cuvier's *Ziphius longirostris.*

Fossil Balænæ.

It is evident, from divers accounts, that more or less considerable portions of the skeletons of the larger cetacea have been found embedded in various places, and among them there are many attributed to the balænæ. But the authors of these accounts have rarely given drawings or descriptions sufficiently precise to enable us to determine the species. Such is the case with some bones found in Clackmannan, in Scotland, in a mutilated state, but which indicated an individual of considerable size. As they were found, however, only at eighteen inches depth in a recent alluvion, it is more than probable that they belong to a living species.

But two skeletons of a whale, of the sub-genus of the *rorquals*, were discovered by M. Cortesi, in Lombardy. The first in 1806, on the eastern side of Monte Pulgnasco, six hundred feet below the summit, which is itself elevated twelve hundred feet above the plain. In this part the hill is formed of regular strata of bluish argilla, inclined towards the north, and filled with marine shells, exactly like those on the opposite hill, where the same naturalist discovered the skeleton of the dolphin which we have already mentioned.

Excepting some ribs a little scattered, the bones of this skeleton were found in their natural connexion. It was surrounded by innumerable shells, and especially by a small species of the oyster. There were also many teeth of the *squalus* there. The head, &c., presents all the sub-generic characters of the rorqual, with specific ones that are incontestable. These consist principally in the dimensions and conformation of the lateral parts of the frontal bone, and in the more speedy union of the transverse crests at the anterior part of the same bone into a middle and longitudinal crest.

The entire length of this skeleton might be about one-and-twenty feet; but a small rorqual, if adult.

The other skeleton of the same species was discovered in 1810, in a neighbouring valley. It was much less preserved, and in its actual state was only twelve feet five inches long. Its characters were precisely similar to those of the other.

A considerable fragment of the head of a balæna was found in the centre of Paris, in 1799, by a wine-merchant of the Rue Dauphine, making excavations in his cellar. The proportions were different from those of the living balænæ, and there is great probability that it belonged to a species hitherto unknown even among the fossils.

Having finished this last order of the mammalia, we shall say a few words on it, by way of recapitulation.

The fossil bones of cetacea which have been collected or described, are far more numerous than those to the notice of which our limits have of necessity confined us. But, in fact, it must be observed that, even were those limits much more extended, we should only be spinning out our observations, and fatiguing our readers to little purpose, by entering more largely into the detail of accounts which have neither sufficient authority nor sufficient precision to confer upon them interest or importance. In most cases the remains themselves are not in the state of preservation which can enable a naturalist to appreciate their forms, and, consequently, to determine the species to which they belonged. But even were it otherwise, the determination of the fossil cetacea must be still attended with difficulties of no common kind. Fossil remains can never be properly determined but by a minute and critical acquaintance with the osteology of living species, an acquaintance which we are yet far from possessing with the animals of the order in question. We find, in general, that the departments of natural history which relate to the large animals of all species, are precisely those in which error and confusion more especially predominate. The reason is, that over such animals man can seldom exercise a sufficient control. It is not possible to know and distinguish any species but those which we are able to examine nearly, and compare carefully one with another. Many of the larger quadrupeds of the earth disdain our sway, and will not submit themselves to our inspection. The wild and fearless tenants of the desert, the mountain, and the forest, are often alike intractable by force or kindness. The puny power of man cannot always cope with their giant strength, or soften their indomitable ferocity; nor can his flimsy stratagems escape their penetration, or deceive their vigilance. We find, therefore, in many instances the accidental possession of a single individual sufficient to overturn many of our received notions respecting the habits, and even the conformation, of a species. A nearer view serves to falsify the accounts of careless, ignorant, or inventive travellers, which may have been long entertained by philosophers as well as fools with easy credulity. A case in point has occurred almost at the very moment I am writing. The presence of a cameleopard in the capital of France has at once dispelled a variety of illusions connected with the structure and habits of this singular quadruped. Animals must be under immediate and constant inspection, and under the inspection of the skilful naturalist, before almost any thing can be predicated with certainty of their peculiarities. Before such inspection the fine-spun cobwebs of theory, and the fairy fictions of imagination, vanish "like the baseless fabric of a vision." The progress of science, while it unfolds to us the real wonders of nature, destroys the marvels and monstrosities of man's creation.

If such observations are applicable to the larger animals of the earth, with how much more force will they not apply to the gigantic inhabitants of the deep? They more particularly hold good respecting the larger cetacea. These have excited universal astonishment by the enormity of their dimensions, and have given rise for ages to the most unparalleled exertions of activity and courage. Yet, except when, by a felicitous chance, their bodies have been cast ashore, in the neighbourhood of some enlightened man, they have been scarcely ever described with exactitude, or compared with accuracy.

Myriads of sailors have caught and divided whales, who perhaps have never had the opportunity of properly contemplating one in its entire state. Yet naturalists have deemed themselves able to compose the history of these animals, from the vague descriptions and the ruder figures given by such uninstructed observers. No critical accuracy, no correct deduction, could exist in such compilations for want of the proper basis of well-authenticated facts. Consequently we find the history of the cetacea, on the one hand, meagre in the extreme, and, on the other, swarming with contradictions, and confusions of nomenclature.

Furnished with such imperfect materials of truth, and perplexed and encumbered by such abundance of falsehoods, the most expert naturalist must encounter incalculable difficulty in separating the one from the other, and reducing the chaotic mass to any thing like harmony and order. He must be aware of attaching too much importance to these vague and contradictory accounts, as to establish on them alone the distinction of species, and still less those of genera and sub-genera. It is no doubt easy, from rude figures drawn from imagination or memory—from confused and mutilated descriptions—from the accumulation of synonimes, which are but copies of each other, to produce a long catalogue of species which have no reality, and which the slightest breath of criticism is sufficient to destroy; but a line of conduct precisely the reverse of this is necessary to be pursued, if natural history is ever to be freed from absurdity and disorder, and established on the basis of truth.

One of those causes which have most contributed to embarrass the history of the cetacea is, that the people of the North, from whom our knowledge of them is chiefly gleaned, as it is in their latitudes that they most abound, designate them all by one common generic name. Thus, *wall* in German, *whale* in

English, *huval* in Swedish and Danish, *qual* in Norwegian, and *hwalune* in Islandic, is applied to all cetacea without discrimination. In French, too, this word, which probably has some relation with *φαλαινα* and *balæna*, has constantly been translated *baleine*, even when it simply signified the dolphin, and has led naturalists, who did not understand the full extent of its acceptation, into the most serious errors*.

We may, however, in summing up this account of the fossil cetacea, speak with tolerable certainty of the following remains.

There is, as we have seen, a collection of vertebræ in the Paris Museum, from the basin of Anvers, which approach the form of the corresponding vertebræ in the dolphins, but the body of which is more elongated in proportion to their diameter, and which appear to have belonged to two or three species of different sizes, the largest of which may have been double the size of the grampus. There are also among them some flatted ones, almost similar to those of the dugongs and lamantins.

From the environs of Havre, and some other places, there are more of these bones in the same collection, the locale of which has not been well described, and which do not appear, in what remains of them, to differ from the existing balænæ and cachalots: but the apophyses are too much fractured to furnish characters that can be distinctly appreciated.

The same may be said of a certain number of entire or mutilated ribs, which were found in various situations. One, for example, from the valley of l'Authie, near Montreuil-sur-

* The word *wall,* imported by the Normans, was used on the French coasts in the middle ages. In several charters of the eleventh century, when mention is made of an association of whalers, they are designated *societas,* or *communio walmannorum.* The cetacea are also called in these charters, *crassus piscis, grassus pesius,* and generally *piscis ad lardum.* It would appear that those animals were estimated more highly in those days than at present. The flesh was a common article of food, and sent in great quantities to Paris. From *crassus piscis* comes the French *graspois,* and the English *grampus. Graspois,* for a long time, signified the fat of the cetacea generally.

mer. This rib entirely resembles that of a small whale; it was found two leagues from the sea, in a sandy stratum, at a depth of twelve feet.

An enormous shoulder-blade was taken from the lake of Geneva, which, to all appearance, had belonged to a rorqual.

A radius, disinterred in the neighbourhood of Caen, though destitute of its epiphyses, exhibits the proportions and forms of the radius of a whale.

But all these last-mentioned fragments, though giving additional demonstration to the existence of the cetacea among the fossils, afford no positive indication of the species to which they belonged, whether living or extinct.

It is sufficient to draw the reader's attention to the better-determined specimens we have already noticed, and the incontestable result which flows from them. This result is, that all the marine mammifera collected in the strata, and whose species it has been possible to characterize, do not differ less from their existing congeners in our seas, than the fossil land animals, of which we have previously treated, do from theirs. Nay, we may go farther and say, that they differ very sensibly from all the cetacea that have been observed up to the present time in every sea.

Thus, the lamantin of the neighbourhood of Angers is not only of a genus foreign to our climates, but of a different species, both from the lamantins of Africa and America, and still more so from those animals of the Indian and Pacific oceans, which had hitherto been deemed as approximating to this genus.

The dolphin, again, with long symphyses, disinterred by M. de Borda, is entirely unknown among the numerous species of this genus described by naturalists. The dolphin with narrow muzzle, of the environs of Angers, and the dolphin with broad muzzle, discovered by M. Cortesi, in Lombardy, though much less remote from the living congeners, are yet distinguished from them by characters of a nature perfectly specific.

The same may be said of the rorqual of Lombardy, for the discovery of which we are also indebted to the researches of M. Cortesi.

But what is far more singular than all this is, that there are among the fossils three or four species so utterly dissimilar to all the other cetacea, that it has been found necessary to form them into a separate genus.

The ZIPHIUS, as we have seen, comprehended animals neither altogether balænæ, nor cachalots, nor hyperoodontes. They hold in the order cetacea a place analogous to that which the mastodons, the palæotheria, the anoplotheria, and the lophiodons hold among the pachydermata, and the megatherium and the megalonyx among the edentata. Like these, they are, in all probability, the relics of a destroyed creation, whose living types it would be vain to seek for in our present world.

These researches on the lost cetacea tend more and more to confirm the opinion to which the examination of the fossil shells had already conducted naturalists. From them it is still more evident, that not only have the productions of the earth changed with the revolutions of the globe, but also that the sea itself, the principal agent of these revolutions, has not preserved the same inhabitants: that when it formed those immense calcareous beds, replete with shells, almost all unknown in the present day, the large mammifera which it supported were different from its modern tenants of the same class. It appears that their gigantic size and tremendous force did not avail them better in resisting the catastrophes of their native element, than did the robust proportions of the mastodon, the elephant, the rhinoceros, or any other of those monstrous quadrupeds avail their possessors in resisting the revolutions of the land. Whether it is in the power of man, or not, completely to extirpate any race of animals, is doubtful; but it is evident that where man was not, nothing could have destroyed these numerous and powerful tribes, but a grand and general convulsion of nature.

FOSSIL BIRDS.

NATURALISTS are agreed that, of all animals, the birds are those whose bones or other debris are most rarely to be met with in the fossil state. Some have even gone so far as absolutely to deny their existence. It is, indeed, a remarkable fact, and one of the many singularities attached to the gypsum strata of Paris, that there are scarcely any other fossil bones of birds, well authenticated, except those which they contain; and it is even but a short period since the true nature of these fossils has been clearly ascertained.

We shall give a rapid sketch of the various testimonies which have been given by writers, from time to time, concerning true or pretended ornitholites.

Walch had pretty early made a considerable collection of these, to which Hermann added many others. But the first of these writers was frequently deceived from the want of due precaution. Even Gesner has declared that stones named after certain birds, supposed to be petrified therein, such as *hieracites*, and *perdicites*, have no other relation with them, than some resemblance of colour.

Neither do the rude figures of birds, traced accidentally on some coloured stones, appertain to the ornitholites, nor yet the stones or flints figured by chance into a likeness to certain parts of birds. The cock of Agricola, and the hen of Mylius, imprinted on a slate from Ilmenan, are of this description.

Many authors have very gratuitously considered certain fossil bones as belonging to birds, merely because they were light and slender; but a little attentive examination soon proved them to be parts of fishes, of small quadrupeds, and sometimes even nothing but shells or crustacea. Thus, the *sulculata littoralis rostrata* of Luid seems to be nothing but the extremity of the dentelated spine of the fin of some fish. Romé Delille, in his catalogue of the collection of Davila, mentions a beak found in the neighbourhood of Reutlingen, and a bone from Cronstadt, which he thought belonged to a chicken; but this pretended beak turns out to be nothing but a bivalve shell, which shows itself obliquely on the surface of the stone. If it were a genuine beak, it would differ most prodigiously from any thing with which we are acquainted in existing birds; as for the bone, there is neither figure nor description of it in the work.

Scheuchzer speaks of the head of a bird, in a black schistus from Eisleben; but he subsequently adds, that it might be taken for a pink-flower—this is quite sufficient.

Many writers quote the description of the environs of Massel by Hermann, as if he had spoken there of the bones of birds; but the fact is, he only mentions small bones, without specifying to what they belong.

The error of compilers respecting the petrified *cuckoo* mentioned by Zannichelli, is still more glaring, and positively ludicrous. That author, in fact, speaks of a fish so called, and which belongs to the genus *Trigla (Trigla culusus* of Linnæus, in Italian *pesce capone)* and not of a bird.

There are other testimonies on this subject, wholly unsupported by details, figures, or descriptions. Such are those in the *subterranea Silesia* of *Wolkmann*, and those of many other systematic mineralogists. It is utterly impossible to establish any thing on such vague indications.

It is quite evident that what are termed incrustations, have nothing to say to the subject in question. We are not inquiring whether birds, exposed in particular situations to waters charged with mineral substances, may be enveloped in those substances, but whether there have been any remains of birds arrested and inclosed in the grand strata which occupy the external surface of the globe.

Thus, the examples of birds, of eggs, and of nests, incrusted in gypsum, tufa, salt, or other minerals, related by Volkman, Lesser, Bruckman, Baccius, Butner, Dargenville, and Bock, even if true, would prove nothing for the existence of ornitholites.

After all these exclusions, nothing remains but some debris contained in certain schists, as those of Œningen, of Pappenheim, and of Mount Bolca, which have any pretensions to a serious examination, or, in fact, which have even been considered as ornitholites by any real naturalists.

Now, all that is cited on this head is either more or less equivocal, or, at all events, unsupported by sufficient figures or descriptions. These schists swarm with the bones of fishes and other marine productions. How can it be supposed that, in such situations, it is always possible to distinguish the bones of fishes from those of birds? What mode is there of judging when no entire limb, when no part of sufficient importance, remains?

The authority of M. Blumenbach must be considered as having the greatest weight on a point of this kind. But all he says is, that at Œningen were found the bones of water-fowl. He refers us to the Memoirs of the Academy of Manheim for an account of the bones of Pappenheim; but all we find there is a notice of a singular reptile (the *pterodactylus*) and not of a bird, as M. Blumenbach says, of the order palmipedes.

It seems, in short, from a careful examination of the testimonies given by various writers on this subject, that no well-authenticated bones of birds have been found any where, except in the gypsum of Paris.

It is not very long since these remains have been properly authenticated. Lamanon, indeed, as long ago as 1782, described the impression of an entire bird found at Montmartre, by M. Darcet. Were we to trust to the figure which he has given, no doubt could remain on the subject, for it represents a bird completely. He has even put feathers on the wings and tail. Unluckily, however, the aid of imagination was summoned, in the drawing of which we speak, and the picture is exceedingly unlike the original.

Fortis, who had conceived strong prejudices against the existence of ornitholites, examined afresh the specimen which Lamanon had described. He also gave a figure of it according to his own notions. This is a striking example of how differently the same object may appear, according to the notions of the observers. In Fortis's figure the head is placed below, all the inequalities of the stone are exaggerated, and the osseous impressions weakened, and the author declares that he can see nothing but a frog or a toad in the fragment in question.

The stone, however, turns out to be a genuine ornitholite. Yet this point might still have remained doubtful, but for subsequent discoveries of similar specimens better characterized.

Peter Camper takes notice of one, but without describing it, in an article on the fossil bones of Maestricht, inserted in the "Philosophical Transactions for 1786." This was a foot found at Montmartre; a second specimen, also a foot, from the same place, was described by the Baron as early as the year 1797. At the same time he learned that two other specimens were in the hands of a person in Abbeville, who had received them from Montmartre: these were the body of one bird, and the leg of another. It was easy to observe that the leg did not belong to the same individual as the body, for even the stone which incrusted it was derived from a different stratum.

Here, then, were four specimens of ornitholites, perfectly well authenticated, as long back as the year 1800. Since that period the Baron has continued his researches, and collected so great a number, that no manner of doubt could remain that the gypsum quarries contained the debris of birds in great abundance.

The feet are by far the most remarkable part in all the ornitholites, even to the most inexperienced eye. The foot, in fact, of every bird is composed in a very peculiar manner, and resembles that of no other animal whatever. The birds are the only class in which there is but a single bone to answer the purposes of the tarsus and metatarsus. In the horses, the metatarsus consists but of a single piece, whereas the tarsus contains many. In the jerboa and alactaga there is also but a single bone of the metatarsus, which supports the three principal toes. But the bones of the tarsus are distinct. In the tarsiers and galagos, the scaphoïd and calcaneum are elongated, so as to give as much length to the tarsus as in certain birds. But the other bones of the tarsus and metatarsus do, nevertheless, equally exist. Frogs, toads, &c., also have the tarsus considerably elongated. But it is always composed of two long bones and several small ones.

In the number of toes and of articulations on each toe in the birds, characters are found no less striking than those furnished by the tarsus.

The birds are the only class in which the toes are all different in the number of their articulations, under which nevertheless this number and the order of the toes are fixed. The thumb, for instance, has two articulations, the first toe of the internal side three, the middle toe four, and the external five.

To this rule there is but one exception, and that is in the case of birds which have no thumb; but the other toes preserve the usual number of articulations.

This rule is never completely observed, excepting in this class.

The quadrupeds have two articulations on the thumb and three on the other toes, be the number of the toes what it may. The tridactyled sloths alone have but two, because the first phalanges are cemented with the bones of the metatarsus. Some few toes, indeed, that are concealed under the skin, want the usual number.

In the reptiles the number of articulations is less equal, but scarcely ever do we find them exactly the same as in the birds.

The following table shows the number and order of articulations in the various reptiles:

Land tortoise		2.	2.	2.	2.
Marine tortoise . . .	2.	3.	3.	3.	2.
Crocodile	2.	3.	4.	5.	
Lizards of all species . .	2.	3.	4.	5.	3.
Cameleons	1.	2.	3.	3.	2.
Seps tetradactylus . . .	2.	4.	5.	2.	
Seps tridactylus . . .	2.	3.	4.		
Frogs, toads, &c.	2.	2.	3.	4.	3.
Salamanders	2.	3.	3.	2.	

Thus we see that in the crocodile alone the number of articulations are the same as in the birds; but as each of their toes is supported besides on a particular bone of the metatarsus, there can be no mistaking the foot of one for that of the others.

Now, all the different feet in the ornitholites examined by the Baron had precisely the characters we have just instanced as belonging to the birds. In one of the feet we mentioned above, the thumb was wanting, but the little supernumerary bone which supports it in many birds was observable. In another specimen these characters were still more complete; the femur was wanting, but the tibia was more entire, and the thumb and three other toes are very complete, and provided with their entire number of articulations.

In another specimen belonging to the Baron, the tibia and tarsus were rather longer than in the last. Feet of a similar description are very common in the gypsum, and seem to belong to a different species.

A third species is indicated by a foot about the same size as the last, but in which the bone seems thicker, and the tarsus more arched in its length. In other respects it has all the characters of the genuine foot of a bird, except that the external toe having left but a single print of its upper part, the three articulations of which it should be composed are not well distinguished.

To a fourth species belonged the foot, which we mentioned

before as accompanying an impression of the body of a bird. That it could not have belonged to that body, is proved by the latter having its femora, while there is another femur with the foot, which is moreover much too large in proportion. The foot itself is also too large, and the bones too thick, to suffer it to be confounded with any of the preceding. The characteristics of the bird are all perfect in it.

From the discovery of some other feet, into a detailed description of which it is quite unnecessary to enter, the Baron concludes from these alone, that there were at least nine species incrusted in the gypsum. From such a number of fragments attesting, by their assemblage, the existence of ornitholites in the regular rocky strata, it was impossible that any doubt could remain on this subject; and, consequently, the negative arguments of Fortis, and other naturalists, against their existence, must fall to the ground before the evidence of facts.

Several other little isolated bones proved, on examination, to be referrible to the class of birds. A portion of tarsus, divided below into three apophyses, each terminated by a demi-pulley, for the articulation of the first phalanges of the three front toes, proved to be of this description.

Among the quadrupeds, none but the jerboa and alactaga exhibit any thing similar. But as there is no other indication in all the plaster-quarries of such animals, the remains in question must not be referred to them.

The femora of birds have also a distinctive character, which belongs to the peculiar nature of the knee. This articulation in birds is provided with a sort of spring, analogous to the hinges of a clasp-knife. The blade, we know, has but two points on which it can rest steadily; namely, when it is completely open or completely shut, because there are only those two points on which the spring is not removed from its natural position.

Now, the birds having but two feet on which to find a solid seat, have received an articulation of this kind with two fixed points; namely, that of the greatest flexion, and that of the most perfect extension. Those are the only points in which the ligaments are not stretched, and in which the bones are preserved in their proper places by the simple action of these ligaments, unless the bird makes an effort to displace them. The head of the peroneum produces this effect, by its mode of catching in a particular fossa of the femur.

This head enlarges very much from front to rear, and its upper edge is nearly a straight line, which mounts obliquely behind, thus rendering its posterior extremity more elevated than the other. The femur rests upon this right line by a projecting line drawn over its external condyle, the middle of which forms an almost semicircular convexity, while the two ends, on the contrary, are a little concave, and the two bones are attached in this place by an elastic ligament which goes from one to the other, crossing almost perpendicularly the line by which they touch.

It is clear, then, that this ligament will be more stretched, according to the degree in which the femur shall touch the peroneum by the convexity of the projecting line which we have described: that is, so long as the leg is neither completely extended, nor completely bent. But in these two extreme positions the peroneum will re-enter into one of the concavities placed at the two ends, and will be retained there by the elastic contraction of the ligament.

The femur, then, of birds is distinguished from that of quadrupeds in this,—that its external condyle, instead of presenting a simple convexity behind, for the external fossa of the head of the tibia, presents two projecting lines: one stronger, which is the true condyle, and which corresponds to the upper external facet of the tibia, and the internal facet of the peroneum; and another more exterior, descending less, and resting on the upper edge of the peroneum. The external condyle, therefore, of birds is forked, or hollowed into a canal more or less deep behind.

The only quadrupeds in which any thing analogous may be suspected, are those which, like the birds, rest and leap on their hinder feet, with the body in an oblique position; such are the kangaroos and jerboas.

We do find, in fact, in the various kangaroos a slight indention in the rear of the condyle to which the peroneum corresponds, but only by a tubercle. The jerboas have not this conformation. In the helamys of the Cape, indeed, there is a particular osselet, which establishes a connexion between the peroneum and the femur, but not in the same manner.

There are moreover many traits which must prevent us from ever confounding the femur of a bird with that of a kangaroo, as well as with that of any other quadruped. Such, for example, is the breadth of the great trochanter from front to rear, &c.

By the observation of these characters, two femora found in the gypsum were clearly recognised to belong to birds. Their cavity was filled with gypsous matter; they had not been crushed by the weight of the strata deposited upon them, and their forms were preserved perfectly entire.

The tibia of birds is doubly characterized, by an upper head corresponding to the form of the femur just described, and by a lower one in the shape of a convex pulley, with a concave neck, on which is articulated the bone of the tarsus. Many such bones are found in the gypsum. The humerus of birds is not less distinguished than the femur and tibia.

The characters consist in the two extremities. In the upper one the head is always oblong from right to left, to play like a hinge in the articulation to which the shoulder-blade and clavicle contribute.

This part of the bone is singularly enlarged by two lateral crests. The upper, or rather the external one, which is angular, and the edge of which is trenchant, and a little re-curved in front, serves to give sufficient attachments to the great pectoral muscle, the powerful action of which constitutes the *primum mobile* of flight. The opposite crest is not so long, its edge is rounded, and a little curved behind, where it forms towards the head of the bone a small crook. Under this crook is the hole by which the air penetrated into the cavity of the bone. In quadrupeds this head is always round, the crests small, and that part of them near the head forms tuberosities.

Even the bats do not resemble birds in the humerus. The mole alone exhibits any analogy with them in this respect, because the manner in which that animal throws the earth backward in burrowing, equally demands a great force in the pectoral muscles. But it is useless to dwell on this exception, as the rest of the humerus of the mole is distinguished by such extraordinary forms, that it is impossible to confound it, not only with the humerus of birds, but with that of any other known animal.

The character of the lower head of the humerus of birds is not less striking than that of the upper. The articulating pulley is divided into two parts, one internal, or inferior, almost round, for the cubitus; another external, or superior, for the radius. This last is oblong, in the direction of the length of the bone, and thus re-ascends a little obliquely over its anterior face. In this manner the radius has a greater arch to describe than the cubitus, and the motion of the fore-arm is not made on a plane, perpendicular to the anterior face of the humerus. The lower part of this radial face enlarges behind, and rests on an external articulary facet of the cubitus.

There is nothing similar in the quadrupeds. The cubital pulley is always concave, and the radial is also hollowed in a furrow, in such animals as have not the power of supination in the fore-arm.

All these distinctive characters of the humerus of birds are found in certain specimens from Montmartre. The characters of the radius are also found in certain specimens from the same place; all have the upper head round, and a little concave—the smallest has the inferior extremity more enlarged, precisely as in the birds.

As these quarries had furnished such a number of separate bones, obviously appertaining to the class "Aves," it was natural to expect that some skeletons, in a state of greater or less completeness, and belonging to the smaller species, might be found there also. This accordingly was the fact. One, in the possession of a M. Elluin, the Baron gives a drawing of, executed by himself. This, though one of the worst preserved, is still perfectly recognizable as the relics of a bird. Though no bone is entire, and the forms of the articulations are lost, yet the position and proportions of the bones are sufficiently visible to enable us to recognize the bill, the head, the neck, the body, the two wings, the two thighs, and a part of the two legs of a bird.

This body appears to have been crushed by the superincumbent strata, and entirely flattened. It has left no impression but a brown lamina, the thickness of which can be scarcely appreciated. Neither the bones of the head, the vertebræ, the ribs, nor the sternum, are distinguishable. Some vestiges alone of the pelvis are visible. There is not the slightest impression of the plumage.

The one found by M. Darcet was in a still more imperfect state; and, as we have already seen, occasioned considerable doubt and discussion. Still, however, one wing is almost completely characterized; and the fore-arm, the metacarpus, and commencement of the great toe, are distinctly visible. The other wing and the bill are also sufficiently traceable, but what remains of the feet and of the bones of the body has lost every description of character.

Three skeletons obtained by the Baron were in a state of preservation far superior to that of the two last mentioned.

The first is an almost entire skeleton of a bird, flattened in the same manner as are all the skeletons of small animals found in the gypsum. This one, when the stone was cut which contained it, was divided into halves, each of which remained adherent to the piece of stone on its own side. The bird had fallen on its belly on the bed of gypsum which was already formed. Previously to the deposition of sufficient gypsum completely to envelope it, it had lost, either by the movement of the water, or by the agency of voracious animals, the principal part of the head, and all the left leg, which were no where to be found in the neighbourhood. Part of the bones remained in their place where the stone was cut, and another part fell in shivers, and left nothing behind but an impression.

The under part of the bill is very distinct in the impression, the left branch remaining almost entire. There are the remains of the two sides of the basis of the cranium, which, like that of all birds, was cellulous.

The vertebræ of the neck are very distinguishable as far as nine.

The clavicle on one side is very well preserved, and there are some remains of that of the other. A small remnant of the shoulder-blade is visible, but the greater portion of this bone has disappeared. The clavicle, however, is so peculiarly conformed in birds, that this bone is alone sufficient to identify the fossil in question as belonging to that class.

The sternum is very much crushed and disfigured. The remnants or impressions of ribs are visible here and there, some of which are covered, or rather interrupted by debris of the sternum, and others by the clavicles.

The pelvis has left an impression not remarkably perfect, in consequence of its being mixed with that of the coccyx, but the two points formed by the ischia and the ossa pubis are very distinct.

All the parts of the two wings are very well preserved in this ornitholite, and exhibit osteological characters eminently classical.

The humerus of one side is almost entire, the cubitus and radius in the two wings have also lost but little, and even one of the small osselets of the carpus is visible.

The metacarpus, which in birds has a very peculiar form,

being composed of two branches cemented together at their two extremities, is very visible; as also is the little osselet which stands for a thumb.

The bone of the first phalanx of the great toe is also formed of two branches in this ornitholite, as in birds in general. The osselet representing the little toe is preserved on one side of it, and that of the last phalanx at its extremity.

The posterior extremities are not nearly so well preserved as the anterior: there are, in fact, but some parts of one only, but notwithstanding this, it is impossible to mistake them for anything but portions of the leg of a bird.

There is a lower half of the femur, and a tibia almost entire, with a small remnant of the peroneum engrafted on its upper part, as in all birds.

The impression of the second fossil skeleton on the stone is so extremely faint, that it requires eyes well habituated to such examinations to recognize it. Yet, when the idea is once conceived that it is the skeleton of a bird, the different parts soon develop themselves very sensibly.

The sternum, cast a little aside on the stone, and an extended left wing, are visible.

The humerus, the fore-arm with scarcely any traces of the radius, the hand, the last phalanx, or little end of the wing, have left their impressions.

There are some traces of the left shoulder-blade, some better marked of the right, and the humerus of this side is almost entire.

Of the neck, but three vertebræ remain.

The body has left but a shadowy impression between the shoulder-blades, but the last ribs on each side are distinguishable enough. The pelvis and the coccyx are but barely visible.

All the right posterior extremity is perfectly to be recognized, as is likewise the lower portion of the left.

The third of these skeletons is, perhaps, the most complete, and the best characterized of any ornitholite that has ever yet been discovered. It came from Montmartre, and there is a minutely exact representation of it in the "Ossemens Fossiles."

The head, neck, trunk, pelvis, the two extremities of the right side, and a part of the left wing, are nearly perfect. The form of the head and the beak, which appears to have been tolerably long and strong, are quite distinguishable. There are the remains of a furca, a part of the right coracoïd bone, and a part of the shoulder-blade on the same side. There are the left coracoïdian and shoulder-blade, the last of which is almost complete. The humerus, cubitus, and radius on both sides, the thumb, or bone of the bastard wing, the bone of the metacarpus, the first phalanx of the termination of the wing, are all visible. Some of the ribs are so well preserved, that the vertebral and sternal part of the rib are quite distinguishable, as also is the recurrent apophysis, a part so characteristic of the ribs of birds, that it alone would be sufficient to identify this ornitholite. The pelvis is not less characteristic, from its general form and the direction of the pubis, than from the holes and emarginations observable in it. The femur, tibia, tarsus, three entire toes, and two phalanges of a fourth, constitute the leg of a bird thoroughly well characterized. The neck and the coccyx have left impressions rather more confused, in consequence of the more complicated forms of the bones which compose them; but the whole is most clearly to be recognized by every person who has ever cast his eyes on the skeleton of a bird.

It would appear that this and the last mentioned ornitholite are precisely of the same species, but the first of the three belonged to a larger species.

It remained only (after having thus clearly ascertained the class) to determine the genera to which those different ornitholites belonged. But this the Baron confesses to be a problem excessively difficult, if not absolutely impossible, to resolve.

The inter-resemblances of birds are considerably greater than those of quadrupeds. The extreme limits of the class are more approximated to each other, and the number of species contained within those limits is much more considerable. The differences, therefore, between two species must at times be utterly impossible to decide, on inspection of the skeletons. Even the genera do not always possess osteological characters of a sufficiently discriminative importance. They are almost all distinguished according to the form of the bill, which is not preserved entire in the ordinary skeleton, and still less so in the fossils, crushed and fractured as they have been in the gypsum of Paris. All, therefore, that can be said on the specific characters of ornitholites amounts to little, and scarcely passes the bounds of mere conjecture.

It is, however, certain that the number of complete feet found in the gypsum furnishes proof of the existence of at least nine species, which M. Cuvier has arranged according to size, beginning with the largest and ending with the least. The other skeletons and fragments he refers to certain of these feet, as belonging to those species, with the exception of the skeleton of M. Darcet. To this skeleton there are no feet, nor any portions of feet remaining; but as the wings and neck are shorter than those of the skeletons above cited, and as in those two the first is smaller than any found, the Baron concludes this bird to have been of a different species from all the others, and to have constituted a tenth.

Several feet and bones, not referable to the skeletons we have mentioned, have been found, nor in all instances referable to the feet which we have noticed as designating distinct species.

A shoulder-blade was found, resembling that of the genus pelican; a foot resembling that of the sea-lark; a metacarpus of a bird of prey, of the magnitude of *Falco haliaetos*, and which M. Cuvier thinks may indicate an eleventh species: there was also a femur which had the strongest resemblance to that of the preserved skeleton of the celebrated *Ibis*, though evidently not exactly of the same species.

But all these ideas must be considered as conjectures, and by no means possessing the decided certainty which attaches to the proposition which we have laid down, concerning the bones of quadrupeds.

It is, however, a very great point, to have proved the existence of the class of birds among the fossil remains of former worlds; to have proved that at this remote era, when the species of animals were so different from those which we behold at the present day, the general laws of co-existence, of structure, in short of all which is elevated above simple specific relations, all which belongs to the nature of organs, and to their essential functions, were the same as at present.

We find, in fact, that when the proportions of the parts, the length of the wings and of the feet, the articulations of the toes, the forms and number of the vertebræ in birds, as well as in quadrupeds, and among the latter, the number, form, and respective position of the teeth, are subjected to the great rules established by the nature of things, almost as certain consequences are to be deduced from reasoning as from actual observations.

It is vain, then, any longer to insist on variations of organic structure being the result of habits or circumstances. Nothing has been elongated, shortened, or modified, either by external causes or internal volition ; all that has been changed has been changed suddenly, and has left nothing but wrecks behind it, to advertise us of its former existence.

Fossil Reptiles.

In describing the fossil remains of extinct animals, M. Cuvier found it necessary, in his great work, to give a preliminary and very detailed description of the fossil osteology of the living species. Through this it was not to be expected that we should strictly follow him, as the necessity of so doing has been in some measure anticipated by the preceding portions of "The Animal Kingdom," and as we should exceed all reasonable limits by dilating much on so extensive a subject. Occasional reference to it, however, has been, and will be necessary, to enable the reader to appreciate the distinctions between the present and the past animal population of the earth. It will, indeed, be necessary to treat a little more formally of specific characters in reptiles, though our notices, of course, must be confined to osteology: the other attributes of this class will be found in their proper places in the present work.

Every research on the differences of the productions of nature must also infallibly conduct us to a consideration of their relations. It is easy for the reader to perceive, that notwithstanding the very varied proportions of the bones we have hitherto surveyed; notwithstanding the very singular external forms which have often resulted from these varied proportions, that, nevertheless, there exists among all the mammifera a sort of common plan, a composition or combination, nearly similar; so much so, that every bone may be recognized by its uses and position through every metamorphosis which it undergoes, and in spite of all the augmentations and diminutions which it experiences. Thus, in all the heads of this class, from man to the balænæ, we trace the frontal bones, the parietal, the nasal, in a word, all the constituent parts of the cranium and face, with such few and trifling exceptions as the absence of the lachrymals in some species, and perhaps that of the interparietals in some others. The other apparent differences in the number of the bones are referable, generally, to the greater or less promptitude with which these bones unite, and the sutures which distinguish them disappear. Thus the parietal in the adult sometimes appears simple, sometimes double, and even triple or quadruple, counting the interparietals, which always finish by uniting together. But when we consider the animal nearer to its birth, these anomalies disappear, and in the fœtus, or generally at the time when all the bones are still distinct, we find a regular number, the same for all species, with indeed some very rare exceptions.

It was a curious question to find out whether this analogy was sustained in the other classes of the vertebralia, and if the differences which they exhibit might not depend upon the periods in which their bones became inter-cemented; if the reptiles, for instance, which always preserve many more sutures in the head than the mammifera, might be considered in this respect as mammifera, in a state analogous to that of the fœtus; if the birds, which, in their early age, have as many as the reptiles, but which, when they approach the adult state, often exhibit less than the mammifera, might be considered, on the contrary, as mammifera passing rapidly from one state to the other, and even going beyond this last, in reference to the union of their bones.

M. Geoffroy St. Hilaire was one of the first to consider this interesting problem; and in many particulars he has treated it with signal success. The Baron has also considered it several times in the course of his lectures, and on other occasions. But many other profound anatomists, especially certain of the German school, have made it the object of more consecutive researches and detailed examinations.

These writers have not only endeavoured to assign to each bone in the oviparous vertebralia its correspondence with some bone, or determined portion of a bone in the mammifera, but, following the ideal and pantheistic system of metaphysics, termed the philosophy of nature, which has for some time prevailed very much in Germany, and the language of which has been adopted in the positive sciences, they have endeavoured to find in the head a representation of the totality of the body, as in general, according to the principles of this philosophy, each part, and each part of a part, should always represent the whole.

Thus M. Oken, in his *programma on the signification of the head*, has proceeded from the analogy which exists, in many respects, between the kind of rings which the bones of the cranium form, and those of the vertebræ, to consider the cranium as composed of three vertebræ. Thus the body of the anterior sphenoid represents the body of the first vertebra, its alæ orbitales the lateral parts of the ring, and the frontals the spinous apophysis: this is called the *ocular* vertebra. The second, or the *maxillary*, is represented in the same manner by the body of the hinder sphenoid, by its temporal alæ and by the parietal bones; and the third or *auricular* vertebra, by the basilary bone, the lateral and upper occipitals. Thus seeking in different parts of the head the representatives of the different parts of the entire body, he has seen in the cranium, taken separately, the *head of the head*, in the nose the thorax of the head, in the jaws, the upper and lower extremities, or the arms and legs.

It is easy to conceive that, with a little imagination, applications very different from these, and very different *inter se*, might be made of a principle so abstracted from common sense, and elevated to such a soaring pitch above mere matter of fact. Accordingly, we find that in 1811, M. Meckel, in his "Materials for Comparative Anatomy," takes the ethmoïd for the body of a vertebra, of which the frontals will be the annular part, and represents the temporals as another vertebra, the body of which is divided into two parts (the petrous portions) by the forced introduction of the body of a third (the basilary bone).

The ethmoïdal vertebra was afterwards adopted as a fourth, and added, under the name of *olfactive* vertebra, to the three of M. Oken, by M. Bojanus, in 1818, in the third number of the *Isis*, and in 1821, in the *Parergon* of his great work on the anatomy of the tortoise.

M. Spix, in his work on the composition of the head, entitled *Cephalogenesis*, published in 1815, sticks to the three vertebræ of the cranium, but departs widely from the views of M. Oken, relative to the bones of the face.

Representing the hyoïd bone, the shoulder, and the pelvis with the extremities thereto attached, as three circles, or groups of osseous pieces of similar nature, he finds them again in the face attached to the three vertebræ of the cranium. The bones which compose the nose represent the hyoïd and laryngian apparatus, and those of the two jaws represent the two extremities of the body, but with an arrangement of relations quite different from that of M. Oken. Thus the nasal bones, properly so called, represent the sternum, the cartilages, and the xyphoïd; the shoulder-blade corresponds to the posterior frontal bone, the clavicle to the cheek-bone. The small bones of the ear represent the pubis, the condyloïd apophysis, the femur, the coronoïd, the tibia, &c. The teeth, according to M. Spix, are only claws, and the alveoli represent the phalanges!

This strange fantasy, of finding a representative of the body in this way, has obliged some authors to give denominations to certain bones of reptiles and fishes, which they would have never thought of otherwise. The wish of constantly finding the number of osseous pieces to be the same has led others into deviations not less strange. When they found it difficult to support their theory on those bones where it appeared most natural to seek a foundation for it, they betook themselves to the neighbouring bones. Sometimes they found themselves obliged to admit the most singular metastases, without even ever thinking of the number of organs and soft parts which it was necessary to displace, and assign a different agency to, before they could transport a bone from one place into another, adjoining. For instance, they would insert (to make out their system) a piece belonging to the sternum between two pieces belonging to the hyoïd bone, or make some similar transposition, as if such a process was quite simple, and that none but osseous parts entered into the composition of the animal.

Such theoretical reveries have been carried to a great extent by these writers in their discussions on the reptile class, and still farther in treating of the fishes, especially regarding the osseous pieces composing the opercula and the hyoïd bone.

But in matters of inductive science, it is absolutely necessary to steer clear of the errors attendant on preconceived theories. This rock, the Baron, with his usual judgment and caution, has particularly avoided. He neither pretends to find a permanent recurrence of the same number of osseous pieces, nor representations in the head of the other component parts of the corporeal system. Neither does he attempt to maintain that the bones of the head are absolutely the same in all the genera, but he endeavours to ascertain how far their correspondence extends, and what are the precise limits of its progress. For this purpose, he commences with that oviparous quadruped, which, in the head at least, affords the most sensible analogies with the mammifera: this is the Crocodile. He points out the bones which correspond with our own, and, in establishing this correspondence, he consults not merely their position, but also the muscles which are attached to them, the nerves, vessels, &c. He candidly instances the bones which escape this analogy; shews where a bone, a foramen, a facet, a suture, appear to be wanting, and where some new one is to be found.

On this principle he proceeds through all the genera. Having no necessity to represent things otherwise than they are, he neither employs vague propositions nor figurative expressions, which have proved so fertile of self-illusion to the ablest writers. Thus, though the results he arrives at are not so calculated to dazzle the imagination as theirs, they satisfy the judgment, from their consistency with facts.

A development of the bones of the head, in treating of the crocodile, becomes of the greatest importance, for when this is done, the study of those of the tortoises, lizards, and crocodiles, is comparatively easy. With the batracian reptiles, indeed, the case is a little different.

The bones of the shoulder and the sternum in the lizard require considerable attention, as there is a greater degree of complexity attached to them.

The hyoïd bone in the batracians is of the greatest importance, because its study enables us to form clear ideas on that of the fishes, concerning which numerous and discordant systems have been proposed.

Relative to this point, we shall find that the successive simplification and final disappearance of the auricular apparatus in the batracian reptiles, and also the gradual development of that of the hyoïd, notwithstanding the existence of a larynx and sternum, will lead us to the idea that the bones of the ear do not re-appear in the osseous fishes under the form of opercula; that the bronchial apparatus has no need to complete its complicated formation by the intercalation of sternal, laryngian, or costal pieces; and, in short, that the operculary apparatus is one specially peculiar to the animals to which nature has accorded it.

As to the bones which compose the other parts of the body of reptiles, the pieces which constitute them are so far from being multiplied, like those of the head, that, in youth, they not unfrequently want those portions of their extremities called epiphyses.

In the crocodiles and tortoises, the extremities of the bones and their principal eminences are clothed with cartilages of greater or less thickness, which harden and ossify with age. They do not, however, as in the mammifera, form an osseous nucleus, separated for a time from the body of the bone or diaphysis by a suture. This is singular enough, as the saurian reptiles, and particularly the monitors, exhibit epiphyses strongly marked.

Without a thorough investigation of the osteology of existing reptiles of every genus, and without reducing this osteology to general rules, it was impossible for the Baron to conduct his researches on the fossil reptiles in a satisfactory manner. The necessity of this led to views and details much more extensive than those which the bones of mammiferous animals had suggested. We shall accordingly (in proportion to our limits) follow him more extensively into this department.

The mammifera were the last, as they evidently are the most perfect productions of the Creative power. Their organization is in all points more complicated and complete, fitting them for the performance of a greater variety of functions, and placing them altogether in a more extended range of existence. There is among them a more decided separation of all the organs which concur to the purposes of life, and nature appears to have employed more art, and less economy, in the process of their conformation.

The reptiles commenced their existence long before the mammalia. Their debris are found in formations of greater antiquity, and the naturalist is compelled to seek for their remains in the deeper strata of our globe.

We have already seen, in the brief review of fossil mammifera, which we have presented to our readers, that the greatest number, beyond all comparison, of viviparous quadrupeds, have left their bones in the latest diluvial strata, in caverns, or in the clefts and crevices of rocks. The sea which passed over them had scarcely sufficient time to deposit many traces of its transitory inundation: at all events, it has not covered them with compact, solid, regular strata. Some local formations alone, and which appear to be of a more ancient date, inclose principally the remains of non-existing genera, and are covered in certain places by marine depositions. But in the coarse limestone, the *calcaire grossier* of the French, in the limestone with *cerithia,* none but marine mammifera have been found, such as phocæ, lamantins, and cetacea. In one instance alone, which may, after all, be attributable to some mistake, has an exception to this law been found. We allude to the molasse, the lignites contained therein, and certain contemporary lignites, in which incontestable bones of mammifera have been observed, and where the Baron found his antracotheria, and palæotheria, accompanied, as in the Parisian gypsum, with trionyces and crocodiles, and where he also met with some bones and teeth of mastodon, and a jaw-bone of castor. These molasse and lignites are said to be invariably lower than the coarse limestone; but though this point were as well established as it appears to be doubtful; though it were true, that the molasse and lignites of two different eras had not been confounded together, still it must be remembered that the strata resting on the chalk are universally allowed to be the most ancient in which the debris of mammifera are observed; that the chalk itself positively contains none, and still less can they be supposed to exist in strata of an anterior state; while, on the other hand, the chalk, and most part of these anterior strata, even to the great pit-coal formation, abound, in certain places, with tortoises, lizards, and crocodiles, species which are very rare in the superficial strata.

We now come to the consideration of another age of the world, an age in which the earth was inhabited only by the cold-blooded reptiles, and in which the sea abounded in ammonites, belemnites, terebratulæ, and encrinites, and in which all these genera, now of such prodigious rarity, constituted the basis of its population. This age is termed by geologists that of the secondary strata.

We shall begin with the crocodiles, in our review of the remains of reptiles.

Fossil Crocodiles.

The bones of crocodiles are more abundant than those of other reptiles, and more easily recognized. They exist in a great number of strata, both in those of a comparatively middle antiquity, as in the gypsum of Montmartre, or those of an antiquity more remote, such as the limestone of the neighbourhood of Caen, from which the frieze-stones are taken, and the blue calcareous marl of the environs of Honfleur.

It will be necessary, before we proceed to the consideration of their fossil remains, to say a word or two concerning the distinctive characters of the existing species of crocodiles, and to give a sketch of their osteology. Without this, the reader would but imperfectly understand the account which we have to furnish concerning the fossil debris. If we did not do this to the same extent in the case of the mammifera, it was because their species and osteology are now in general better known.

The precise determination of species and of their distinctive characters constitutes the first and grand basis on which all the researches of natural history should be founded. The most curious observations, and the newest views, lose nearly almost all their value when deprived of this support. Nor should the dryness of this sort of labour prevent those who are desirous of arriving at satisfactory results from undertaking it.

For a considerable period of time, the larger animals were precisely those concerning whose species naturalists were in possession of the least accurate information. Their size, the difficulty of killing, of transporting, of preserving them, and the remoteness of the climates which they inhabit, rendered it impossible to assemble and compare together, at once, a sufficient number of individuals.

It has been only, for instance, of late years, comparatively speaking, that it became known that many species existed of the elephant and the rhinoceros; and though it was maintained more anciently that the species of the crocodiles were numerous, yet so vague and variable were the characters assigned to them, and so little in conformity with truth, that those who maintained the negative of this question could scarcely be blamed.

The ancients, who might, if they pleased, have compared the crocodile of India with that of the Nile, have entered into no such detail. One author alone, Ælian, has casually noticed, in a word, the *gavial*, and the common crocodile of the Ganges. A little more attention was paid by them to the crocodile of Egypt, of which, however, their knowledge was imperfect enough, though more accurate than their knowledge of the hippopotamus. Herodotus has given a tolerably exact description and history of this crocodile, and even his errors have some foundation in truth. Aristotle reduced to their just value many assertions more or less erroneous of the Father of History on this subject; and added, to the external and internal description of this animal, many particulars equally correct and valuable. The successors of these two great writers did little more than either copy them implicitly, or add to their relations circumstances of doubtful authority, or idle superstitious tales.

There are few things in history more remarkable than the utter incuriosity of the Romans respecting all subjects of natural science. No other nation ever possessed such advantages as they did for the study of zoology, and scarcely any other civilized nation ever paid less attention to the subject. With opportunities infinitely more numerous than ever occurred to any other people, of observing the rarest animals, they took no pains to transmit to us anything valuable concerning them. With respect to the crocodile, they are liable to the same reproaches which have been made with so much justice against them on the score of the hippopotamus. These two animals, they saw, for the first time, under the edileship of Scaurus, when five crocodiles were exhibited. On another occasion these animals were brought to Rome by certain inhabitants of Dendera, who played a variety of tricks with them. One of the most astonishing spectacles of this kind was given by Augustus, in the year of Rome 748, seven years before the Christian era. The Flaminian circus was filled with water, and thirty-six crocodiles were exhibited and destroyed there. Crocodiles were also shown by Antoninus and Heliogabalus, according to the accounts of Julius Capitolinus and Lampridius, and it is highly probable that they were exhibited on many other occasions, which the authors, whose works have descended to us, have not thought proper to notice.

Nevertheless, the Romans, and the Greeks, who lived under their dominion, availed themselves of all these opportunities of seeing the crocodile, only for the purpose of giving some exactness to their figures of this animal. It is certainly very well represented on their medals and monuments. The mosaic of Palestrina, the plinthus of the statue of the Nile, various medals of the different emperors, and sundry engraved stones, clearly prove that the artists were sufficiently familiarized with the external conformation of the crocodile. But we have no reason to believe that any of the ancient naturalists recognised more than one species. The fact that some of them have spoken of a crocodile called *suchus* or *suchis*, does not, as we shall have reason to observe hereafter, invalidate this assertion.

Nor shall we find that even the moderns themselves, before the attention of M. Cuvier was drawn to the subject, have carried their exacter modes of observation and arrangement in natural history into this department of the animal kingdom. The most enlightened naturalists of the eighteenth century have confounded together, contrary to all rules, distinct species of the crocodile, and mixed with them certain of the larger lizards, which have no claims to such an arrangement. Thus, Linnæus, in the editions of the *Systema Naturæ* published during his own life, admitted only a single crocodile, without even distinguishing the species of the Ganges with elongated beak. His contemporary Gronovius separated from the *crocodile*, properly so called, *the cayman*, or *American crocodile*, *the crocodile of the Ganges*, (to which last he joined the *black crocodile* of Adanson,) and a fourth species which he named *the crocodile* of Ceylon, and which he distinguished by this accidental character, and one peculiar only to the individual which he described, namely, that the two external toes only are completely palmated. Laurenti has established two particular species besides the crocodile and the cayman, founded upon indifferent figures from Seba, (*Crocodilus Africanus et C. terrestris*,) but he has totally forgotten the *gavial* and the *black crocodile*. Lacepede, like the two preceding writers, admits four species, but combines them differently. His first species is the *crocodile*, under which he includes, after the example of Linnæus, the common crocodiles of the old and new world as one and the same species. 2. The *black crocodile*, which he merely slightly notices after Adanson. 3. The *gavial*, or long-beaked crocodile of the Ganges, of which he was the first to give a good description. And lastly, an animal which he terms *fouette-queue*, because he judges it to be the same as the *lacerta caudiverbera* of Linnæus. Its description was taken, however, from an altered figure of a crocodile given by Seba.

Gmelin reduced the crocodiles to three species, 1st, by joining the *common crocodile* and the *crocodilus Africanus* of *Laurenti*, under the name of *lacerta crocodilus;* 2ndly, uniting the *gavial*, *crocodilus terrestris* of Laurenti, and the *black crocodile*, under his *lacerta Gangetica;* and 3rdly, by separating the *cayman* under the name of *lacerta alligator*. Finally, Bonaterre returned to the quaternary number, by adding the *fouette-queue* of M. de Lacepede to Gmelin's three species, and neglecting the black crocodile. These differences, however, in the establishment of species were nothing in comparison of those which existed in their characters and synonymy.

Without entering into an account of these inconsistencies, which produced a perfect chaos in this department of natural history, we shall notice, in a few words, the characters which circumscribe the genus.

All the lizards or saurian reptiles which have the tail flattened at the sides, the hind feet palmated, or semi-palmate; the fleshy tongue attached to the floor of the mouth, almost as far as its edges, and by no means extensible; sharp, simple teeth on a single range, and a single penis in the male;—these are CROCODILES. The three first-mentioned characters determine the aquatic nature of these animals, and the fourth constitutes them voracious carnivora.

All the animals of this genus hitherto known also unite the

following characters, which future discoveries may, perhaps, prove to be less general and less essential:—

Five toes before; four behind; three toes only armed with claws on each foot; thus there are two before and one behind without a claw: all the tail and the upper and under part of the body covered with square scales; the greatest part of those of the back raised by longitudinal crests or ridges, projecting more or less: the flanks furnished with small rounded scales only: similar ridges forming on the base of the tail, two crests denticulated like a saw, which unite in a single one for the rest of its length: the ears closed externally by two fleshy lips: the nostrils forming a long narrow canal which opens internally in the throat: the eyes having three lids: two small pouches opening under the neck, and containing a musky substance.

Their anatomy also presents characters common to all the species, and which clearly distinguish their skeleton from that of other saurian reptiles. These are:—

1. The vertebræ of the neck have certain kinds of false ribs, which, touching each other by their extremities, hinder the animal from completely turning the head aside.

2. The sternum is prolonged beyond the ribs, and has false ribs of a description altogether peculiar, which do not articulate with the vertebræ, but serve for the purpose of guarding the abdomen, &c.

The crocodiles with all these characters constitute a very natural genus. This is a truth which was sufficiently perceptible to various systematic writers, but they were wrong in uniting to this genus certain species which in truth possessed the character assigned by their system, but which differed from the genus crocodile in every other respect.

It is not unimportant nor uninteresting to trace the steps by which the Baron Cuvier was led to a just classification of these animals. To arrive at the distinction of species, he commenced by putting out of the investigation the long-beaked crocodiles, vulgarly termed gavials, or crocodiles of the Ganges, they being allowed, by universal consent, to constitute, if no higher a subdivision, at least a species perfectly distinct from all others.

There remained all those animals known under the names of *crocodile*, and of *cayman*, or *alligator*, and so frequently taken one for the other. These animals are very numerous in the collections of natural history in France, in consequence of the relations of that country with Egypt, Senegal, and Guyana, which, with the East Indies, are the climates where crocodiles chiefly abound.

The Baron, after an examination of about sixty individuals of both sexes, from the length of twelve or fifteen feet down to the very young which had but broken the shell, seemed to think that they might all be reduced to two species, which he thus defined:—

1. *Crocodile.* Oblong muzzle; upper jaw notched on each side, to let the fourth lower tooth pass; hinder feet entirely palmated.

2. *Cayman.* Obtuse muzzle; upper jaw receiving the fourth lower tooth in a particular hollow which conceals it; hind feet semi-palmate.

All the individuals of the first form, whose origin the Baron could then learn with certainty, came from the Nile, Senegal, the Cape, or the East Indics. All of the second from America, either from Cayenne or elsewhere.

The Baron thus at first established two very distinct species of crocodile, without reckoning the gavials with the long muzzle; and he deemed himself justified in assigning the Old Continent as the country of the one, and the New as that of the other. Moreover, at the time of which we speak, he gave an indication of a third species, that of North America, of which he then possessed but a single individual, and the distinction has been since completely confirmed.

Such were the results of the earliest labours of M. Cuvier published in the year 1801. But during the ten subsequent years very important researches took place on the subject of the crocodiles—researches, prosecuted by various naturalists of different countries, and by the Baron himself. These researches occasioned certain modifications in the results of his former labours.

They proved, in the first instance, that what he had regarded merely as two species were in reality two subdivisions of the genus, susceptible of being themselves divided, by means of secondary characters, into several different species. They also showed, in the second place, that the two subdivisions are not entirely peculiar to the two continents to which he had respectively attributed them: for the crocodile of St. Domingo, for example, though certainly forming a species apart, does yet much more closely resemble the crocodiles proper, of the Old Continent, than those which are most commonly found in the New, and to which the Baron had restricted the appellation of *cayman*. From this the possibility was deduced that a reciprocal discovery might be made in the old continent of some species appertaining or approximating to the subdivision of the caymans.

The Baron, in enumerating the names of those naturalists to whom we are indebted for the augmentation of our knowledge on this genus, expressly excepts the editors of Buffon. They have given nothing original. Their figures even are but copies from other figures badly chosen. Daudin alone has given some slight indication of a new species. Neither has Shaw, in that part of his General Zoology which treats of reptiles, done anything to elucidate the subject. He admits but two species with short muzzles, namely, the *common crocodile* and the *alligator*. To represent the latter, he takes the altered figure of Seba above mentioned, which had been used for Lacepede's *fouette-queue*, and his two figures of crocodiles are caymans.

M. Faujas de Saint-Fond wrote especially on this subject in his geological essays and other works. He had numbers of individual specimens under his inspection, and a full opportunity of marking the distinctive characters above assigned to the crocodile and cayman. But he preferred pronouncing, without an examination, "that the cayman so nearly approaches the African species, that some naturalists, and I (adds he) am of the number, regard it only as a simple variety appertaining to climate."

A strong proof that this gentleman gave no proper consideration to this question, is that, in another work of tolerable circulation, he has given the figure of a crocodile, which he believed to be taken from an African individual preserved in the Museum of Paris. This, however, turned out to be the figure of a very different species from the East Indies, and of a species of which there was no specimen in the Museum. Another proof is an assertion in his Geological Essays, where he says, that supposing caymans to exist in the fossil state, the semi-palmation of the hinder foot must disappear, nor would their *second character* be more stable. Now, as this second character consists in the form of the osseous head, it is quite obvious that it would be as stable as any other that could be met with in fossil bones.

M. Schneider was the first who endeavoured to throw some little light on this confused subject. For this purpose he collected carefully all the passages of ancient authors respecting the crocodile, that a clear idea might be formed of the crocodile of the Nile; and also everything that modern writers have alleged on the same subject. He compared this recomposed description with that of the crocodile of Siam made by the missionaries, and of an American crocodile from Plumier, whose manuscript is preserved at Berlin.

But as the differences which M. Schneider deduces from this comparison, result only from the terms, or the peculiar views employed by authors who had no intention of giving distinctive characters, and as the American species dissected by Plumier happened to be that which has an exact analogy with the crocodile properly so called, the labours of this gentleman have led to little in the illustration of the genus except in the case of the crocodile of Siam, the peculiarities of which are well distinguished in his comparison. The species of the Nile is so badly authenticated by him, that most of the characters attributed to it belong in reality to the cayman. Even the cranium, of which he gives a figure, is not that of the crocodile, but one belonging to a cayman of a particular species.

Still there are many passages in M. Schneider containing useful and true indications respecting the multiplicity of species in America. He has also the merit of having been the first to recognize the distinction of that species, called by the missionaries the crocodile of Siam, and which certainly seems to be distinct. His *porosus* seems to be the same as the *biporcatus* of Cuvier, which we shall notice presently. The *pores* to each scale, which M. Schneider makes a specific character, are more or less to be found in all the crocodiles, properly so called. He also enumerates as distinct the *longirostris* or *gavial* universally known to be so. That which he names *sclerops*, and very erroneously gives as the crocodile of the Nile, is the most common cayman in Guinea. His *crocodilus trigonatus* appears to be another cayman, which is named *palpebrosus* by Cuvier. His *crocodilus cazrinatus*, *oopholis* and *palmatus*, all belong to the crocodile division, but his characters are so insufficient, that it is impossible to refer them to one species more than another. Lastly, his *crocodilus pentonix* is an imaginary being. It may be added that all his figures are made with singular carelessness and inaccuracy; neither do the descriptions of the text always accord with them.

M. Geoffroy St. Hilaire rendered a very eminent service to this branch of natural history by bringing from the Thebaïs a well-authenticated crocodile of the Nile. He has informed us that the fishermen of that country pretend to be acquainted with two other species. He brought back the mummy of a cranium from the catacombs, which induced him to examine analogous individuals in the Paris collections; and as this cranium and these individuals differed in some points from the common crocodile, he concluded them to belong to one of the species mentioned by the fishermen. To this species, M. Geoffroy is disposed to refer the crocodiles which were held in such peculiar veneration by the ancient Egyptians, and to apply to it the name of *suchus*, mentioned by Strabo and Photius. His numerous observations on the habits of the crocodile perfectly explain everything, doubtful or obscure, on this subject in the writings of the ancients, and form very valuable additions to the natural history of this reptile. He has also given a comparative description of the bones composing the head of this animal, embracing new and interesting views on the osteology of reptiles. But one of the most important facts relative to specific classification, ascertained by M. Geoffroy, is the establishment of the astonishing resemblance of the crocodile of St. Domingo with that of the Nile; and, consequently, of the great differences which distinguish the former from the common cayman of Cayenne.

The Baron, in his grand work on "Fossil Osteology," has finally divided the genus Crocodile, the characters of which we have already laid before our readers, into three sub-genera. This division was not made until he had personally inspected every new specimen that it was possible to procure;—taken a second review of all that he had already seen, and again perused all the most ancient writers on the subject—it was impossible, in fact, by any means, to make closer approaches to the truth, or arrive at an enumeration more complete and distinct. The genus itself became so clearly circumscribed and determined, that it could not be confounded with any other of the reptilia; and M. Cuvier's first division of it (already mentioned) in a general way, was fully confirmed by his subsequent observations. The general form, thus determined, became modified in its details into three particular forms, which we shall now lay before the reader.

The first sub-genus is named CAYMAN or ALLIGATOR. The name of *Cayman* or *Caïman* is almost generally employed by the Dutch, French, Spanish, and Portuguese colonists, to designate the crocodiles which are most common in the neighbourhood of their establishments. Authors are not agreed on the origin of this name. According to Bontius it came from the East Indies: his words are "*per totam Indian* CAYMAN *audit.*" Marcgravê makes it come from Congo—"JACARE *Brasiliensibus*, CAYMAN *Æthiopibus in Congo;*" and Rochefort says, it was original with the native islanders of the Antilles. The opinion of Marcgrave was confirmed to M. Cuvier, by the report of a very enlightened inhabitant of St. Domingo, who declared, that the slaves on their arrival from Africa, at the sight of a crocodile, instantly gave it the name of *cayman*. The word *alligator* has been more peculiarly employed by our own countrymen, but rather in a loose way. Notwithstanding the Latin air of this word, it has no relation with its apparent etymology. Some say that it is derived from *legateer* or *allegater*, which is the name of the crocodile in some parts of India. This, however, is not a well authenticated etymology: it seems more probable that it is a corruption of the Portugese *lagarto* which, itself, is clearly derived from *lacerta*.

This first sub-genus has the head less oblong than the crocodiles. Its length to its breadth, taken at the articulation of the jaws, is usually as three to two: never more than double. The length of the cranium makes more than a fourth of the total length of the head. The teeth are unequal: they are at the least nineteen, and sometimes two-and-twenty on each side below—nineteen at the least; and often twenty above. The first teeth of the lower jaw pierce the upper at a certain age; the fourth, which are the longest, enter into hollows in the upper jaw, where they are concealed when the mouth is closed. They do not enter into notches. The legs and hind feet are rounded, and have neither caruncles nor indentations at their edges. The intervals are at most but one half filled by a short membrane. The foramina of the cranium, in such species as have any, are very small: one species is entirely destitute of them.

The second sub-genus is that of the CROCODILES *properly so called.*

The name of *Crocodile* was given originally to the Nilotic species, according to Herodotus, by the Ionians, because, says he, it resembled the *crocodiles which, among themselves,* (the Ionians,) *inhabit hedges.* These were probably the lizard, named *Stellion* by Linnæus, and which is still called in modern Greek, by a slight change, *koslordylos.* In its primitive acceptation Κροκοδειλος signified "*that which fears the bank or shore.*" The true crocodile of the Nile was formerly, according to the historian above cited, named *Chamses* by the ancient Egyptians, and at the present day, in Egypt, is called *temsach*, according to all travellers.

In the individuals of this sub-genus the head is oblong, the length is double, and sometimes more than double the breadth. The length of the cranium is less than one-fourth of the total length of the head. The teeth are unequal. There are fifteen on each side below, and nineteen above. The first teeth of the lower jaw pierce the upper at a certain age; they pass into notches, and are not lodged in hollows of the upper jaw. The hind feet have at their external edge an indented crest or caruncle: the intervals of the toes, at least of the external ones, are entirely palmated. In the cranium, behind the eyes, are two oval foramina, which may be observed through the skin, even in dried individuals.

The GAVIALS constitute the third sub-genus. The muzzle is narrow, cylindrical, extremely elongated, and a little swelled out at the end. The length of the cranium is scarcely one-fifth of the entire length of the head. The teeth are nearly equal—twenty-five to twenty-seven on each side below; twenty-seven to twenty-eight above. The first two and fourth two of the lower jaw pass into notches of the upper, and not into hollows. The cranium has large foramina behind the eyes and the hind feet are indented and palamated like those of the crocodiles proper. The slender form of their muzzle renders them, though superior in size, much less formidable than the two other sub-genera. They usually content themselves with fish.

A brief notice of the species of the sub-genera will be necessary.

The first is the *Pike-muzzled Cayman (Crocodilus lucius)*, so called by M. Cuvier, who was under the necessity of new-naming all the species, to prevent the confusion which must have resulted from retaining the popular appellations. Several individuals, all from North America, came under the inspection of the Baron; who, however, hesitates to say that this is the only cayman found in that part of the globe. Dr. Leach has given a figure of it in the "Zoologist's Miscellany," t. ii. p. 117. CII.

It has all the characters common to the Caymans. Its muzzle is very flatted, the sides are almost parallel, and they unite in front by a parabolic curve. These circumstances give it a striking resemblance to the muzzle of a pike. The internal edges of the orbits are very much raised; but they are not united by a transversal crest or ridge. The external apertures of the nostrils are from the earliest age separated from each other by an osseous branch, which does not take place at any other age in the other species. The cranium has two oval, oblique fosses, of no great depth; in the bottom of which are small holes. The nape of the neck is armed in the middle with four principal plates, each of them being raised by a ridge. There are besides, two small ones in front, and two behind. On the back are eighteen transverse ranges of plates, each raised with a ridge. The number of the ridges, or of the plates of each range is as follows:—

One range with two ridges, two with four, three with six,

six with eight, two with six, and the rest with four. The unequal ridges sometimes found on the sides are not reckoned here.

These ridges are tolerably elevated, and nearly equal; but on the tail, as in all the crocodiles, the lateral ridges predominate, until they unite. There are nineteen transversal ranges of them as far as the union of the two crests, and as many after. It is to be observed, however, that these two numbers are more subject to variation than those of the ranges of the back.

One individual of this species measured but five feet; and the longest seen by M. Cuvier was not more than from six to seven. According, however, to the accounts of travellers, this species grows as large as any other; and Catesby, in particular, says, that he has seen some fourteen feet in length. The breadth of the cranium, at the articulation of the jaws, makes one half of its length; by which it will be found that the muzzle of this species is both wider and longer than those of the following.

This species seems adapted to a northern climate, and has been found as high as $32\frac{1}{2}°$ north latitude, even in the depth of a rigorous winter. In Louisiana, they throw themselves into the marshes when the cold commences, and fall into a lethargic sleep without being frozen. Their lethargy is so complete, that they may be cut in pieces without being awakened.

The Spectacled Cayman (Crocodilus Sclerops) is the most common in Cayenne. Its muzzle, though wide, has not the edges parallel. They proceed, approximating throughout their entire length, and forming a figure rather more triangular than in the preceding species. The surface of the bones of the head is very unequal, and appears throughout as if rotted, or gnawed into small holes. The interior edges of the orbits are greatly raised. From their anterior angle springs a projecting rib-like bone, which extends in front, and a little outwards, branching towards the teeth in aged individuals, especially in males. Another very marked projection proceeds from the anterior angle of one orbit to that of the other transversely. This is the most remarkable character of this species, and the one from which its name is derived. Behind the eyes, only, are two rather small holes.

Besides some scales, spread behind the occiput, and which form there in certain individuals a transverse range of tolerable regularity, the nape is armed with four transverse bands, very robust, which touch each other, and proceed to join the series of the bands of the back. The first two consist each of four scales, and, consequently, four ridges, the middle ones of which last are sometimes very much effaced. The two others have most frequently but two of these scales with their ridges.

Of the transverse ranges of the back, two have two ridges, four have six, five have eight, two have six, and four have four.

But with age, the lateral scales, not much marked at first, assume the form of the others, and two must be added to the number of ridges in each range. In general, it is rare to find two individuals perfectly similar in this respect.

All these ridges are not much raised, and are nearly equal with one another. Even the lateral ones of the base of the tail predominate but little over the others. It is only at their juncture that they become very projecting.

There are eleven, twelve, or thirteen ranges before this union, and twenty-one after. But these numbers vary.

This species grows to a considerable size; some being eleven, and some fourteen feet in length.

The *Yacaré* of M. D'Azara (which is the present species) proceeds no further south than the 32d degree—precisely the same limit which bounds the northern extension of the preceding species. In the flat island of Marajo, or Johannes, at the mouth of the Amazon, these animals remain in summer in the marshes; and when these dry up, what remains of water in the bottom is so completely filled with them, that no liquid is to be seen. It is thought that, at this time, the large devour the smaller. They cannot re-ascend the river, because the island is surrounded with salt water. Those of Guiana, according to Laborde, sometimes remain almost dry in the marshes, and are then considered most dangerous.

The *Cayman, with bony eyelids, (C. Palpebrosus,)* is the third species of M. Cuvier. In this he has established two varieties. Of the first, of which he received specimens from Cayenne, the following is a summary of the characters:—

The muzzle is a little more elongated and less depressed than in the preceding species. The surface of the bones is, however, equally vermiculated. No osseous band unites the orbits, but the upper lid is entirely filled by an osseous plate, divided into three pieces by sutures. The nape is armed with a range of four small scales; then come four transverse bands, each being provided with two projecting ridges, and they join those of the back. The cranium is not pierced, nor are foramina observable there at any age. There are twenty-one teeth below on each side, and nineteen above. The interval between the two external toes behind is obviously less palmated than in the preceding species, which renders the animal, according to M. Cuvier, more terrestrial. Cayenne is assuredly the country of this variety, to which the Baron appears to restrict the epithet *Palpebrosus.*

From four other specimens this eminent naturalist forms his second variety. The characters are:—

Osseous lids like the preceding: a ridge proceeding from the anterior angle of the orbit. A small emargination at the posterior edge of the cranium. The second band of the nape wider than the others, with two or three small ridges irregularly arranged; the large ridges cut into scalene triangles, greatly raised, which give a very bristling appearance to the nape. The ridges of the back and tail also project very much.

The Baron is very much in doubt whether these two be merely varieties, or distinct species. The crocodile of St. Domingo differs little more from that of the Nile than they do from each other. The country of the second is not determined, and a difference of continent would strengthen the supposition of a distinction of species. Seba, who has evidently figured the last one we have described, makes it an animal of Ceylon; but as the Baron observes, there is no more dependance to be placed on this assertion than on many other erroneous ones made by the same author relative to the origin of the specimens in his collection. One, however, of the individuals, on which the Baron founds the variety, was ticketted thus, in the Paris Museum, the letters being half effaced—*Krokodile noir du Niger*, which is the orthography and handwriting of M. Adanson. This naturalist tells, in his *Voyage*, that there are two crocodiles in Senegal, and M. de Beauvois says that there have been seen in Guinea a crocodile and a caïman. But an embarrassment still remains, for Adanson says that his black crocodile has a more elongated muzzle than the *green*. Now the latter is the crocodile of the Nile, and it happens unluckily that the variety of which we are now treating has the muzzle much shorter than that of the Egyptian species. The Baron therefore leaves these two animals we have described, provisionally, as varieties, giving to the latter the epithet *trigonatus*, which M. Schneider appears originally to have bestowed upon it.

The difficulty with the sub-genus of the CROCODILE is of a different nature from that which is attached to the investigation of the caymans. The species most easily authenticated resemble each other infinitely more. In the numerous varieties of age and sex, of which multifarious specimens have arrived in Europe, so many various shades, yet graduating towards each other, have been found, that it is impossible for a naturalist to know where to stop in their determination.

The *Common Crocodile of the Nile (Crocodilus vulgaris*

Cuv. *Lacerta crocodilus*, Linnæus. *Crocodilus Niloticus* Daud.) has, notwithstanding its ancient celebrity, been almost continually mistaken by those modern naturalists who have attempted to distinguish the species of this genus. Laurenti, and even Blumenbach, have taken for it the *palpebrosus;* and Schneider, for the *sclerops.* This has chiefly proceeded from the very detestable figures generally given by Egyptian travellers of this animal, and the slender authentication of most of the specimens dispersed through the different cabinets of Europe. To the researches of Geoffroy we are indebted for the first establishment of precise notions on this subject.

In the true crocodile of the Nile, the length of the head is double the breadth. The sides are in a general direction, nearly rectilinear, and the head thus represents an elongated isosceles triangle. The foramina with which the cranium is pierced are large, and broader than they are long. The muzzle is rugged and uneven, especially in aged individuals, but has no particular projecting ridge. Immediately behind the cranium, on a transverse line, are four small isolated scales, with ridges. Then comes the large plate of the nape, formed of six scales with ridges; then two scales a little separated; and afterwards the transverse bands of the back, almost always fifteen or sixteen in number. The first twelve have each six scales and six ridges; the three bands between the thighs have only four each.

All these ridges are nearly equal, and moderately projecting. There is, moreover, on each side, a longitudinal range of seven or eight ridged or carinated scales, not so much united to the assemblage of the others. The lateral ridges of the tail only commence to become prominent on the sixth band, and to form two crests; these unite on the seventeenth or eighteenth band, and there are yet twenty-eight more to the end of the tail.

The equality of the scales, of the ridges, and of their number in each band, and their position on two longitudinal lines, give, says M. Cuvier, the crocodile of Egypt the appearance of having the back regularly paved with squares, or diamonds with four angles.

The scales of the back and nape, especially those of the two longitudinal middle lines, are wider than they are long. Those of the belly have a pore more or less marked towards their lower edge. The colour of the upper part is a bronzed green, more or less clear, marbled with brown; that of the lower a yellowish green.

The Nilotic species also inhabits Senegal. It is probable also that it is found in the Zaïre, in the Ioliba, and other rivers of Africa. It certainly exists in Madagascar, as appears from a specimen sent to France by M. Havet.

Among the crocodiles referable to this species, there are some which have the head rather more elongated in proportion to its breadth, and a little flatter, or rather less unequal, at its surface. Beside the differences in the form of the head, these individuals exhibit some in the shades of their colours. These differences, joined to the testimony of the fishermen of the Thebais, would seem to authorize the distinction admitted by M. Geoffroi, if not of another species, at least of a particular race of the crocodile inhabiting the same country. M. Geoffroi has given the name of *Suchus* to this variety, a name which we find in Strabo.

It was the opinion of Jablonsky and Larcher, that the *Suchus*, or *Souchis*, was a particular species of the crocodile, and that which, in preference to the other, the Egyptians used to rear in their temples.

This opinion, however, is liable to great controversy. It appears certain that neither Herodotus, nor Aristotle, nor Diodorus, nor Pliny, nor Ælian, had any idea of two species of the crocodile in Egypt. Herodotus, after telling us that the inhabitants of Elephantina used to eat crocodiles, informs us they are named *Champses*, and this he states in a general way, without applying it to this particular district, or to any particular species. He says, Καλέονται δὲ ὀ κροκόδειλοι ἀλλα χάμψαι. He does not mean that they are marine *crocodiles* in the rest of Egypt, and *champses* only in Elephantina, for he afterwards informs us that the word crocodile is Ionian.

As for the passage in Strabo, it will not be found, on examination, to prove anything in favour of a distinct species, and appears to apply only to the individual particularly consecrated. His words are, Κάι ἔςιν ἱερὸς (κροκοδέιλος) παρ᾽ἀυτοῖς ἐν λίμνη καθ'αὑτὸν τρεφόμενος, χειροήθης τοῖς ἱερεῦσι, καλεῖται δὲ Σοῦχος.

The proper translation of this is—"They have a sacred crocodile, which they rear by itself in a lake, which is tame (or gentle) to the priests, and which they name *Suchis*." Thus, in the same manner the sacred bull of Memphis was named *Apis*, that of Heliopolis *Mnevis*, and that of Hermonthis *Pacis*, these names not being intended to designate particular races of the ox, but consecrated individuals.

Strabo, in his account of the sacred crocodile, which he himself presented with food, speaks but of a single individual. Herodotus, in the same way, attributes but to one individual the ornaments and honours which he describes. Diodorus speaks of the crocodile of Lake Mæris, of the he-goat of Mendès, in exactly the same style as he does of *Apis* and *Mnevis*, obviously meaning only individuals.

Plutarch is very express upon this subject. He says, "Though some Egyptians reverenced the entire species of dogs, others that of wolves, and others that of the *crocodile*, they only rear one individual respectively: some bring up a dog, others a wolf, and others a *crocodile*, for it would not be possible to rear them all."

Ælian, indeed, in the history which he relates of one of the Ptolemies, who used to consult them as oracles, seems to countenance the supposition that there were several "Quum ex *crocodilis*, antiquissimum et prestantissimum appellaret;" but Plutarch, in relating the same history, speaks of but one, "the *sacred crocodile*."

It is very true, that the entire species was spared in those places, where an individual was reared and consecrated. It is also true that these consecrated individuals, fed and well treated by the priests, in the course of time grew tame. But this, so far from being a peculiar character of the species, was constantly cited by the ancients, as a proof that there was no animal, however cruel, that might not be tamed by the care and assiduity of man; and especially when food was abundantly provided. Aristotle expressly concludes, from this familiarity between the *priests* and the crocodiles, that the most ferocious animals might live peaceably together as long as there was no lack of nourishment. Whether this illustrious philosopher considered that the priests on their side furnished an illustration of this fact with the crocodiles, we cannot pretend to determine.

There are, moreover, very sufficient proofs, that the crocodiles most common in the districts where they were brought up and reverenced, were not an atom more gentle than those of the rest of Egypt. On the contrary, they were more cruel, because they were less timid. Ælian relates, that among the Tentyrites, who destroyed these animals as much as they possibly could, one might bathe and swim in perfect safety in the river. While at Ombos, Coptos, and Arsinoë, where they were revered, it was not safe even to walk upon the bank, much less to wash the feet, or draw water. He adds, that the inhabitants considered it an honour, and rejoiced when these animals devoured their children.

Whatever might have been the origin of so besotted a worship as that of the crocodile, we have sufficient evidence that the Egyptians did not attribute it to the gentleness of the species thus held in adoration. On the contrary, many were of opinion that it was the ferocity of these animals which raised them to the rank of deities, for by that was arrested the progress of the Arab and Libyan robbers, who, but for the crocodiles, would have been passing and repassing the Nile and its canals incessantly. Diodorus especially mentions this, among other reasons, and Cicero, before him, states it in the plainest language: "*Ægyptii nullam belluam nisi ab aliquam utilitatem consecraverunt; crocodilum, quod terrore arceat latrones.*"

A strange passage of Damascius, a Greek writer of the Lower Empire, quoted by Photius, has given rise to the opinion of Jablonski and Larcher. His words are:

Ὁ ἱπποπόταμος ἄδικον ζῶον; ὁ Σοῦχος (or more properly Σοῦχις) δίκαιος. Ὄνομα δὲ κροκοδείλου καὶ ἐιδος ὁ Σοῦχος; οὐ γὰρ ἀδικεῖ ζῶον. "The hippopotamus is unjust; the SUCHIS is just: it has the name and figure of the crocodile: it hurts no animal."

The explanation of this is very simple. In the time of Damascius, who lived under Justinian, the Pagans were persecuted, and sacred animals were no longer reared in Egypt. Nothing remained of the ancient worship but in tradition and the reports of books. Damascius was, from all that is known of him, evidently a very ignorant and credulous man. He had read or heard that the *suchis, or sacred crocodile* of Arsinoë, was harmless, and he immediately sets it down as an innocent and peculiar species. This explanation is sufficient, if we translate the word ἐιδος species; but the signification of this word is ambiguous, and the mode in which it is used by this author is not calculated to fix the sense. We have thought proper to translate it in its original and natural acceptation, in which we are fully convinced it was employed by Damascius. It is, moreover, evident that the *suchis*, even supposing that it was a weaker kind of crocodile, must have been carnivorous. To say, therefore, that it *did no injury to any animal*, is false and absurd, and such an error ought to deprive the passage in question of all credit.

The second species established by the Baron is the *double-*

crested crocodile (crocodilus biporcatus), which is the *porosus* of Schneider. M. Cuvier had the good fortune to possess specimens of this species at all ages, from its issuing from the egg to its attainment of the length of twelve feet. This enabled him not only to distinguish the characters with the utmost certainty, but also furnished him with the most useful knowledge concerning the variations of form which the crocodiles in general undergo in proportion to age.

The head differs from that of the common crocodile only by two projecting crests or ridges which proceed from the anterior angle of the orbit, and descend in almost a parallel line along the muzzle, disappearing by degrees. The scales of the back are more numerous than in the preceding species. The first range has four; the following have six. The eight which come after have each of them eight: then there are three with six, and three with four. The entire number of ranges is seventeen. These scales, instead of being square, and wider than long, are, on the contrary, oval, and longer than wide. In young individuals there are pores to all the dorsal scales, and at the triangular intervals left between them. The ventral pores are also very obvious in this species.

This crocodile is the most common in all those rivers which flow towards the Indian Ocean. It is found in Java. Peron has observed it in Timor and in the Sechelles Islands. It has also been taken in the Ganges, and M. Cuvier received from Calcutta a skeleton seventeen feet in length.

The next species is the *lozenge crocodile (crocodilus rhombifer.)* The country of this crocodile is unknown, and the Baron had only seen two individuals.

The chaffron is more gibbous than in the other species, its transverse section representing at least a semicircle. From the anterior angle of each orbit proceeds a blunt rectilinear ridge, which promptly approaches its correspondent one, and forms with it and the internal edges of the two orbits an incomplete lozenge at its posterior angle. The four limbs are clothed in stronger scales than in the preceding, each elevated in the centre with a bulky projecting ridge. This gives them the appearance of being more vigorously armed.

The fourth species is the *crocodilus galeatus*, Cuv., Siamensis, Schneider. Its distinguishing character consists in two triangular osseous crests implanted one behind the other on the central line of the cranium. It resembles almost in everything the common species of the Nile, attains the length of more than ten feet, and inhabits the rivers of Siam.

It is only known by the description given by the French missionaries at Siam.

The *crocodilus biscutatus* of Cuvier, black crocodile of Adanson, is the fourth species. The muzzle is more elongated than in the preceding species, but less so than in that of St. Domingo. The two middle longitudinal lines of the crests are less projecting than the lateral ones, and these last are somewhat irregularly disposed. The nape is armed with only two large pyramidical scales on its centre, and two small ones in front. The number of transverse scales, as far as behind the thighs, is only fifteen. The scales of the two longitudinal central lines are broader than they are long.

This species has been found in the river Senegal by Adanson. This naturalist says that it is blacker and more ferocious than the common species, which is also to be met with in the same river.

The *slender-muzzled crocodile* of St. Domingo, *crocodilus acutus*, Cuv., is the fifth species. The muzzle is more slender than that of all the other crocodiles properly so called.

The breadth of the head, taken at the articulation of the jaws, is comprised twice and a quarter in the length. The length of the cranium is little more than a fifth of the total length of the head. The males, however, have all these proportions a little shorter than the females, approximating in this respect to the females of the common crocodile, especially when young. On the middle of the chaffron, a little in front of the orbits, is a rounded convexity, more or less sensible. The upper face of the muzzle presents no projecting lines. The edges of the jaws are still more sensibly festooned than in the Egyptian species, taking individuals of the same age.

The plates of the nape are nearly the same as in the crocodile of the Nile, but those of the back, and this is a distinctive character, properly form but four longitudinal lines of crests (as in the preceding), the middle ones of which are more raised, and the external very projecting. These are placed more irregularly, and some of them scattered along the external side. This armour of the back does not resemble in number of pieces, nor in equality, that of the Egyptian crocodile. The middle pieces are wider in proportion. There are but fifteen or sixteen transverse ranges as far as the origin of the tail. In the tail are seventeen or eighteen ranges before the union of the two crests, and seventeen after. The middle ridges cease at the eighteenth or nineteenth range.

The feet do not differ from those of the common crocodile. The under scales are each provided with a pore. The head is to the length of the body as one to seven and four-tenths. The upper part of the body is of a deep green, spotted and marbled with black: the under of a paler green.

The males have all the proportions of the head a little shorter than the females.

Another species of living crocodile, with the nape of the neck armed *(cuirassée)* has been marked by M. Cuvier, who names it *crocodilus cataphractus.* The specimen was seen by the Baron in this metropolis, in the museum of the College of Surgeons. The muzzle is still more long and narrow than in the crocodile of St. Domingo. It has not the peculiar convexity on the chaffron belonging to this last species, nor any other remarkable mark.

There are seventeen teeth on each side in the upper, and fifteen in the lower jaw. The foramina of the cranium may be seen through the skin, as in the crocodiles. But what characterizes it particularly, more than the muzzle, is the armour of the nape. After two oval isolated plates, and a range of four others, smaller, come five scaly bands continuous with each other and with the scales of the back, formed each of two large square scales. The two first pair are very broad. The following three diminish gradually, and altogether they form on the nape a cuirass as solid as that of any cayman or gavial. The scales of the back are careened and disposed in transverse ranges of six each, except the two first, which have but four.

This species, says the Baron, is obviously distinct from the preceding, but unluckily no note respecting its origin has been preserved.

We now come to the species of the Gavial.

The great gavial; crocodilus longirostris, Schneider; *lacerta Gangetica*, Gm.; has been rather improperly named the *crocodile of the Ganges*, which would lead us to suppose that no other crocodile existed in that river, which is not the fact.

The muzzle is nearly cylindrical. It swells a little at the end, and widens at the root. The head is singularly enlarged, especially behind. Its transverse dimension is comprised twice and two-thirds in its total length. The upper table of the cranium behind the orbits, forms a right angle one-third wider than long. The length of the cranium, taken from between the anterior edges of the orbits, is comprised four times and a third in the total length. The orbits are more wide than long. The space which separates them is wider than themselves. The foramina of the cranium are larger than in any other species; larger even than the orbits, and like them, more wide than long. They do not grow narrow scarcely even towards the bottom.

There are twenty-five teeth on each side below, and twenty-eight above, in all one hundred and six.

The length of the beak is to that of the body as 1 to 7½.

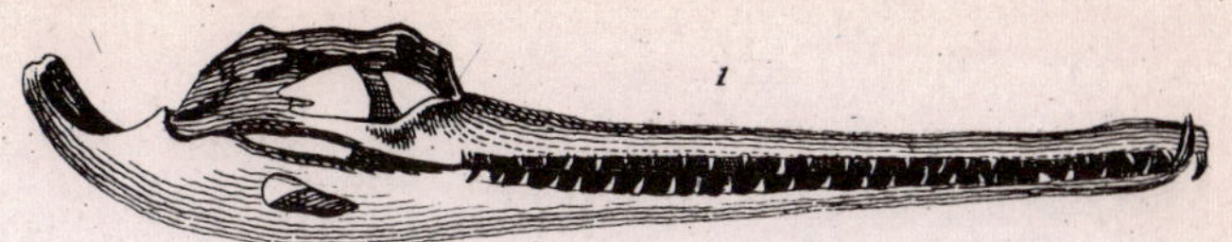

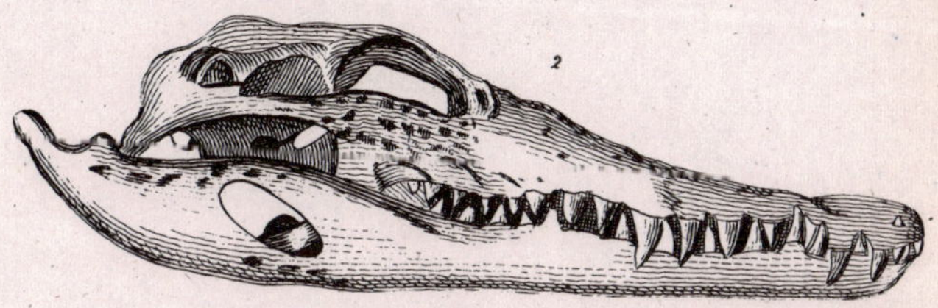

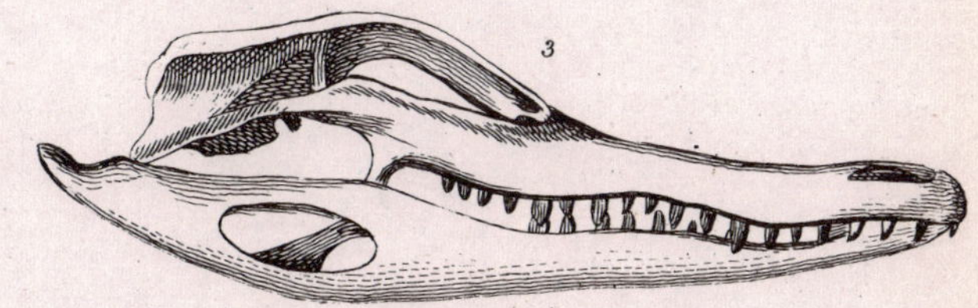

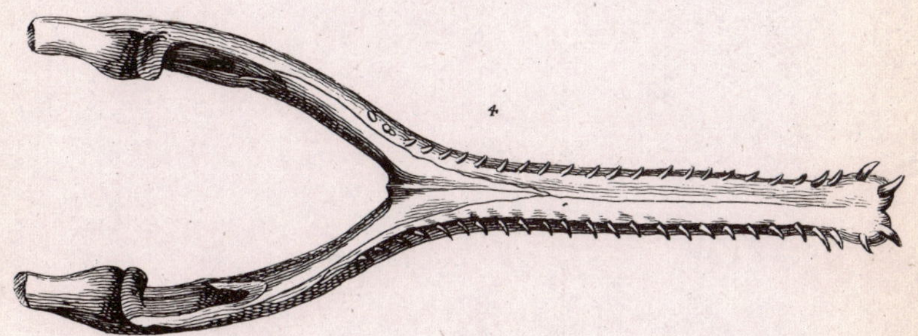

1. Great Gavial. 2. Common Crocodile.

3. Caiman. 4. Inner view of lower Jaw of Gavial.

There are only two small shields behind the head, followed by four transverse ranges, continuous with those of the back.

This gavial attains to a gigantic size. The Baron received one whose skeleton measured seventeen feet; and M. de Lacepede observed in the Paris Museum the portion of a jaw, from the size of which he concluded that the individual to which it belonged must have been thirty feet ten inches in length.

The little Gavial (*Crocodilus tenuirostris*, Cuv.) has the cranium less long and less wide in proportion to the muzzle than the last noticed species. The length of the cranium, taken from between the anterior edges of the orbits, is comprised three times and one-third only in the total length.

The upper table of the cranium behind the orbits forms a square of equal length and breadth. The orbits are more long than broad, and larger in proportion to the head, and separated by a space one-half more narrow than each of them. The foramina of the cranium are more long than wide, and very much narrowed in the bottom. The number of teeth appears to be the same as in the preceding.

The length of the beak is to that of the body as one to seven; and, therefore, is a little longer than in the last.

The nape is armed behind the cranium with two pairs of oval escutcheons, then with four transverse ranges. The first of these has two large scales; the two following, two large and two small; the fourth, two large ones, of which the bands of the back are a continuation. They all have four large squared scales, and two very narrow ones on the side. All these scales have equal but not very elevated ridges. The number of dorsal bands is eighteen. The crests of the tail are double as far as the nineteenth band.

If this little gavial be anything but a young one of the great, of which the Baron seems doubtful, it is certain that it also inhabits the waters of the Ganges, for M. Cuvier received from Calcutta a specimen exactly according to the above description.

That the reader may have clearly before him the number of living crocodiles, which it is absolutely necessary he should be in full possession of, for the thorough discrimination of the fossil species;—we will copy here M. Cuvier's tabular synopsis of them from the "Ossemens Fossiles."

CLASSIS. AMPHIBIA.
ORDO. SAURI.
GENUS. CROCODILUS.

Dentes conici, serie simplici; Lingua carnosa, lata, ori affixa; cauda compressa, supremè carinato serrata; plantæ palmatæ, aut semipalmatæ; squamæ dorsi, ventris, et caudæ, latæ subquadratæ.

* ALLIGATORES.

Dente infero utrinque quarto in fossam maxillæ superioris recipiendo, plantis semipalmatis.

1. *Crocodilus lucius.*

Rostro depresso parabolico, scutis nuchæ quatuor.
Habitat. in Americâ septentrionali.

2. *Crocodilus sclerops.*

Porca transversa inter orbitas, nuchæ fasciis osseis quatuor cataphracta.

(*Seb.* I. *t.* 104. *f.* 10. Mediocr.)

Habitat. in Guyanâ et Braziliâ.

3. *Crocodilus palpebrosus.*

Palpebris osseis, nuchæ fasciis osseis quatuor cataphracta.
Habitat. in Guyanâ.

4. *Crocodilus trigonatus.*

Palpebris osseis, scutis nuchæ irregularibus carinis elevatis trigonis.

(*Seb.* I. *t.* 105. *f.* 3.)

Num varietat. præced. ?
Habitat. in Guyanâ.

** Crocodili.

Dente infero, utrinque quarto, per scissuram maxilla superioris transeunte, plantis palmatis, rostro oblongo.

5. *Crocodilus vulgaris.*

Rostro æquali, scutis nuchæ 6, squamis dorsi quadratis, sexfariam positis.

(*Ann. Mus. Paris* x. *t.* 3.)

Habitat. in Africâ.

6. *Crocodilus biporcatus.*

Rostro porcis 2 subparallelis, scutis nuchæ 6, squamis dorsi ovalibus, octofariam positis.

Habitat. in insulis Maris Indici, in Gange, &c.

7. *Crocodilus rhombifer.*

Rostro convexiore, porcis 2 convergentibus, scutis nuchæ 6, squamis dorsi quadratis sexfariam positis; membrorum squamis crassis, carinatis.

Habitat.

8. *Crocodilus galeatus.*

Crista elevata bidentata in vertice, scutis nuchæ 6.

(*Hist. An. Paris, t.* 64.)

Habitat. in Indiâ ultra Gangem.

9. *Crocodilus biscutatus.*

Squamis dorsi intermediis quadratis, exterioribus irregularibus subsparis, scutis nuchæ 2.

Habitat.

10. *Crocodilus acutus.*

Squamis dorsi intermediis quadratis, exterioribus irregularibus subsparsis, scutis nuchæ 6, rostro productiore, ad basim convexo.

(*Geoff. An. Mus. Paris,* II. *t.* 37.)

Habitat. in Magnis Antillis.

11. *Crocodilus cataphractus.*

Nucha fasciis 4 osseis cataphracta, rostro productiore.
Habitat.

*** Longirostres.

Rostro cylindrico, elongato, plantis palmatis.

12. *Crocodilus gangeticus.*

Vertice et orbitis transversis.
(*Faujas. Hist. Mont. S. Petri, t.* 46.)
Habitat. in Gange fluvio.

13. *Crocodilus tenuirostris.*

Vertice et orbitis angustioribus.
(*Faujas. loc. cit. t.* 48.
Habitat. in Gange fluvio.
Num pullus præced. ?

It will now be necessary to enter into a few details respecting the osteology of the living crocodiles. This was as imperfectly known previously to the researches of M. Cuvier, as the osteology of many of the large viviparous quadrupeds. It is not necessary to trouble the reader with the labours of his predecessors; suffice it to say, that he came to the task provided with materials, the extent and value of which could scarcely be surpassed. Nor, as he himself observes, is the greatest abundance of specimens superfluous in labours of this kind. It is necessary to examine almost all the variations that the osteology of a species can undergo in different individuals, before the naturalist can venture to establish a new species on any isolated bones; on this account, it was the constant endeavour of this most eminent contributor to human knowledge, for many years, to multiply, as much as possible, skeletons of all interesting species, but more especially of those to which approximating ones existed.

We shall first proceed to the determination of the bones of

the head, in the crocodiles properly so called, and their comparison with those of the mammifera.

One advantage in studying the osteology of the crocodile, and indeed that of many other reptiles, is, that the sutures do not become effaced. It is not so easy, however, to refer each bone to its analogous one in man and other animals, as to reckon them, and on this subject anatomists have differed much.

The muzzle of the crocodile is elongated and depressed. The external aperture of the nostrils, placed near its anterior extremity, is directed upwards, pretty nearly as it is in the Lamantin. There is but a single incisive foramen, because the intermaxillaries have no middle apophysis, which is likewise the case in the animal just mentioned. The intermaxillaries surround the external nostrils, excepting one place, very narrow where the point of the nasal bones is placed between them. This is the case, more or less, with the majority of mammalia.

On each side the maxillary supports the jugal behind, which goes to form the external edge of the orbit. This edge would be the lower one in most animals, but here it is external, in consequence of the opening of the orbit being directed towards the top.

Underneath, the palatine bones prolong the roof, furnished to the mouth by the intermaxillaries, and by the maxillaries; but while they prolong it they also render it more narrow, because they leave a void between themselves and the elongations of the maxillaries, which support the jugals, and which void serves for the passage of the crotaphite muscles.

The lachrymal bone occupies on the cheek an oblong space, between the nasal, the maxillary, and the jugal. It reenters the orbit by a sort of plane contiguous to the jugal and to the maxillary, in which plane is pierced the lachrymal canal: all this is exactly an arrangement similar to that in the mammifera; but we must now observe the differences.

In the mammalia, the frontal bone would commence immediately at the external edge of this lachrymnal bone, and would occupy all the space between the two lachrymals above the nasal. It would descend into the bottom of the orbit to articulate broadly to the palatine, and the anterior sphenoïd; and in such genera as the quadrumana and ruminantia, which have the frame of the orbit completely osseous, it would give an apophysis behind, which would unite to the jugal to surround the orbit.

But this is not the case with the crocodile. There is, indeed, a frontal bone, covering, as in the mammalia, the interval of the orbits, furnishing a roof to them, or rather here, in consequence of their direction, an internal border descending almost to the root of the nasals. This bone even exhibits, in the individuals just broken from the egg, a remain of a longitudinal suture, as there is one in the mammalia, and which is speedily effaced. But a suture, which never exists in the mammalia, and which always, on the contrary, continues in the crocodile, separates in front of the frontal on each side, a bone, which is thus interposed between the lachrymal and the chief frontal, and descends from the edge of the orbit to the root of the nasals. It re-enters the orbit like the lachrymal, and descending lower, there unites itself by an apophysis with the palatal bone.

Between this apophysis and the palatal on one side, and the maxillary on the other, and under the lachrymal, is a large aperture which penetrates into the nasal cavity. It at once fills the place of the sub-orbital canal, and of the pterygo-palatine, and spheno-palatine foramina; but it is especially filled, in the fresh animal, by certain motive muscles of the lower jaw, muscles which we shall find to be peculiar to the ovipara.

The principal frontal bone does not descend into the orbit under an osseous form, and all the space between it and the palatine, as far as the sphenoïd, or what may be called the inter-orbital partition, is simply cartilaginous or membranous in the fresh animal, which leaves it entirely vacant in the skeleton.

Some traces of this arrangement are observable in certain of the mammifera. In the *saimiri* for example, and some of the *moschi*, in which the interorbital partition is reduced to a single lamina; it has membranous spaces.

The bone above mentioned, placed between the lachrymal and frontal, M. Cuvier first considered as a second lachrymal; but a more attentive examination clearly proved it to be a portion of the frontal bone, that which in man is named the *internal orbital apophysis*, or in the mammalia the *antiorbital apophysis*, which here is constantly detached from the body of the bone; M. Cuvier names it the *anterior frontal*.

It is sufficient to place the head of a mammiferous animal, of a ruminant, for instance, by the side of the head of a crocodile, to be assured that in the latter this dismemberment of the frontal bone has taken place; one might, without deranging anything, design on the frontal of the mammiferous animal, the suture which exists in the crocodile, and thus detach, in the first, an anterior frontal which would have the same position, almost the same figure, and absolutely the same office as in the crocodile.

This theory is entirely confirmed by the observation of the fresh head. There we see the frontal, conformably to its ordinary part, covering the anterior portion of the encephalon;—separating the orbits—giving a point of attachment to the levator muscles of the eye—allowing the olfactory nerves to pass under its part situated between the orbits. We find that it is expressly between its two dismemberments, called anterior frontals, that these nerves proceed from the cranium, after having been swelled into ganglia, and divided into numerous threads,—that these threads cross a cartilaginous sieve, placed between the two anterior frontals, as in the mammifera is placed the sieve-like plate of the ethmoïd. We find that it is under this sieve that the anfractuosities or cornets commence, over which the pituitary membrane is spread, and where the threads in question are distributed; but that these cornets remain cartilaginous like the sieve, and like the vertical lamina, which separates the orbits under the passage of the olfactory nerves, a vertical lamina, which, were it ossified, would probably appertain to the anterior sphenoïd, as it does appertain to it in such of the raminants as it exists in, that is, in the *Moschi*. In the *Saimiri*, its lower front part is of the ethmoïd, and its hinder part of the anterior sphenoïd. All the upper part is membranous. The nature, then, of these anterior frontal bones is clearly established, though their existence has been questioned by some comparative anatomists. It is no business of ours to meddle with the controversy in an humble compilation like this. It is sufficient for the satisfaction of our readers to know that we follow the best of all authorities.

Behind the orbit is again another separated bone, which completes the frame, by proceeding to join, by an apophysis, a corresponding apophysis of the jugal. Inspection alone will prove that this piece answers to that part of the frontal which gives the postorbital apophysis; and even the connexion of this part with the jugal, in the ruminantia, is entirely similar to that which takes place in the crocodile. This is named by the Baron the *posterior frontal.*

In fact, this piece is virtually nothing but the postorbital apophysis. It performs the same functions, for it closes the orbit, and is placed before the fossa temporalis and the crotaphite. It has the same position and connexions, being situated over the junction of the frontal and parietal bones.

Behind the chief frontal and the two anterior frontals is a large unequal bone, which covers all the middle and hinder part of the cranium, and affords, by its sides, an attachment to a part of the crotaphite. This presents no difficulty: it is the *parietal.* It is simple in the crocodile, as it is in a vast number of quadrupeds when adult; but it is double in a great number of other ovipara. It is even probable that it may be so likewise in the fœtus of crocodiles in no great degree of advancement, but it is simple in them on their coming out of the egg.

There is no dispute respecting the four parts of the occipital which constitute the hinder portion of the head of the crocodile. They are obviously the same as in the young mammifera—only the single condyle, placed under the occipital foramen, appertains entirely to the basilary bone. The upper occipital and the two lateral pieces have also a more important part to play than in the mammifera, because they are hollowed with cavities for the internal ear, for which the os petrosum is insufficient. The same arrangement prevails for the birds, and probably for all the ovipara.

No difficulty remains but for those parts which, in man, are termed the temporal and sphenoïd bones, and for the different pieces into which these bones are dismembered.

The *alæ temporales* of the sphenoïd are easily recognizable by their position, figure, and function of supporting the central lobes of the cerebrum. It is not surprising to see them forming distinct bones, for the same is the case with all the fœtuses of mammalia.

It must be remarked here, that this osseous piece encloses at the same time, and in a single mass of ossification, the ala temporalis, and a great portion of the ala orbitalis. When we examine a fresh crocodile, we find that, if the olfactory nerve and the optic pass between this ala and its corresponding one, the nerves of the third, the fourth, the sixth pair, and the first branch of the fifth, pass through foramina formed in the body itself of the ala, and the assemblage of which, if they were continued, would represent the spheno-orbital cleft.

The internal pterygoïd apophyses of the sphenoïd are very obvious, especially when we consider that they not only remain distinct from the body of the bone in many of the mammalia to an advanced age, for which reason they have been named *ossa pterygoidea,* but that in some ant-eaters they come underneath, uniting one to another, in concert with the palatine bones, to prolong the nasal tube as far as the basilary region.

In the crocodile, even from the fœtus state, these pterygoidean bones are united one to another under the body of the bone, to form the roof of the hinder nostrils. They unite also below by a suture to form the floor of this same tube, and they extend horizontally into a large wing, or broad surface, in which the pterygoidean muscles are inserted above, and which the membrane of the palate doubles below.

One ridge or process from their roof, corresponding to another of their floor, divides the nasal tube in two. Their upper lamina, or plate, proceeds forward in the form of two demi-cylinders to form again the roof of the double tube of the back nostrils over that part where the palatines constitute its floor, as far as the descending apophysis of the anterior frontals, and even by the internal face, a little in front into the cavity of the nose.

The body of the sphenoïd gives rise to no difficulty. It is situated at the centre of the floor of the cranium, is slightly concave, supporting the part of the brain placed behind the optic tubercles, articulating by its sides with the temporal wings in front, behind with the petrous processes, and by its posterior extremity with the basilary or lower occipital bone, descending between this occipital and the ossa pterygoidea, so as to shew itself externally only by a small surface below the lower occipital. An open canal in this surface traverses the entire body of the bone, and opens in front by two branches into a wide funnel, where the pituitary gland is lodged. In front of this funnel the sphenoïd gives out a vertical truncated lamina which enters into the composition of the interorbitary partition, and which is the only osseous part of it.

Above this lamina is an empty space, the sides of which are formed by the temporal wings, and the vault by the frontal bone. In the fresh subject, the membranous and cartilaginous interorbitary partition ends at the middle of this space, and bifurcates for the purpose of closing it.

It is through the upper part of this space that the olfactory nerves pass; the optic pass through the middle. Some vessels pass through the two sides of the osseous vertical plate of the sphenoïd. The nerves of the third and fourth pair, and the first branch of the fifth, pass through particular foramina (before mentioned) of the temporal alæ. Those of the sixth through a canal of the body of the sphenoïd.

This vertical lamina in front of the lodgement of the pituitary gland evidently corresponds to a part of the anterior sphenoïd in the mammalia, which, in the same manner, assumes the form of a vertical lamina in the species whose interorbital partition is slender, such as the *saïmiri* and the *musk*. At the same time it is clear that there is no particular orbital wing, since the nerves which, in the mammalia, pass through the spheno-orbital cleft or foramen, or in other words, through the interval of the orbital and temporal wings, pass here through particular foramina in the temporal wing; and also, that the optic nerve, the essential function of whose orbital wing is to surround its passage, passes into a foramen of the membrane, or of the cartilage.

In the fœtus, indeed, a small point of ossification has been found above the place through which the optic nerve passes, which, however, is soon enveloped in the growth of the temporal wing. This is the only vestige of an orbital wing in the crocodile; but it by no means fulfils the functions of that process, for it is not between it and the rest of the temporal wing that the nerves of the spheno-orbital foramen pass.

It is only in the small vertical lamina that we might look for an osseous representation of the anterior sphenoïd; but no suture is to be found distinguishing this lamina from the rest of the sphenoïd.

To complete what relates to the sphenoïd, we must speak of a bone common to almost all reptiles, but which is never found separate, either in the mammifera or birds. This is a large bone with three branches, which proceeds from the *os pterygoideum*, or its internal apophysis, to the union of the jugal, the maxillary, and the posterior frontal bones. This, says the Baron, if not altogether a new bone peculiar to the animal, is at least a decided dismemberment from the sphenoïd, as the fore and hind frontals which we have mentioned are dismemberments of the frontal bone. It can by no means be compared to any of the bones naturally distinct in the fœtus of the mammalia. M. Cuvier has, therefore, bestowed upon it a particular name, calling it the *transverse bone.*

In the Crocodile, as in the other ovipara, many parts of the ethmoid remain cartilaginous. Four alone become osseous. The first, or lower two, are articulated to the internal edge of the palatines in front of the anterior frontals, and of the vaulted portion of the ossa pterygoidea. Between them and the neighbouring part of the palatines, commences on each side the double canal of the back nostrils, which proceeds to terminate at the posterior edge of the ossa pterygoidea. These pieces are analogous to the lower and canaliculated part of the vomer in quadrupeds.

The two other ossified pieces of the ethmoïd adhere to the roof of the nostrils, between the nasal, the lachrymal, the anterior frontal, and the chief frontal bones. Nothing of them is to be seen externally in the caymans or the gavials. But they are easily distinguishable in part outside, between the frontal and nasal bones in the crocodiles proper. They are manifestly analogous to some portion of the upper cornets.

We must now speak of the *temporal* bone, and determine the analogy of its parts.

In the fœtus of mammalia this bone is divided into four pieces:—1. The squamous and zygomatic, which, as we descend in the scale of quadrupeds, becomes more and more foreign to the cranium, so that in the ruminants it appears rather pasted or glued on above, than entering into the composition of the parietes of the skull.—2. The tympanic, having at first in the fœtus no part ossified, except the frame of the tympanum, and extending itself successively so as to form an *os tympani,* and a *meatus externus.*—3. The petrous portion which envelopes all the membranous labyrinth.—4. The mastoïd portion which covers the *os petrosum* behind the squamous portion and the *os tympani*, but which is so soon soldered to the *os petrosum*, that it can scarcely be recognized as distinct in the youngest fœtuses, in which it is sometimes double.

In the crocodile we find an *os tympani*, and three other bones, two of which are external to the cranium, and one altogether internal.

The *os tympani* is easily recognized, for it gives an attachment to the membrana tympani, forms a lodgment for the osselet of the organ of hearing, and contributes to form, in a great measure, a cavity in front of the two fenestræ, from the bottom of which cavity proceeds the Eustachian tube.

The *os petrosum* is equally observable by its internal position, and by its lodging in a great measure the labyrinth, and essentially contributing to the formation of one of the fenestræ; but, in the crocodile, neither the *os tympani*, nor the *os petrosum*, is sufficient to lodge the cavity of the tympanum and the labyrinth.

The *os tympani* communicates with some large cells analogous to the mastoïd, or mammillary cells in man, some of which extend into the lateral occipital piece, and others into the upper occipital. These are common to the ossa tympani of both sides, and unite the two cavities.

The same is the case with the Eustachian tube. It commences in a sinking of the bottom of the cavity of the *os tympani*, descends almost vertically, passes between the basilary, the sphenoïd, and the lateral occipital, and terminates in the skeleton, at the point in which these three bones unite: but it is afterwards continued by a membranous tube, and approaches to its correspondent part to arrive, by a common aperture, into the hinder mouth, behind the back nostrils.

The labyrinth, like the tympanic cavity, and the tube, is surrounded by many bones. Its principal part, the vestibulum, is lodged in a cavity, to the formation of whose parietes the *os petrosum*, the upper, and lateral occipitals, concur. The upper and lower semi-circular canals wind in narrow tubes hollowed in these same parietes, and, consequently, in these three bones.

The portion of these parietes which separates the vestibulum from the cavity of the cranium is very slender, and divided by a suture with three branches, which marks the limits of the three bones.

On the side of the *os tympani* the paries is pierced by two transversely-oblong fenestræ, separated by a thin division. The upper one, which answers to the *fenestra ovalis* in man, and which is closed by the osselet of hearing, is formed partly by the *os petrosum*, and partly by the lateral occipital. And the other, which is analogous to the *rotunda*, is altogether in the lateral occipital to which the separating division belongs.

These two fenestræ are elongated from front to back. They open into the same osseous cavity, which is pretty large; but a slender ridge, proceeding from the bottom and anterior partition of this cavity, and continued in the fresh subject by a membrane, divides it into two parts, of which that which is lower and anterior, and communicates with the under fenestra (*rotunda* in man), contains a small lenticular mass, of a substance resembling hardened starch, and quite analogous to what is found in the sac of the ear of thornbacks and squali. This external and anterior part evidently represents the cochlea. But it is far from being so much developed even as it is in birds, in which it is yet considerably less than in the mammalia from its trifling inflexion. Still in the aves, especially in the ulula, there is found in it a demi-osseous partition, sensibly tending to a spiral curve. The internal and upper part, into which the upper fenestra (*ovalis*) opens, as do also the semi-circular canals, is the vestibulum.

This extension of the two cavities of the auditory organ, in the different bones, is found more or less in all the ovipara. That of the tympanic cellulæ, in particular, is much larger in certain birds.

The *os tympani* exhibits a large concave surface underneath, which articulates with the sphenoïd, the pterygoïd, and the great ala temporalis. Between this last and the *os tympani* is pierced the foramen, through which the fifth pair passes. This is the same with many mammalia in relation to the foramen ovale.

The hinder free edge of the *os tympani,* which projects behind, supports almost entirely the articulary facet for the lower jaw. This is nothing very anomalous, for in several mammalia, and even in man, the bone of the tympanic closet begins to contribute to form the posterior edge of the articular cavity.

All these functions of the tympanic bone in the crocodiles are performed in birds by the bone which has been named *os quadratum,* and the latter is distinguished from the former only by its mobility. And we may consider the os quadratum a true tympanic bone.

This bone, however, in the crocodile, does not include the whole bony cavity of the tympanum, even abstracting the mastoïd cellulæ, but neither does it do so in the mammalia. The os petrosum, the temporal squamous process, and frequently the sphenoïd, contribute also to the formation of the same part. This grand cavity is formed of many bones. Its internal paries is always the os petrosum. The os tympani forms, in general, a great portion of this cavity, constituting the entire of its external and lower paries.

The squamous temporal piece is entirely separated from the cranium, an approach to which arrangement we find already commencing in the ruminantia and cetacea.

The mastoid piece of the crocodiles proper and gavials has this peculiarity, that it advances laterally to unite itself with the posterior frontal, and to surround with it, and the parietal, the foramen of the upper face of the cranium which communicates with the fossa temporalis. In some caymans it even unites with these three bones to cover this fossa altogether above.

What we have now given respecting the head of the crocodile, is taken from the "*Anatomie Comparée*" of M. Cuvier, and extracted by himself into the "*Ossemens Fossiles*." We shall borrow from the latter work, to which we must of necessity be indebted for all details of this kind, a summary recapitulation of the comparison of the bones of the head and face with their analogues. But it will first be necessary to mention the osselet which represents, or rather is substituted for, the four little bones of the ear of mammalia, namely, the malleus, the incus, stapes, and os orbiculare. It consists of a small, long, narrow, elliptical plate, applied over the upper fenestra, and from which proceeds a sort of handle, long and slender, and which is fixed to the membrana tympani. There it is curved, and assumes a cartilaginous consistence. From the hinder paries of the long cavity proceeds a muscular filament, which is attached to the handle of the bone, at about a third of its length, and a doubling of the internal tunic of the tympanum forms a triangular ligament which extends to the same point, and thus contributes to fix this handle to its recurved and tympanic part.—Now to the recapitulation.

The intermaxillary, maxillary, nasal, lachrymal, jugal, and palatine bones, belong to the head of the crocodile equally with the mammalia, occupy the same positions, and fulfil the same offices.

The ethmoïd is similarly formed, of a sieve-like plate, lateral wings, upper cornets, and a vertical plate; but it remains, for the most part, cartilaginous. Two pieces seem to represent the lower portion of its vertical plate or *vomer*, two others some portion of its upper anfractuosities.

The frontal bone in its position and functions is the same as in the mammalia, but its ante and post-orbital apophyses are distinct bones.

The same is the case with the occipital, and it remains divided into four portions, as in the mammiferous fœtus.

The same may be said of the body of the sphenoïd, but it is

not separated from the anterior sphenoïd. Its great wings are similarly situated, and perform the same functions as in mammalia; but they always remain separated from the body of the bone, as in the fœtus of that class. They embrace a great part of the space and functions of the orbital wings, of which last no vestiges remain, except some little points of detached ossifications in the membrane which closes this part.

The pterygoïd wings are similarly situated, and perform similar functions as in the mammalia. But they always remain separated from the body of the bone, and unite below to prolong the nasal tube, as in the ant-eaters.

The situations and functions of the tympanic bone are the same as in mammalia, but it gives the facet for the articulation of the lower jaw. So are those of the mastoïd, but its processes are rather more extended than in mammalia. The place and offices of the os petrosum are the same, only that the labyrinth extends into the neighbouring bones.

Between the osseous box of the tympanum and the jugal bone, is one which can only correspond with the zygomatic of the temporal bone; and between the pterygoïd wing and the jugal and maxillary bones, is another, which corresponds, but feebly, to an external pterygoïd apophysis of the sphenoïd, which would be entirely detached from its principal bone, a thing which never occurs in mammiferous animals. Thus we see that all the essential differences are reduced to this distinction and division of the frontal bone.

In consequence of the importance of this head, we have been a little more extended in our remarks upon it. In our subsequent anatomical details we shall be as brief as is consistent with perspicuity, noticing what is of chief import only.

We have before mentioned, that an apophysis of the tympanic bone forms a facet for the articulation of the lower jaw. This facet forms a gynglymus of no great depth. In the mammalia, even in the fœtus state, as soon as the lower jaw has acquired any consistence, it presents but one bone on each side. The crocodile, like most reptiles, has six. These are the *dentary*, in which are hollowed the alveoli of all the teeth. The *opercular*, thus named by M. Adrien Camper, which covers the entire internal face, except all in front, where it is formed by the dentary. The *angular* and *subangular*, placed one above the other, and extending to the hinder extremity of the jaw. They leave a space between them in front occupied in its anterior part by the end of the dentary, and afterwards by a large oval foramen. Between the angular and opercular, on the external face of the jaw, is another oval foramen smaller than the last, and above it an empty space. The anterior point of this space is bordered by a particular little bone, of a crescented form, which the Baron names *complementaire*.

The condyle, all the upper face of the posterior apophysis, which gives an attachment to the digastric muscle, and all the internal face of this part appertain to a special bone called by M. Cuvier *articulaire*.

Many interesting particulars may be remarked concerning the teeth. Their number does not vary according to age. The crocodile just broken from the egg has as many as the animal of twenty feet in length. Their internal solid part is never completely filled up, though, like other teeth, they are formed by superposition of laminæ. At whatever age the teeth of the crocodile may be pulled out, there is found, either in the alveolus, or in a cavity of the tooth itself, a small tooth in a greater or less state of advancement, and ready to occupy the place of the old one as soon as it shall have fallen. This succession would seem to take place many times, and to continue during the life of the animal. Thus the teeth of the crocodile are always observed to be fresh and pointed, and not more worn in the old than in the young subject. The mode in which this replacing of the teeth is performed is very curious. The teeth of the crocodile being generally perfect cones widening towards the root, could not fall out of their alveoli, whose entrance is narrower than the bottom, but that the new tooth, as it develops, and fills the cavity of the old one, compresses its substances against the sides of the alveolus, destroys its consistence, splits it and disposes it to detach itself to the level of the gum at the slightest shock. The fragments which remain in the alveolus are afterwards easily expelled by the forces of living nature. In the crocodiles that change their teeth rings are often found thus formed in the alveolus by the remains of the old and broken teeth, and through which the new ones begin to shoot; and the same is observed, as we shall see presently, in the fossil jaws of true crocodiles.

Very frequently, the basis of the cone of the tooth is not entire, and a notch is observable in it on the side next the inside of the jaw. This proceeds from the germ having been formed a little on the inside of the alveolus, and commencing from this side to hinder the continuance of the old tooth. Sometimes there are two of these, for a second germ will sometimes be found before the fall of the tooth which is in place.

The hyoïd bone in crocodiles is very simple. Its body consists of a large and cartilaginous plate, convex below and concave above. The contour of its anterior part is semicircular, and the posterior part, more narrow, is terminated behind by a concave edge. The lateral angles of this edge ossify by small degrees, but always continue embodied with the rest of the cartilage, so that they cannot even be considered as vestiges of posterior cornets. The anterior semicircular part has two small notches, filled by a membrane. Behind this semicircle, where contraction begins, the anterior horn is articulated on each side. This is osseous, a little square, and goes off obliquely behind and towards the top. There it is terminated by a small cartilaginous appendage, which is neither articulated, nor suspended to the cranium by a ligament, but only by certain muscles which have some analogy with those of birds. The anterior edge of the plate is raised a little at the base, and forms a slight representation of an epiglottis, which would be very broad and very low. On this cartilaginous plate the larynx reposes, composed only of a cricoïd cartilage, and two annular arytenoïds, so that the plate performs at one and the same time the function of epiglottis, corpus hyoïdis, and thyroid.

The number of vertebræ is subject to some variation, but M. Cuvier considers the average number to be sixty. In the young the number is generally more complete, as their tail has not been mutilated by any accident. He reckons in some seven cervical, twelve dorsal, five lumbar, two sacral, and four-and-thirty caudal. All these vertebræ, including the axis, have the posterior face of their body convex, and the anterior concave. An important remark, as will be seen hereafter. Both of these faces are circular. In some young individuals, forty, and even forty-two caudal vertebræ have been found. The last M. Cuvier regards as the *normal* number, for the tail.

The *atlas* is composed of six pieces, which, as it would appear, remain distinct during life, and are retained only by cartilages. The *axis* has five. With respect to the rest it would be tedious to enter into details, and unimportant, at present. Any variations that may occur in the fossil specimens shall be duly noticed and explained.

The ribs are twelve in number on each side, without reckoning the appendages of the cervical vertebræ, which might very well be named false ribs. The first, and sometimes the first two ribs, have no cartilage to unite them to the sternum. The following eight or nine have each a cartilage or sternal part, which quickly ossifies, but which unites to the vertebral part by an intermediate portion, which for a long time, perhaps always, remains cartilaginous.

The sternum, even in the oldest individuals, has but one piece, which is osseous. This is flat, elongated, pointed in front and behind, the anterior part of which goes under the neck in front of the coracoïd bones, and the posterior part is enchased in a cartilaginous, rhomboidal, or elliptical disk, at the anterior lateral side of which is a groove into which the coracoïd bones are articulated.

The *omoplate* is very small in proportion to the size of the animal. Its plane forms a very narrow isosceles triangle. Its neck grows cylindrical, is curved internally, and widens to present a face to the coracoïd bone. On the external edge of this is an apophysis, which, with a corresponding apophysis of the clavicle, contributes to the formation of the fossa which receives the head of the humerus.

The head and body of the coracoïd resembles in form those of the omoplate. It has a thick and arched neck, and a plane portion which unites with the lateral edge of the sternum. This bone alone in the crocodile performs the office of buttress against the sternum, for in this genus there is no true clavicle.

The *humerus*, in front, behind, above, and below, is curved in two directions. Its upper part is a little convex in front, and the lower concave. Its upper head is compressed transversely, so is the lower, and it is divided in front into two condyles.

The *cubitus* has no olecranon, nor sygmoïd facet. Its upper head is articulated to the external condyle of the humerus by an oval facet wider on the radial side. Its body is narrowed and compressed. The lower head is smaller, and descends a little lower on the radial side.

The *radius* is shorter and more slender than the cubitus, and almost cylindrical.

There are but four bones in the *carpus*,—a radial and a cubital; a third which may be regarded as a sort of pisciform, articulating to the cubital osselet and to the cubitus; and a fourth of a lenticular form, between the cubital and the metacarpians of the index and the medius.

The metacarpians pretty nearly resemble those of the mammifera; the differences are very minute. To the thumb are two phalanges, to the index three, the medius and the annular have four, and the last digitus three.

The *os ilium* is vertical, concave without and convex within, where it receives the transverse apophyses of the sacral vertebræ. The *ischium* is nearly formed like the coracoid bones. The *femur* is a little longer than the humerus. Its upper head is compressed. It has but one trochanter, which is a pyramidal blunt eminence. The *tibia* approaches the usual form in mammalia. Its upper head is gross and triangular. The lower is crescentwise and placed obliquely, and its surface is concave. The *peroneum* is slender and cylindrical. Nothing in the *calcaneum* is worth remarking as different from what is found in quadrupeds. But the figure of the *astralagus*, as in all the lizards, is very singular and anomalous. The contour of its anterior face is determined by four faces; one upper, which is small and square, for the peroneum; one internal, for the tibia, oblique, and elongated; another is external, and crescented, the upper and lower parts of which only bear against the internal side of the peronean prominence of the calcaneum. All the lower part of the astragalus is occupied by an irregular surface, very convex, whose external posterior part rests in the astragalian apophysis of the calcaneum, and the rest of which supports the first two metatarsians.

There are three bones more which must be reckoned among those of the tarsus; one analogous to the cuboïd, placed between the calcaneum and the last two metatarsians; another, cuneiform, which answers to the second and third metatarsian; and one supernumerary, flattened, triangular, with a point slightly crooked, which is attached to the external side of the cuboïd. This holds the place of the fifth toe.

A few words respecting the differences in the skeletons of the *caymans* and *gavials*, from what we have now described, will finish all we have to say respecting the living crocodiles.

In the head of caymans, there are these differences from that of the crocodiles. The anterior frontal, and the lachrymal, descend much less upon the muzzle. The holes pierced on the upper face of the cranium, between the posterior frontal, the parietal, and the mastoïd bones, are much smaller, and often disappear altogether, as in *palpebrosus*. A portion of the vomer is visible in the palate, between the intermaxillaries and the maxillaries. The palatines advance farther, and widen in front. The posterior nostrils are wider than long.

In the head of the gavial the differences are much more sensible. The enormously long muzzle is formed underneath, one third by the intermaxillaries, and two thirds by the maxillaries ; the palatines advance in a point which occupies only a sixth of its length.

The bones of the nose above, do not by any means come so far as to end at the aperture of the nostrils. They terminate in a point nearly towards the upper fourth of its length. The intermaxillaries surround the external nostrils, and mount in a point towards the under fourth of this length. All the intermediate part is formed solely by the maxillaries. All this muzzle has pretty nearly the form of a depressed cylinder ; towards the middle its height pretty nearly equals two thirds of its breadth.

The cranium of the gavial is much broader in proportion to the muzzle, and in proportion to its own length than that of the crocodile. Its length is a fourth of the total length to end of muzzle, and is less than the breadth by about one-tenth. The orbits are more wide than long.

The lachrymal descends in a sharp point along the nasal, much more forward than the anterior frontal.

The foramina, between the parietal, the posterior frontal, and the mastoïd, are enormous, larger even than the orbits, and more wide than long in the adult individual, which very much contracts that part of the cranium which covers the parietal.

The foramina of the lower face, between the palatines, the maxillaries, and the bones which unite these last to the pterygoïds are shorter in proportion than in the crocodile.

The *ossa pterygoidea* form, above the palatines, a kind of gross bladders, swelled out and oval, about the bulk of a hen's egg, instead of a simple cylindrical vault as in the crocodiles and caymans. These communicate with the nasal canal only, by a hole of moderate size. These vesicæ are not observable in the little gavial, and the Baron considers them the result of age, as he has found this part much more swelled in the old than in the young individuals of *Gangeticus*. In this respect it presents an additional analogy with the sphenoïdal sinuses.

The lower jaw of the gavial, independently of its elongation, which corresponds with that of the muzzle, has this peculiarity, that its symphysis predominating as far as the last tooth, the bone called *opercular* is comprised within a little more than one-third of the length of this suture.

As for the rest of the skeleton, no other difference exists between that of the cayman and crocodile, but in total length. The bones of the cayman are almost all a little broader in proportion.

The form of the bones of the gavial, also, have a prodigious resemblance to those of the crocodile, only that the spinous apophyses of the vertebræ are more squared. M. Cuvier considers the normal number of vertebræ for all the living crocodiles of each sub-genus to be, seven cervical, twelve dorsal, five lumbar, two sacral, and forty-two caudal, in all sixty-eight.

We now come to the fossil remains of this remarkable genus of reptiles.

The Fossil Crocodiles appear to be by no means of rare occurrence in the ancient secondary strata; and what is remarkable is, that, though they all belong to species different from each other, yet they are almost all referable to the sub-genus with elongated muzzle, namely, the *Gavials*.

We shall first notice an account of two fossils, discovered in this country many years ago, though it is doubtful whether they are really referable to this genus. The account of one was published in 1718, by Dr. William Stukely; and the other in 1758, by Messrs. Wooler and Chapman.

A description of Stukely's specimen will be found in the thirtieth volume of the "Philosophical Transactions." It was the impression of a skeleton which was found at Elston near Newark, in Nottinghamshire.

The stone which contained it had served for a long time near a well, for the purpose of resting on it the vessels of those who came to draw water. The impression was on the side next to the ground, and was accidentally discovered by the turning of the stone. The stone was a bluish argile, and probably came from the quarries of Pulbeck, which belong to the western declivity of the long chain of hills which extends throughout the whole of Lincolnshire, and contains an abundance of coquillaceous remains, and even of fishes.

As usual, this skeleton was supposed to be human; but Stukely quickly perceived the contrary, and declared it to be that of a crocodile, or a *porpus*. This was certainly giving himself a sufficient latitude. His first conjecture, however, was the only one that could be sustained, as, according to his own account, the remains of a large pelvis were visible, which could not have belonged to a cetaceous animal.

The resemblance of this stone to those of Honfleur, where animals of this genus were assuredly found, disposed M. Cuvier to adopt the opinion that this was the impression of a crocodile; but the subsequent discovery in similar strata, of ichthyosauri and plesiosauri, threw some doubts about this conjecture.

A portion of the spine remained, containing sixteen vertebræ, the spinous apophyses of which are a little oblique, cut squarely, and nearly equal; the anterior six have large ribs. There are, more forward, the fragments of three ribs which were attached to some vertebræ lost by the breaking of the stone. The five vertebræ, which succeed those bearing the ribs, had long and narrow transverse apophyses, or, probably, false ribs, of no great elongation. The following four had but small ones. The *os ilium*, or, at least, an impression which seemed to have some relation with this bone in the crocodile, comes after the last of these four, which is the sixteenth. But it is difficult to say if it has not been displaced, and it might easily be believed that it was originally behind the fifth of those vertebræ with large transverse apophyses, which would then be the lumbar vertebræ. Then come twelve traces which might have been the marks of the bones formed like a V, placed under the vertebræ of the tail.

On the sides are two bones which Stukely took for femora; but by their form they might be judged to be ossa ischia, tolerably like those of the crocodile. There are besides, on the left side, two short and broad impressions, which might have been the top of the tibia and peroneum.

The plesiosauri and ichthyosauri have long bones much resembling the ischia just mentioned; and, on the whole, M. Cuvier is of opinion that the specimen in question is as likely to belong to either of these new genera as to the crocodile. Our respected countryman, M. Conybeare, has decided that it does belong to the plesiosaurus.

The other specimen was found on the sea-coast, near Whitby, in Yorkshire, in a blackish slate, called aluminous rock, and which comes off in exfoliations. Ammonites are observable in it, the interior of which is filled with spathic concretions.

As the flood tide used to cover this skeleton with five or six feet of water, it was considerably damaged by the sand and pebbles cast upon it. As it was at no great distance from a steep shore, very much elevated, and which the sea is incessantly undermining, there is no doubt but that it was formerly covered by the entire of this cliff. When a drawing of it was taken, a part of the vertebræ, and the slenderest bones of the head, had already been washed away by the sea, or carried off by virtuosi.

The spinal column was nine feet long, but probably not complete. There was also a head a little displaced, two feet nine inches in length.

But twelve vertebræ of the tail remained in their place, and a series of ten other vertebræ, which appeared to have constituted the lumbar, the sacrum, and the basis of the tail. Those of the neck, back, and middle of the tail, left nothing but their impressions. The space which they occupied does not appear to have been sufficient for more than eight, so that the tail could have had no more than twenty-two or twenty-three vertebræ, unless it had been truncated at the end. We might also believe that this spinal column was not complete in front when it was incrusted in the stone; for there is by no means sufficient room for the usual number of vertebræ in the crocodiles.

The head is turned, presenting its lower face. The occipital condyle is visible behind. On the two sides we find the zygomatic arches which terminate, as in all the crocodiles, in two broad condyles for the lower jaw, and which are placed on the same transverse line as the occipital condyle.

The cranium occupies but a narrow space, and the interval between it and the arches was furnished only with very thin small plates, coming doubtless from the pterygoïdean laminæ. The head grows narrow in front, not suddenly, but by degrees, as in the crocodiles of Altorf and Honfleur, ending in a pointed muzzle, covered in certain places by the remains of the lower jaw. In these places, in the two jaws are observed large pointed teeth, placed alternately, and crossing each other narrowly. But in those places where the lower jaw had been removed, the teeth of the upper were also taken away, and nothing was visible but their deep alveoli, placed at the same respective distances as the teeth themselves. There were large fangs or tusks towards the point of the muzzle stronger than the others. The enamel of these teeth was well polished.

The vertebræ seem to have been placed on the side. Each was three inches long. Near the place where the pelvis should have been, was found, in digging into the stone, a portion of the *os femoris*, three or four inches long, and a very small portion indeed of the ossa innominata, to which this femur was articulated. Some fragments of the ribs were also found near the dorsal vertebræ.

On the whole, the Baron's opinion is doubtful respecting this relic. He seems to hesitate between referring it to the new genera before mentioned, or to the crocodile, but yet leans more to the latter. Of its belonging to the class of reptiles there can be no doubt, yet M. Adrien Camper hazarded the assertion that it was a *balæna*, though the balænæ have no teeth, and M. Faujas St. Fond confidently pronounced it *physeter*; the teeth in the upper jaw, the femur, and portion of pelvis completely refute the last notion. The physeters have teeth only in the lower jaw, and the vestiges of pubis are very faint in all the cetacea.

The crocodiles of Franconia are better ascertained than the preceding. The deposition in which they are found is very similar to that in which the undoubted crocodiles of Honfleur are found. It is described as a calcareous stone, or bad sort of marble of a grey colour, full of ammonites and other ancient shells. The quarries are near the little town of Altorf, which was formerly subject to that of Nuremberg, and which has passed along with it under the domination of the Kingdom of Bavaria. The position, as well as nature of the strata, leads geologists to consider them as belonging to the middle layers of Jura.

The first head of this genus was discovered by a burgomaster of Altorf, and described by M. Walch in 1776, in the *Naturforscher*, a German periodical. He considered it to belong to the gavial. With the exception of the muzzle, the rest of the head adhered in such a manner to the stone, that it was impossible to design it distinctly. Some other discoveries of the same kind were subsequently made in the same place, which, though some controversy and confusion prevailed concerning them, are undoubtedly referable to the genus crocodilus.

According to Schrœder, at Erkrode, half a league from Brunswick, was discovered, in 1755, an entire skeleton of a crocodile, whose head, one foot long, with all its teeth, is said to be in the Ducal collection in Brunswick.

The bones found in the Vicentine territory were not situated in depositions precisely resembling the foregoing; but they belonged, however, to the limestone of Jura.

Considerable portions of jaws have been discovered in a mountain near Rozzo, on the confines of the Vicentine and the Tyrol, in calcareous stone of a reddish yellow colour. This stone is the ammonitiferous limestone of Jura, covered by the other kind which is destitute of shells.

There was found there the anterior portion of a muzzle, and two halves of the lower jaw detached from each other, but remaining nearly in their natural position. The lower jaw is twenty-five inches and a half in length, and eight inches broad. Many of the teeth had fallen, but became engaged in the stone, where they still surround the maxillary bones. The alveoli are visible in their proper places, and even a part of the roots. M. Sternberg, who gives this description, assures us that no little tooth was found in the cavities of the large ones. These bones appear to have belonged to a crocodile, but by no means to the common gavial, as M. de St. Fond confidently asserts. The posterior portion of the jaw would not be in a right line with the anterior, that is, with the part which belongs to the symphysis, but would make an angle with it, and thus remove more from its correspondent part on the other side, if this specimen was a relic of the common gavial.

This is a sufficient character for distinguishing this head, and especially the lower jaw, from that of gavial, and to approximate it to the head found at Altorf, and one of those of Honfleur. The Baron would assign them all to one and the same species, if he could depend on the drawings.

There are few countries more remarkable for petrifactions, than that which extends along the banks of the Altmuhl, one of the streams of the Danube, towards Pappenheim and Acihsted, where numerous quarries of a whitish calcareous slate, in great esteem, continually present impressions of fish and crustacea, entirely unknown in Germany at present; and, in all probability, in living nature itself. Some reptiles, also, of very curious kinds, have been found there; among which, the *pterodactyli*, which shall be described hereafter, are particularly to be remarked.

These schists belong to that prolongation of the chain of Jura which, after the fall of the Rhine at Schaffhouse, extends into Germany, over the borders of the Mein, near Cobourg. The sides of the valley of the Altmuhl are very precipitous, and it is easy to see, over two hundred feet of height, the strata which compose them. The calcareous schists, so abundant in fishes, in crustacea, in reptiles, and even in asteriæ, but which contain scarcely any other shells but two species of tellinæ, and some small ammonites, occupy the summit. They rest on a considerable mass of magnesian limestone. It is not stratified, and scarcely exhibits any traces of petrifactions. This, and the schists which cover it, do not prevail at all, to any extent, through the chain of Jura. They only begin to appear between Donawert and Noodlingen. The said limestone extends much farther northward than the schists, and the celebrated caverns filled with bones, which we have mentioned, when treating of the fossil mammalia, exist in it. Under it are banks of limestone of a greyish white, abounding in ammonites, and furnishing enormous frieze-stones, and a brownish or greyish sandstone, of a fine grain, which constitutes the bases of all the hills in this district. The most celebrated of these quarries is that of Solenhoffen, in the very valley of the Altmuhl, a little below Pappenheim. Here was found a remarkable fossil, which has been described by M. de Sœmmering.

The two plates containing it are of a marly calcareous schist, yellowish grey, spotted with red and yellow oxide of iron, and mixed here and there with parcels of quartz, with very delicate blackish and crystallized veins. Some impressions have been seen in them which have been referred to ammonites, but are as likely to belong to planorbes. There is also the impression of the tail of a small fish, and some remains of an insect.

The bones themselves are browner than the stone. On being analyzed, they were found not to have lost all their animal matter, and particularly to have preserved a remarkable proportion of phosphoric acid.

The largest of these plates, about three feet long and fifteen inches in breadth, contains the head, trunk, and tail of the animal, from one extremity to the other, and very little deranged, and a hind foot almost entire, detached from the trunk and encrusted at some distance. Scaly parts are also mingled with the bones. M. Sœmmering has published an excellent figure of it. In it is to be seen the lower jaw at its upper face, having twenty-five or twenty-six teeth on each side. The upper jaw is seen at the palatine surface, also the upper paries, and other parts of the cranium together, but a little detached from the muzzle. The condyle for the articulation with the atlas, and the articulary facet of the tympanic bone for the lower jaw, are also distinctly to be recognized. The series of the vertebræ is deranged only towards the end of the tail, and contains seventy-nine. Those of the neck have lost their transverse apophyses. Twenty-three ribs, more or less entire, are all out of place, or nearly so. A fragment of sternum, of the os ilii, an ischium of the left side, and a coracoïd bone, (these three are detached,) and some other bones, not so well determined, are to be seen. The left hind foot is in its place, but detached and disarticulated. The right hind foot, on the contrary, is cast out of its place, but has preserved its parts in their natural connexions.

This figure is as sufficient for determining the characters of the animal, as if the latter were under one's eyes itself. On the first glance, this fossil was found to resemble the little gavial more than any other known animal. The proportions, number of parts, teeth, &c. are similar. To discover the differences required a more attentive examination, but they were all found to be specific. The symphysis of the lower jaw is much less long in proportion. It exceeds only by a tenth the length of each branch. In the little gavial it exceeds it by one-third. In the great gavial it exceeds it by a fourth and more. There must be a corresponding difference of proportion in the upper muzzle, but as it is detached from the cranium it cannot be given so exactly. The teeth of the lower jaw are regularly and alternately longer and smaller, counting from the fourth; so that the fifth is one-half shorter than the sixth, the seventh than the eighth, and thus in succession. In the gavials, great and small, this regular inequality does not take place. The teeth which follow the fourth are nearly equal, except such as appear to have shot forth more recently.

In the upper jaw there are at first, on each side, two small teeth, then a very large one a little back, and the others are nearly equal and short. In the little gavial there is at first, on each side, a small tooth, then at some distance another small one, then one a little larger, and the following are nearly equal, but as long as those below.

If the aperture observable in the figure be that of the external nostrils, it is more broad, less long, and placed more forward than in the little gavial. If it be the incisive foramen, as might be conjectured from the position of the head, the character would be still more distinctive.

What appears to be the foramen which the crocodiles have between the parietal, the mastoïd, and the anterior frontal, is much larger than in the little gavial, although it has the same form. It exceeds the size of the orbit, which does not even take place in the great gavial, in which also this aperture has more breadth than length. The reverse is the case in the fossil.

Seventy-nine vertebræ are to be reckoned in the fossil skele-

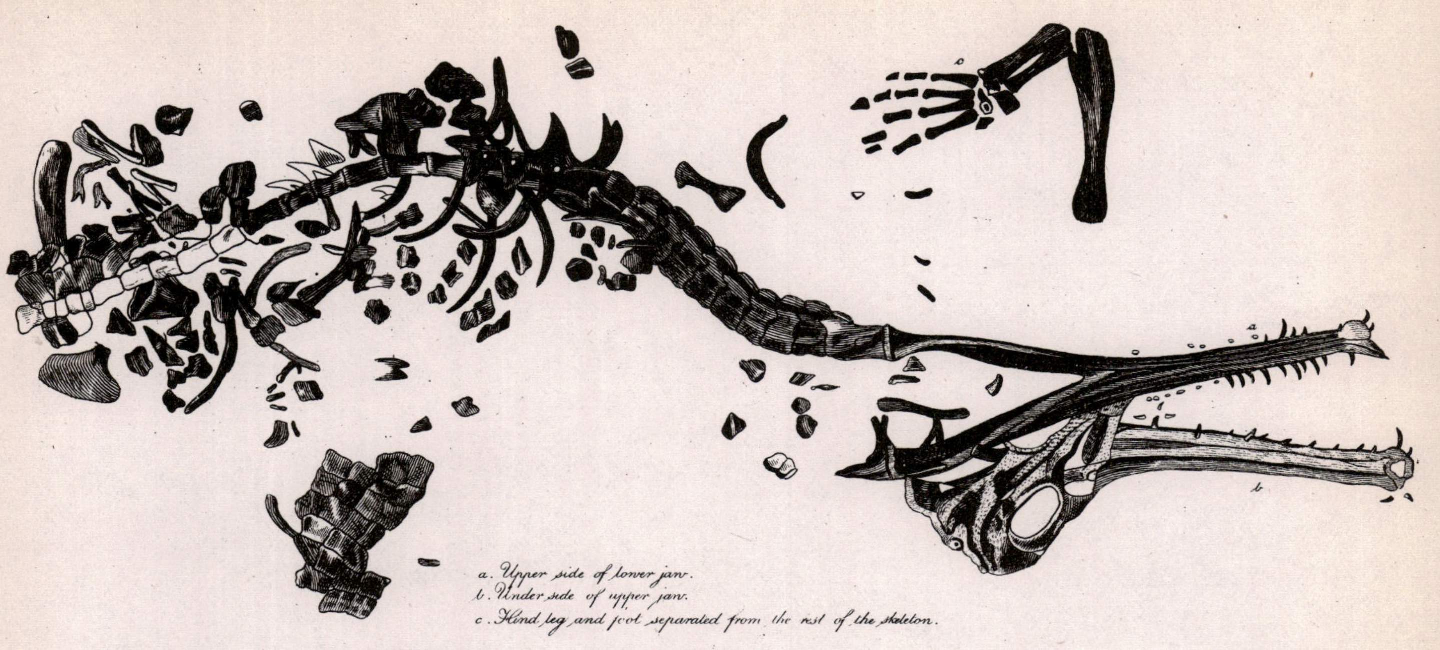

a. Upper side of lower jaw.
b. Under side of upper jaw.
c. Hind leg and foot separated from the rest of the skeleton.

Part of *Crocodilus priscus* of Sœmmerring.

ton. The little gavial has never more than sixty-eight. This difference is especially owing to the tail. It has at least ten vertebræ more than that of any known crocodile.

The spinous processes of the cervical vertebræ are square and touch together, which, in the little gavial, takes place only in the dorsal. The articular apophyses of the same vertebræ are also less advanced beyond their body. This difference continues in the back. There are no traces of the lower spinous processes which exist in the last cervical and first dorsal vertebræ of the little gavial. The vertebræ of the tail, beside their greater number, are sensibly thicker and shorter than those of the little gavial. Their spinous processes, like those of the cervical, are broader, and approximate more together, especially towards the middle of the tail.

The neck of the iliac bones is longer, and the ischium has its widened part much more broad and short than in the little gavial. The length of the femur in the fossil is more than double that of the tibia. In the little gavial it exceeds it only by a fourth. The tibia is thicker in proportion to its length, and the same difference takes place in the metatarsian bones, and particularly in those of the little toe. These differences are assuredly more than sufficient to prove that this fossil gavial is of an unknown species. M. Sœmmering has called it CROCODILUS PRISCUS, and gives these characters: *Rostro elongato cylindrico, dentibus inferis alternatim longioribus, femoribus dupla tibiarum longitudine.*

The entire length of the individual described by M. Sœmmering is two feet, eleven inches, seven lines, French measure. It is very remarkable that the tail should not be longer in proportion than the body, though it has ten additional vertebræ.

In the collection at Dresden is another fossil specimen, found at a place called Boll, in Wirtemberg, remarkable both for its baths and the fossils in their neighbourhood. It is situated between the Wils and the Lindach, two streams of the Necker at the north-west foot of the Albe of Suabia, which is a continuation of Mount Jura. This fossil would appear to have been but ill preserved. Its gangue is a grey, schistous, argillaceous earth, and an impression of ammonite is visible in it. The head, breast, anterior limbs, and hinder part of the tail, are wanting. What remains, however, is sufficient to justify the reference of this specimen to the genus of the crocodiles, and M. Cuvier thinks it probable that it is the same species as that last described. The situation of the two along the two borders of the same chain gives additional weight to this conjecture.

This fragment is forty-five inches and a quarter in length. The two knees are separated three-and-twenty inches and a half. Nothing is very distinct, except five vertebræ of the back, the femora, a part of the leg, and the left foot. But the form of the vertebræ, long, narrow, cut squarely at the two ends, and more contracted in the middle, would suffice to prove it a crocodile, and nct a monitor. The vertebræ in the last-mentioned genus would be wider in front, narrower behind, terminated in front by a concave arch, and behind by a convex. Other minuter particulars go to establish the resemblance between these remains and those of *crocodilus priscus.*

The town of Caen, in Normandy, is surrounded by quarries of a very fine limestone, from which beautiful stones have been supplied for the construction of the city, and for churches and other public edifices throughout the province. It is even said that most of the cathedrals erected in this country, by our Norman princes, were built with stones brought from Caen. The nature of this stone bears some resemblance to that of a hardened chalk, and the geological position of its beds is unquestionably lower than that of the chalk in the neighbourhood of Paris, which extends very far into Lower Normandy, and occupies all the upper part of that province, as well as Picardy and the opposite coasts of England.

The whole soil of this country is essentially composed of four kinds of strata. The upper stratum, immediately above the vegetable soil, but which elsewhere passes under the chalk, is

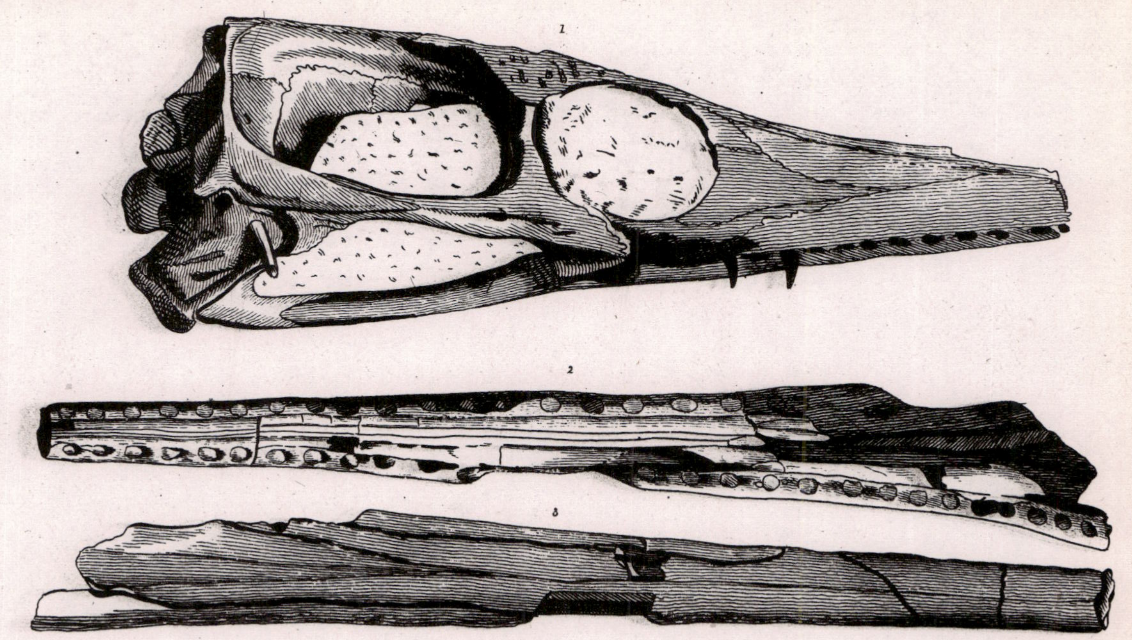

1 . . . Crocodile of Caen. pa. 231.

2 . 3 . Portion of muzzle of Crocodile of Caen. pa. 231.

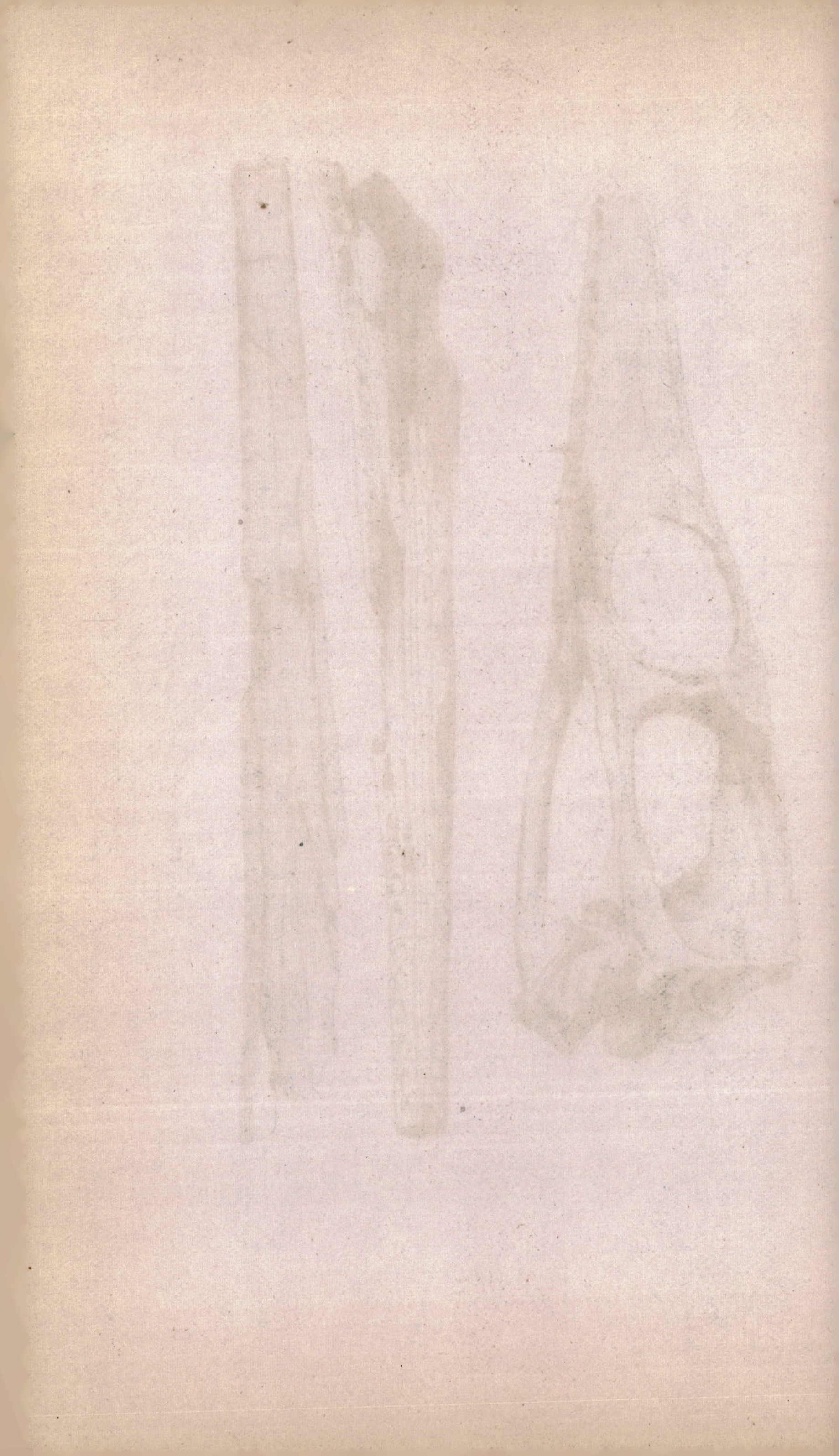

a limestone, with large spathic grains, filled with polypiers, encrinites, terebratulæ, and some ammonites. The second is of that sort of stone especially named Caen-stone, which contains ammonites, pinnites of particular species, and other shells, but in no great abundance. The third is composed of a very hard bluish marle, and often pyritous, and is of very great extent. In this might be found remains of crocodiles, similar to those of Honfleur. It contains abundance of gryphites, ammonites, nautili, oysters, terebratulæ, fossil bones, encrinites, and fossil wood.

Below this is the oolitic limestone, which occupies an immense space in the department, and furnishes almost all the chalkstone of the country. Its strata are horizontal, of various thickness, and separated by potters' clay. They contain oolites, belemnites, nautili, and starred encrinites. A fish has been found there like the *dapedium politum*, taken from the blue lias of Lyme Regis, by Mr. de Labèche. Its lower beds are very thin, and contain siliceous rolled flints. It rests on the red sandstone, schists, diabases, and marbles, which are subordinate to it.

Below the chalk and the green ferruginous sand, which serves for its basis, is a bed of blue marle, which begins to show itself at Havre, and rises more on the other side of the Seine, at Henqville. In this bed near Havre some bones of crocodile have been found.

Under this bed rest some vestiges of the Portland stone, and below this, coral rag. Under this last are often found beds three hundred feet deep, of another blue marle analogous to the Oxford clay, which forms that tract called, in French, the "Vâches noirs," where crocodiles, which shall be described hereafter, have been found, and also the plesiosaurus. Between this marle and the Caen-stone there should be still two banks analogous to those we call cornbrash and forest marble. After this last mould is again a bank of oolite; and finally, the last bank of blue marly stone, analogous to our lias, which rests on the red sandstone. Mr. de Labèche believes that this lias of France contains bones of the ichthyosaurus, as well as that of England. From the ascertainment of these three distinct banks of blue marle, there must be a considerable difference between the ages of the crocodiles there found.

The view we have just given is taken from Mr. de Labèche. M. Prevost, who has very accurately examined the coasts from Calais to Cherbourg, agrees with him to a certain point. According to him, the upper strata are the bluish argilla of the environs of Havre, with lignites. Then comes the limestone of Caen, the upper strata of which contain polypiers, trigoniæ, and cerithia, and the lower contain bones of crocodile. Under this limestone come the lower blue argilla, and the oolites, alternating with the lias, containing, or supposed to contain, the ichthyosaurus, and all would rest on the limestone with gryphytes, and the limestone used for lithographic operations.

Be all this as it may, it is incontestable that the crocodile of Caen, like the last described, and those of Honfleur, and many others, belongs to this great assemblage of strata, which continental geologists have agreed to call the *formation of Jura*, and which holds a sort of middle rank among the secondary strata, being placed below the chalk and above the other secondary formation, which has been named *Alpine*.

This crocodile of Caen does not appear to have been very rare in these environs at the epoch in which it lived; for, within but a few years, comparatively speaking, have been found the remains of at least ten individuals. The specimen which has excited most attention was found at the end of 1817, in part of the banks of the Caen-stone, on the right of the Orne, and in the quarries of a village called Allemagne, a short league to the south of Caen.

The principal piece was composed of from fifteen to sixteen vertebræ, placed on a continuous line, and pretty nearly in their natural position, with some portions of ribs, and a great number of scales still in connexion, and almost such as they were when they formed the armour of the animal.

At the same time, and at no great distance, was discovered a considerable portion of a head, which was presented to M. Cuvier; he also received some incomplete vertebræ and a group of scales, which had been more anciently found in another quarry near the same place.

In 1822, M. Lamouroux procured two considerable blocks, on which was the impression of a head, tail, part of the ribs, and some long bones. The bones which formed these impressions had been lost in getting out the blocks. Some portions of the parietal, however, were preserved, the frontal and the muzzle almost entire, some vertebræ, and some other fragments. The portion of head above mentioned having been carefully disengaged, presented almost every advantage that could be desired for determining this part of the osteology. It was one half of the left side, which had been detached longitudinally from the other half, and only showed, at first, its vertical and longitudinal section, but when it had been disengaged from its stony covering, all the parts were found perfectly preserved, from the occiput to beyond the anterior extremity of the lachrymals. This half being thus complete, it was easy to represent the other half.

On the first view, the Baron pronounced this specimen to have appertained to a gavial equally different from the living species and the fossils discovered before. These are its special characters:—On the upper faces the sides gradually approach each other to form the muzzle. The anterior frontal advances less upon the cheek, the lachrymal advances more, and is broader at its base. The jugal, on the contrary, is more narrow. The edges of the orbits are not raised; the orbits approach each other more, and are of a circular figure. The principal frontal between them is not concave. The foramen of the crotaphite is much larger in proportion, and nearly squared, not round. The posterior frontal, which separates this foramen from the orbit, is much longer and narrower. The parietal surface between the two crotaphidians is more elongated. The occipital crest is not in a right angle, but in a very slender, trenchant lamina, extending from the parietal to the mastoïdean angle. On the occipital face the mastoïdean angle is not uninterruptedly united with the back of the articular apophysis of the os tympani. It is separated by a deep depression, above which advances a trenchant crest, which belongs to the lateral occipital. The mastoïdean bone has a concavity in its descending part, of which there is no trace in the gavial. The emargination of the lower edge of this face between the articular apophysis of the os tympani and the tuberosity of the basilary bone, is much less than in the gavial; and, consequently, this apophysis is less salient towards the bottom, and less detached.

On the lower face of this head the palatines do not close the nasal cavity underneath, but opposite the posterior edge of the great palatine foramen. So that the hinder nasal fossa is very large, and does not open but towards the extremity of the basilary face, where, in the common crocodiles, the back nostrils are situated a little before the foramen of the arteries. The pterygoïdean wing is not widened externally, as in all the crocodiles, but is contracted by a broad emargination in the part where it is about to unite to the bone.

On the lateral face of this head, the orbital edge of the jugal is not raised, and does not leave behind it a deep emargination, as in the gavial. The jugal does not re-ascend to articulate with the posterior frontal, but, on the contrary, this last descends to unite itself with the jugal, at the external edge of the orbit, at its posterior angle. The vacancy between the orbit and the anterior edge of the os tympani is greatly elongated in the fossil, and occupies four-fifths of the temporal fossa; so that the tympanic cavity is much shorter and more thrown behind. The anterior part of this temporal foss is also very narrow and acute, which gives it quite another figure from that of the gavial. The bone which is analogous to the stapes, and which, singularly enough, has been preserved in this specimen, is cylindrical, and much thicker in proportion than in any crocodile or any known reptile whatever.

In this fossil are eleven alveoli, two of which alone preserve their teeth. These cavities are all nearly of the same diameter, and are filled with the matter of the stone, which proves that the teeth had fallen before the incrustations took place.

M. Cuvier has finished the determination of this head from the impression, and some remains of other bones in the blocks before mentioned. These pieces, though belonging to an individual four times the size of that to which the demi-head we have just described belonged, yet appertained to the same species, as the Baron judged from what remained of the frontal, the parietal, and the anterior frontal bones. The narrow crest formed on the parietal by the approximation of the temporal fossæ, he considers as an ordinary effect of age, which enlarges the crotaphite muscles. To the same cause he attributes the curve of the crest in the form of a chevron, which is strongly marked in this crocodile. The frontal just mentioned is singularly flat; a ridge slightly salient traverses the middle of its length, and its surface is rendered a little unequal by vermiculations.

According to the impression, the muzzle of this crocodile was longer in proportion than that of the gavial. It grew gradually more slender towards the end, where it dilated a little, and the total length of the head must have been more than thirty-seven inches.

There was also found a portion of muzzle twenty inches long, which confirmed the characters drawn from the impression. Its depression was stronger than in the gavial; the bones of the nose descended lower, and formed a more acute angle. Along its length was a sort of central rib, slightly projecting, and marked with a longitudinal furrow. From the roots remaining on either side, it appears that, in this length of twenty inches, there must have been at least thirty teeth.

According to the impression, the entire muzzle must have been twenty-nine or thirty inches in length. M. Cuvier thinks it probable that there were forty-five teeth on each side in each jaw. This would make the entire number one hundred and eighty. The gavial has only one hundred and twelve.

In another considerable fragment of the lower jaw, part of the opercular bones was visible; and on the right side three teeth, the middle one of which was double the size of the two others. The teeth are long, narrow, arched, and very pointed, but not trenchant.

From a model of the anterior end of the lower jaw, the Baron observed that it was much depressed, widened a little in front, and emarginated at its extremity. On one side twelve teeth were visible, alternately longer and shorter, but all tolerably long in proportion to their bulk. In the gavial they are not nearly so close.

On the annular portions of two cervical vertebræ, it was observed that they did not differ from those of the common crocodile, but in having spinous apophyses broader from front to back, and more inclined behind. Two sacral vertebræ, a portion of the ossa ilii, and the cotyloïd cavity, three lumbar vertebræ, and ten dorsal, bearing ribs, were found; and also some caudal, &c.; all these presented characters different from the existing species. In the bones of the extremities which remained there was more analogy with those of living crocodiles; but still there were many slight variations easily perceptible to an experienced eye.

Abundance of scales and impressions of scales were found, many of them still adhering to the parts of the body to which they belonged, so as to leave no doubt of their appertaining to the same species. They differ from those of the living crocodile more than any other part of the skeleton, and this crocodile of Caen was beyond all comparison the best provided with defensive armour of any of the genus. The scales are very thick, rectangular, slender towards the edge, and have all their external surface hollowed with little semi-spherical fossets, about the bulk of a pea, and closely pressed against each other. They were arranged, as in the living crocodiles, in regular series, both longitudinally and transversely. The hinder edge of one covered the base of that which followed.

From the largest fragments, the Baron judges that this species might have been twenty feet in length. Some of the remains, however, did not belong to individuals more than thirteen, and ten feet long; so that, notwithstanding the clear distinction of species, this crocodile of Caen did not exceed in dimensions its living congeners.

It resembled *crocodilus priscus*, in the alternation of size in the teeth, and the dilatation of the anterior extremity of the two jaws. However, the more elongated and slender form of the upper jaw in front, and the form of the temporal fossa more wide than long, seem to mark a sufficient difference for the species.

Some remains, very similar to the last, were found in the formation of Jura, and sent to M. Cuvier from Switzerland. Some of the teeth, however, from their greater bulk and obtuseness, seemed to indicate another species. It is most remarkable, as the Baron observes, to find an animal, so especially a native of the fresh water as the crocodile, in the strata of the formation of Jura. It is also worthy of remark, that it is accompanied by abundance of tortoises, equally belonging to the fresh water. This fact, joined to many others, proves that there existed here dry lands, watered by rivers, at some wonderfully remote era, and long before the three or four successions of these kinds of strata which have been observed in the neighbourhood of Paris.

We shall now speak of the bones of two unknown species of gavial found near Honfleur and Havre.

These bones, the Baron thinks, may have belonged to two depositions different from each other, but superior to that of the Caen-stone. But he considers them much more ancient than the immense mass of chalk which rests upon them, and much anterior to those which contain the bones of even the most ancient quadrupeds, such as the gypsum of Paris; for this gypsum rests upon the most common coquillaceous limestone, which last rests upon the chalk.

The substance of the bones is of a very deep brown, and takes a fine polish. It is soluble in acids, and assumes a reddish tint, which shows that it is coloured by iron. It has, however, preserved a portion of its animal nature. The great cavities of the bones, as the box of the cranium, the canal of the nostrils, and that of the vertebræ, are filled with the same hard and greyish marle which envelopes their exterior. But the pores, or small cells, are occupied by a calcareous demi-transparent spath, sometimes tinted with yellow. Each cellule is usually carpeted with pyrite, which envelopes the spath with a thin and brilliant bed. The interior of the shells found here is also, sometimes, furnished with it; and some are found whose substance has been entirely replaced by the pyrite.

The most considerable piece in the collection of which we are now to speak, is a lower jaw almost complete. The articular extremity of the branches appears to be all that is wanting.

This jaw bears the most incontestable characters of the crocodile. The teeth are conical and striated. The majority are broken, but some are entire, and the two trenchant ridges very distinguishable. Many of those which are in their places show in their cavities the little germ which was to replace them.

The sutures which divide this jaw into six bones on each side are easily distinguished. They are pretty nearly in the same positions, and of the same form, as those which compose the jaw of the gavial. Still, an attentive examination soon leads to the discovery of characters which distinguish it very clearly from the last-mentioned species.

The branches are much longer in proportion than the anterior or symphysized part, which they exceed by some thirds of

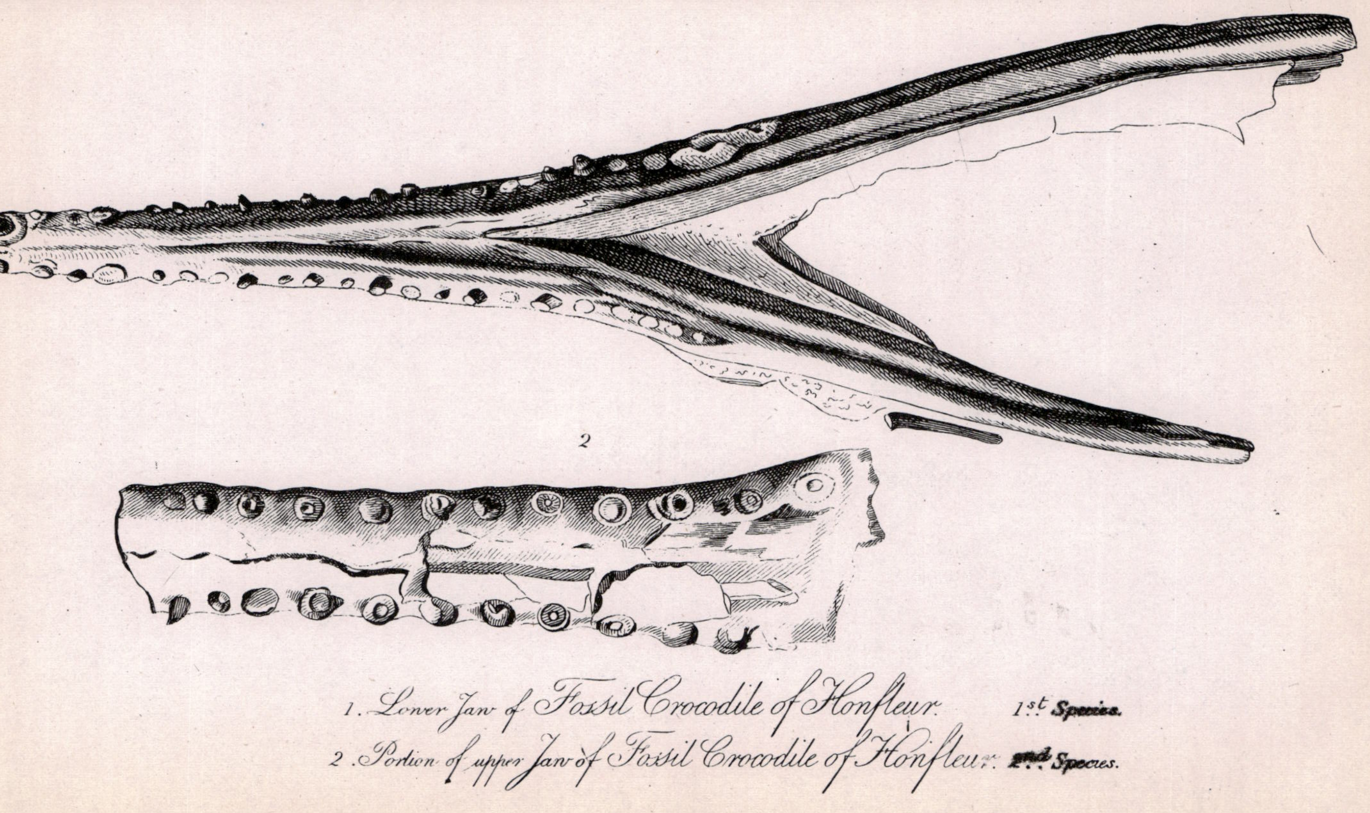

1. Lower Jaw of Fossil Crocodile of Honfleur. 1st Species.

2. Portion of upper Jaw of Fossil Crocodile of Honfleur. 2nd Species.

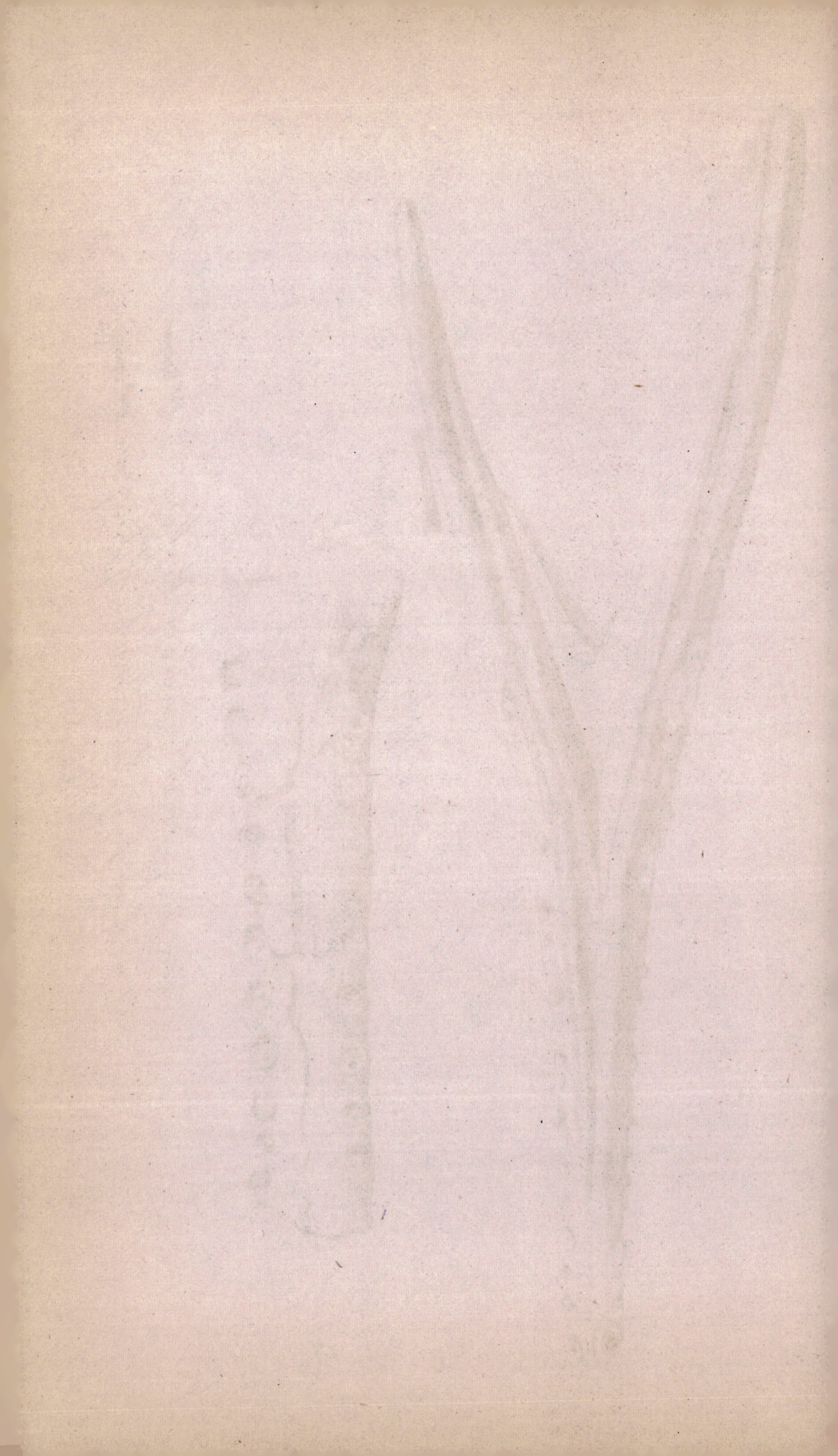

an inch. In the gavial, when the articular part has been removed, as here, they are, on the contrary, shorter by one-third; and even, adding this part, they are shorter by a sixth.

They do not form together so open an angle as in the gavial. Theirs is about thirty and some degrees. That of the gavial is nearly sixty, taking these two angles within and between the lines which the internal edges of the branches form.

For the same reason, they separate less from the external line of the symphysis, and almost appear prolongations of it. In the gavial they remove from it by a much more sensible inflexion.

The upper face of this jaw is hollowed by a middle and two lateral furrows, which approach each other in front. It is smooth in the gavial.

The transverse diameter near the anterior third of the symphysis does not exceed by quite one-fifth the vertical diameter. In the gavial it is almost double. The emargination which separates the branches penetrates more forward between the teeth than in the gavial. There are seven teeth on each branch; in the gavial there are but two or three. Nevertheless, the entire number is less; for there are but two-and-twenty on each side, while the gavial has five or even six-and-twenty. Finally, there does not appear to have been any oval foramen on the external face of the branch.

In the same strata were found fragments of another lower jaw, much more approximating to that of gavial. If not gavial, it is certainly distinct from the last. Another fragment, apparently from the same species, was found near Auxerre. It was remarkable for its size, and the resemblance of its proportions to those of gavial. The Baron thinks the individual to which it belonged might have been seventeen feet and a half in length.

These two lower jaws suggested to M. Cuvier that two species might exist at Honfleur. By great chance he got together six pieces which had belonged to the same cranium, and by means of these re-constructed a considerable portion of the head, containing all the occiput, the greatest portion of the upper face, and of the sides as far as the muzzle. He also got three fragments which had appertained to one and the same muzzle, united them as they had been in nature, and joining them to the partial cranium which he had already formed, he found the muzzle so well adapted to the cranium, that he had no doubt of its having originally belonged to it, and that the nine fragments were from one and the same head. At all events, it was quite certain, that even if the muzzle did not belong to the head in question, it belonged to an individual of the same species and of the same size, which was fully sufficient for all purposes of determination.

In the mutilated state, this head was two-and-thirty inches in length, and allowing four inches for the end of the muzzle which was wanting, it must have been three feet long. The largest head of gavial in the Baron's possession was but thirty-one inches. But this is the least difference existing between the two species. The muzzle of the gavial is both wider in proportion and in actual measurement; and the excess of width is still more marked in the occiput. The cranium of the fossil has an oblong form, quite different from that of the gavial, and unites itself to the muzzle by an insensible narrowing, and not by a sudden contraction.

The occiput of the gavial is limited above by a horizontal right line. In the fossil it is an angular line, whose middle projecting angle corresponds to the sagittal crest.

The crotaphite foramina of the fossil are much wider and very considerably longer than those of the gavial. Their figure is elliptical, and their grand diameter longitudinal. They intercept a long and narrow sagittal crest, and not a short and flat parietal surface, as in the gavial. The arch which is formed by the posterior frontal and the mastoïd, which limits the crotaphite foramen below, is not straight as in the gavial, but forms a convexity externally.

The length of the cranium, and of all that surrounds it, corresponds to the length of the crotaphite foramina, in consequence of which the orbits are carried forward much more in proportion than in the gavial.

The frontal of the fossil is not concave, as in the gavial, but flat. It is less emarginated by the orbits. Their edges are not raised; from which it results, that the bones which form these edges, the anterior frontals, the lachrymals, and the jugals, have a more even form, and by no means concave. The anterior frontals are much larger in proportion in the fossil than the lachrymals.

There is one thing especially worthy of remark in the fossil: instead of the slight emargination of the orbital edge of the anterior frontal which is in the gavial, there is a smooth demi-canal, which descends on the junction of the anterior frontal and nasal bones. The nasal, instead of re-ascending entire along the internal edge of the lachrymal and the anterior frontal, widens to embrace the point of the anterior frontal in an emargination of its base. The external apophysis of this base separates the lower point of the jugal from the lachrymal and the anterior frontal.

The base of the muzzle below, at the spot where the palatines enter into its composition, is much more gibbous, and more high vertically in the fossil than in the gavial. These characters were confirmed by the examination of another fragment unnecessary to be described here.

This distinction of two species shown by the lower jaws, and confirmed by the examination of heads, was further corroborated by many portions of the skeleton. An attentive examination of the vertebræ proved that they form two systems, and would have indicated the existence of two crocodiles in these marly strata, of different species, even if it had not already been recognized by the examination of the jaws. The vertebræ afforded even more positive demonstrations of this fact. In the first specimen examined, the atlas and axis were found soldered together. Of the atlas was preserved only its lower piece, and a part of its lateral pieces destined to embrace the condyle of the occiput. All that had contributed to form the canal had disappeared. The axis is more complete, having lost only the hinder part of its annular piece. Many characters, even in this specimen, indicated a species different from that of the gavial of the Ganges and all other living crocodiles. The tubercle of the axis showed that the false rib of this vertebra had two heads, as well as those of the succeeding cervical vertebræ. In the crocodile and gavial there is but one. The posterior face of the body of the axis is concave; while it is convex in all known crocodiles. From these, and many other fragments found in the same neighbourhood, M. Cuvier has established that there was one species here exhibiting a peculiar vertebral system in which the vertebræ are convex in front. This he calls the *convex system.*

But there were also found in the same places, and mixed up with the former fragments, others demonstrating a very different system, which Baron Cuvier terms the *concave system.* The vertebræ which compose it have not the body narrowed in the middle. Their transverse apophyses do not spring from the union of many salient crests. They much more resemble those of our living crocodiles; but their principal difference, both from our living species and the first fossil species, is, that neither of the faces of their bodies are convex, but both slightly concave. As to the rest, they have the sutures, and all the arrangement of apophyses which characterize, generically, the vertebræ of crocodiles. M. Cuvier is inclined to refer the vertebræ of the first system, or the convex, to the first jaw which we described from the Honfleur remains—that belonging to a shorter and more obtuse muzzle. As that least resembles the living crocodiles, it seems natural to refer to it the vertebræ which least resemble theirs. Be this, however, as it may, the clear distinction of all the fossil fragments found in the places which we have described, from all that characterizes living species, is completely made out. This must be evident even from the necessarily abridged account to which our limits confine us. It is scarcely necessary, as the Baron remarks, to reply to those who would explain the extraordinary differences of those bones from those of the gavial, by the influence of age, of nutriment, of climate, or of the passage to a state of petrifaction. Could all these causes united, continues this distinguished philosopher, place in front the convexity which the other crocodiles have behind in the vertebræ? Or could they change the origin of the transverse apophyses, flatten the edges of the orbits, diminish or augment the number of the teeth? As well might it be said that all our living species are derived from one another.

Having noticed the remains of the most ancient crocodiles which belong to the formations called *Jurassic*, we must now mention such as are a little more recent, and appertain to the age of the chalk. Crocodiles have been discovered in the ferruginous sands under the chalk [in this country—in the chalk at Meudon; and immediately under the chalk in lignites and plastic argile in several places. Unfortunately, there have not been collected specimens in a sufficient state of preservation to determine the species: all, therefore, that can be said respecting them is reduced to mere indications.

A tooth from the chalk of Meudon, in its form, curve, and slight ridge on one side, much resembled that of the common crocodiles. The individual from which it came might have been twenty feet long.

Our countryman, Mr. Mantell, in his "Illustrations of the Geology of Sussex," has described the fossil remains in the different strata of that county very minutely. He has given a very particular account of those which are found in vast abundance in the forest of Tilgate, near Horsham, and has determined the strata of which the soil of this forest is composed. These are such as are immediately below the chalk formation. After the common chalk with silex, comes a lower chalk, without silex, then a sort of grey marlė or tuphous chalk, a bluish argillaceous marle analogous to that of Havre, the green sand, an argile containing beds of limestone, which we call Sussex marble, and, lastly, the ferrugineous sand which is so much extended throughout all this country.

The different strata are situated obliquely, so that the lowest rises on the eastern side, and forms plateaus as high as those of the chalk on the western side; on one of which the forest of Tilgate is situated, so that the ferrugineous sand is immediately under the diluvial or superficial stratum.

Amidst innumerable remains of testacea, crustacea, and fish, this sand also contains an abundance of reptiles' bones, and especially those of tortoises and crocodiles. There are many teeth, vertebræ, and other bones of the latter. Many of the megalosaurus, to be noticed hereafter, and especially some very singular teeth belonging to a reptile, but which are worn by detrition, as in the herbivora.

The vertebræ are a little concave at the two extremities, which approximate them to the crocodile of Caen, and the second of those of Honfleur. They are, however, nearer the first. The teeth are, for the most part, more obtuse than those of the common crocodiles.

In the deposition of lignites and plastic argile at Auteuil, near Paris, some small remains were found belonging to this genus. This formation is much more ancient than that of Montmartre, where there were also found a frontal and mutilated humerus of crocodile. In the lignites of Provence was found the upper part of a femur, which M. Cuvier seems to regard as having belonged to a peculiar species. In the isle of Sheppey, a cervical vertebra was found by M. de Luc, and determined by the Baron.

It is remarkable that crocodiles equally resembling the living species, accompany the palæotheria and lophiodonta in the fresh-water calcareous formation as well as in the gypsum; and what, perhaps, is still more surprising is, that they are found there with trionyx.

Thus, in a collection of bones made at Argenton, there were found seven left femora, in a fragmentary state, indicating the existence of, at least, seven individuals, all of moderate size, not supposed to be more than nine feet long. The number of teeth in this collection was prodigious, and their character differed somewhat from that of the teeth of known crocodiles. There were, also, remains of vertebræ and some fragments of head, from which M. Cuvier concluded the existence of a species different from those of Caen and Honfleur. There have, also, in several other places, been found remains more or less resembling the correspondent parts in living species, but of which it would be tedious to enter into any details here, and, from their fragmentary state, unimportant.

We find, upon the whole, that if the existing crocodiles are more numerous than was formerly believed, the fossil species of this genus also present a sufficient variety. Six, at the least, perfectly distinct, can be reckoned, and which do not differ less from the living crocodiles than they do from each other. These are the *crocodilus priscus*, two of Honfleur, one of Caen, all which four appertain to the sub-genus of the gavials; and the species of Montmartre and Argenton, which are referred by the Baron either to the sub-genus of Crocodile, or Cayman.

Had more considerable parts of skeletons been found in the other places, which presented remains of these animals, M. Cuvier thinks it probable that the characters of some other species might have been determined. However, the knowledge which has been gained on this subject is of great interest. It proves that the crocodiles have undergone the same fate as the mammalia, and that their species have not resisted the catastrophes which have broken and convulsed the external covering of our globe. But there is another truth of the greatest importance, of which they present us with the first indication, namely, that the various classes of vertebrated animals do not date their origin from the same epoch, and that the reptiles in particular are considerably anterior to the mammalia.

We have seen before, that the fossil mammifera of the genera most known, appear to be those which perished only at the latest revolution of this globe; that their debris fill the most superficial strata; that many of them have even preserved some of their soft parts, and that some have been even found entire, having been seized by the ice at the moment of their destruction, and not disengaged from it after. If we ascend higher in the series of ages, or, in other words, if we penetrate more deeply into the strata, we discover mammalia of genera less known, or that have for ever ceased to exist; such as the palæotheria, the anoplotheria, and the lophiodonta. They appertain to stony strata, formed, it is true, by the fresh water, but which cover others equally stony, and of an origin evidently marine. With these singular beings are also found a few species of existing genera, but their number is small, and it is quite clear that they by no means formed the character of the animal population of those remote ages.

Again, earlier than these we find only marine mammalia, dolphins, phocæ, lamantins, and other remains of similar genera. Beyond these, there are no further vestiges of mammalia, or, at all events, there are none whose origin is not more than doubtful. The antracotheria of the lignites, and the other mammalia there, nor those of the schists of Œningen, form any exception to this, for the position of these strata is not well ascertained, and the Baron does not think that these schists, and many strata of lignites, are of such antiquity as is commonly supposed.

In any case, the strata which are regarded as the most ancient of the tertiary, must be the first which would have enveloped the remains of mammalia; and supposing what is said of the small number of places presenting debris of this class to be correct, it would only oblige us to admit of an additional revolution—that is to say, of the existence of some tracts inhabited by mammifera, previously to the invasion of the sea, by which the cerithian limestone was formed.

The crocodiles, however, give rise to no doubts of this description. We see them appear in the very first secondary strata. The monitors of the coppery schistus alone precede them in point of time, but they show themselves immediately after in the lias in this country, the *banc bleu* of Normandy, or in the bluish and pyritous calcareous marle, which has so much analogy with the coppery schistus.

From then, until the epoch of the last catastrophe but one, some species of them have always subsisted, and in considerable abundance. To those of the different strata of the Jurassic formation, succeed those of the chalk. There are some above the chalk in the lignites of Auteuil and Mimet, and in the sandstone of Kent. Above the cerithian limestone, which the French call *calcaire grossier*, some are found in the marly fresh-water formation of Argenton, and in the gypsum of Paris. Finally, there even may be some in the loose and superficial strata, in which so many remains of elephants and other large quadrupeds are buried, if the small number of fragments collected at Brentford were not brought there from some other position.

It must be confessed, however, that they are of exceeding rarity in the last-mentioned depositions. None have been seen in the immense collections of bones made from the valley of the Arno, nor in those from any part of Germany or France. This is the more extraordinary, as the crocodiles of the present day live in the torrid Zone with the elephant, the hippopotamus, and all the other genera which have furnished those remains. Some few, indeed, are said to have been recently found in the superficial strata of the Vale of Arno.

We shall begin with a brief and rapid survey of the osteology of the living species, and then proceed to the fossil.

The sub-genera of the tortoises differ much more in the head than those of the crocodile. In the LAND-TORTOISES, such as the *great Indian tortoise*, the head is oval, and obtuse in front. The interval between the eyes is wide and gibbous. The nasal aperture is large, of greater height than breadth, and inclining a little back. The orbits are large, almost round, enframed on all sides, directed sideways and a little to the front. The parietal region of the cranium sharpens behind into a large and very salient occipital spine, and has on each side two very large temporal fossa, under which are two enormous *ossa tympani*. Behind these last, and a little above, are two bulky mastoïd protuberances, and under them are the apophyses, which serve for the articulation of the lower jaw. These apophyses descend vertically, and do not go backward as in the crocodile. Below, the basilary region is plane, the palatine concave. At the anterior part of this last, the osseous back-nostrils open, the palatines having no palatine floor, and the palatine portion of the maxillaries being grooved as far as the anterior fourth part of the muzzle. This disposition is rendered necessary by the manner in which the tortoises respire, and which as much resembles that of the frogs as it differs from that of the crocodiles.

The occipital region, on the whole, is vertical, although the occipital spine, the mastoïdean protuberances, and the articular condyle of the head, which is a very salient tubercle, render it very unequal.

The first remarkable trait in the composition of the head of tortoise is, that there are no bones of the nose.

In the fresh animal, the external osseous nostrils are contracted by cartilaginous laminæ which represent these bones; but in the skeleton, immediately at their upper edge, is found the frontal anterior bone, which takes its usual place in the frame work of the orbit, is also articulated as usual to the ante-orbital apophysis of the maxillary, forms the anterior partition which separates the orbit from the nose, and articulates below with the palatine and the vomer, leaving between itself, the maxillary, and the palatine, an oblong foramen, which opens into the back-nostrils.

The osseous cavity of the nose is oblong, and formed by the maxillaries, the inter-maxillaries, the vomer, the two anterior, and the principal frontals. The extent of the frontals, and the absence of nasal bones, cause the first to articulate one with the other, and to extend above the orbit, and outside the principal frontals, as far as the posterior frontals in this species, and very near them in some others.

The inter-maxillaries have no ascending apophysis. They form, as usual, the end of the muzzle, and go back into the palate between the maxillaries, and even between the back-nostrils, as far as the vomer. These back-nostrils are two wide apertures, pierced on each side, in the midst of the floor of the nasal cavity, between the maxillaries, the inter-maxillaries, the vomer, and the anterior frontals.

The bottom of the cavity of the nose is covered above, and closed behind by the principal frontals, which leave between them a wide aperture, closed in the fresh subject by a cartilage, which allows the threads of the olfactory nerve to pass.

Lower down, and laterally, there is between the frontal, the anterior frontal, and the vomer, a tolerably large space, closed in the fresh subject by a continuation of the same cartilage which represents the *os planum*.

In the land-tortoise there is nothing, or scarcely anything, of a simple, inter-orbital, cartilaginous partition, which want results from the great depth of the nasal cavities behind, and from the anterior and cartilaginous portion of the cerebral box approaching them considerably. But this is not the case with other sub-genera.

The frontals cover but a small portion of the cerebral cavity, for they are but short, and form, in conjunction, a lozenge more wide than long.

The parietals form together a pentagon, the most acute angle of which unites to the occipital spine. They cover more than half of the cerebral case, and go back by a squamous suture on the occipital and *os petrosum*. On each side the parietal descends very low into the temporal foss. It occupies there almost all the space occupied in the crocodile by the temporal wing of the sphenoïd, and there remains of this wing, in the tortoise, only a very small piece, which unites itself on one side to the descending portion of the parietal; on the other, to the palatine, to the internal pterygoïdean bone, to the body of the sphenoïd, to the *os tympani,* and to the *os petrosum.*

The jugal bone is, as usual, articulated with the external and posterior angle of the maxillary. It is narrow, and extends under the orbit, behind which it meets the posterior frontal which completes the frame in this part, and the squamous temporal, which of itself forms all the zygomatic arch, an arrangement of which there are a multitude of examples in the cetacea.

This temporal widens to unite itself to the *os tympani,* which is extremely large. It forms almost a complete osseous frame for a wide tympanum; and under this it descends as an apophysis for the articulation of the lower jaw. This frame gives entrance into a vast cavity, completed only at its upper angle by the mastoïdean bone. At the bottom of this cavity is a foramen, through which the auditory osselet passes to arrive at a second cavity, formed externally by the tympanic bone, internally by the *os petrosum* and occipitals, underneath a little by the sphenoïd, and closed behind by cartilage. This tympanic bone, moreover, composes a good part of the posterior parietes of the temporal fossa.

Between it and the parietal in this same foss, the *os petro-*

sum shows itself, and the cranium is closed behind by the occipital, which is here divided into six bones, and not into four; for the lateral occipitals are each divided into two parts, the external one of which the Baron terms the *external occipital.*

The fenestra ovalis is situated, as in the crocodile, at the *os petrosum* and the common lateral occipital; but the fenestra rotunda is pierced in the *external occipital,* in the same manner as it is in the lateral occipital of the crocodile.

In this tortoise, as well as in the crocodile, the grand foramen for the passage of the fifth pair is in front of the *os petrosum,* between it and the temporal wing. In the sea-tortoise this foramen is between the *os petrosum,* and the descending portion of the parietal.

The auditory osselet is simple, as in the crocodile, and formed of a slender stem, which widens at its approach to the fenestra ovalis, and attaches itself there by a round and concave face, so that it has pretty nearly the figure of a trumpet. The external edge of this stem is in a great part cartilaginous, and is terminated by a plate of the same substance, and of a lenticular form, which is enchased in the membrana tympani, and which, perhaps, may be considered as analogous to the malleus.

The Eustachian tube is altogether cartilaginous or membranous. It commences in the external chamber of the cavity in the upper part, by a wide emargination of the posterior edge of the tympanic bone, and goes obliquely within, passing between that bone and the depressor muscle of the lower jaw, as far as an emargination of the lateral and posterior edge of the pterygoïdean bone, through which it penetrates into the back part of the mouth, on the side very near the articulation of the lower jaw, and especially very far behind the interior nostrils.

The orifices of the two Eustachian tubes may be seen at the palate, or rather at the back of the roof of the hinder mouth, in the form of two little holes sufficiently remote from each other.

Behind the maxillaries and the posterior frontals, on the two sides of the vomer are the palatines, surrounded behind and externally by the pterygoïdea, which last extend along the external edge of the palatines as far as the maxillaries. The rest of these pterygoïdean bones covers the lower face of the cranium, between the two ossa tympani and the two temporal wings, leaving visible behind only a small triangular portion of the body of the sphenoïd. The palatines have only their upper part, and want the curved part which prolongs the floor of the palate behind the maxillaries. The lachrymal bone is not discoverable in the tortoises, no more than in the phocæ and dolphins.

The olfactory and optic nerves come forth through cartilaginous partitions of the cranium, and have no particular foramen in the skeleton. The same appears to hold good for the third and fourth pair. The sixth passes through a small canal of the body of the sphenoïd. The fifth pair has a large foramen between the *os petrosum* and the temporal wing, divided into two at the exterior. At the external edge of the palatine there is a foramen analogous to the pterygo-palatine.

Internally, the cerebral cavity is more high than wide. The bottom is very even. But in front, in the sphenoïd, is a deep fosset for the pituitary gland, a sort of sella turcica. From the sides of this part originate the cartilaginous partitions, which, joining the ante-cerebral partition of the frontal, close the cavity of the cranium in front, support all the anterior portion of the encephalon, hold the place of the lamina cribrosa, of the orbital wings, and of the greatest portion of the temporal wings; another considerable part of which is replaced by the descending portions of the parietal, so that what remains does not participate in the formation of the cerebral case, except a little in front of the foramen of the fifth pair.

There is no bony vestige of an anterior sphenoïd any more than in the crocodile.

This description, taken from the Indian tortoise, is equally applicable to all land tortoises. In the *testudo Græca* the cranium is less gibbous between the orbits. The principal frontals, more long than broad, reach to the edge of the orbit, between the two frontals, and re-descend into its roof.

In the *emydes*, or common fresh-water tortoises, the head is more flatted. The principal frontals, though more broad than long, do not always reach the edge of the orbit, as, for example, in the *testudo Europæa*. The posterior frontal is broader. The frame of the tympanum is not complete; and instead of a foramen, there is a scissure for the passage of the auditory osselet, from one chamber of the cavity into the other. The basilary region and the palatine make but one plan, the palatines not even being concave. The *test. scripta, picta, scabra, dorsata, centrata, clausa, virgulata*, are all thus distinguished.

Some emydes, as the *emys expansa*, have characters in common with the sea and fresh-water tortoises, and others beside, peculiar to themselves. The head is depressed, the muzzle short, the orbit small and very forward. There is no osseous vomer, so that the two back nostrils form but one foramen in the skeleton. The palatines have not the palatine portion. The frame of the first chamber of the tympanic cavity is complete. This chamber communicates only by a narrow foramen with the mastoïdean cellule, and the Eustachian tube originates there by a cleft which is an extension of the foramen through which the osselet passes into the second chamber.

The temple is covered as in the sea-tortoises by the parietal, the temple, the jugal and the posterior frontal bones. The last of these is very narrow. It has a portion descending into the temple, which, uniting itself to an ascending portion of the palatine, and a re-entering portion of the jugal, forms a partition, which separates the orbit from the temporal fossa, leaving no communication but a large foramen near this descending part of the parietal, which replaces the temporal wing.

The pterygoïd bone unites itself in front to the palatine and jugal, and not to the maxillary, which does not extend so far back. Its external edge recurves with the neighbouring portion of the jugal, and thus forms, in the lower part of the temple, a sort of canal which commences at the foramen of communication of the temple with the orbit. Its posterior angle, on the contrary, is directed a little towards the bottom, descending more than the articulary facet for the lower jaw, and leaving between it and the raised portion of the external edge a broad emargination. Between this angle and the articulary facet is a foss hollowed in the tympanal, the sphenoïd, and pterygoïdean bones.

The mastoïd tubercles are depressed, very salient behind, and pointed. Their point is formed one-half by the mastoïdean bone, and the external occipital. The sphenoïd shows itself underneath on a wider surface than in the land-tortoises, and the basilary appears there less. The lateral occipitals are also very small, and speedily are soldered with the basilary. The tubercle for the articulation of the atlas projects less than the mastoïdean apophyses.

In the *emys serpentina*, at a certain age, no external occipital is found distinct. It is united to the lateral occipital, but in the land-tortoises to the upper occipital. The head is depressed in front, the muzzle very short; the orbits moderate, and approximating to the muzzle; the temple covered only at its anterior portion by a lamina of the parietal, less complete than in the sea-tortoises, and by a widening of the posterior frontal and of the jugal. The palatines have no palatine lamina: the palatine and pterygoïdean region is very flat. The foramina analogous to the pterygo-palatines are very large, and the passage of the auditory osselet is made by a foramen, and not a scissure.

In the TRIONYX, or soft tortoises, the head is depressed, and elongated behind. The muzzle in certain species, as in that of the Nile, is pointed, short and rounded in others. The intermaxillaries are very small, and have no nasal or palatine apophyses. Behind them is a large incisive foramen. The maxillaries unite between them in the palate on a tolerably long space, so that the back nostrils are further back than in the sea-tortoise. The palatines do not unite below to prolong the palate; they are hollowed into a demi-canal in front, and less extended than in the land-tortoises. The body of the sphenoïd reaches to them, proceeding between the two pterygoïdean bones, which are not united one to the other, but go from the lateral occipital, between the tympanic cavities and the basilary bone, and at the sides of the body of the sphenoïd, as far as the palatines and maxillaries, which renders all the basilary and palatine region broad and flat.

The anterior frontals advance between the maxillaries, and in this part occupy exactly the place of the proper nasal bones, without being distinguished by any suture. They even form a point over the external aperture of the nostrils, as the bones of the nose often do in the mammifera.

The principal frontals form almost a square. They reach the edge of the orbit. The posterior frontal is as wide above as it is high. The jugal forms a part of the posterior and lower edge of the orbit, and almost all the zygomatic arch, of which the squamous temporal bone constitutes but a small portion in front of the tympanic cavity. This last has its frame complete. The osselet passes through a foramen into the second chamber, which, as in the other tortoises, is closed behind only by cartilage.

The spine of the occiput and the mastoïdean tuberosities are all pointed, and more salient behind than the articulary condyle. The space occupied by the os tympani at the posterior edge of the temporal foss is very narrow, but it widens in re-descending towards its apophysis for the lower jaw. The temporal wing is placed below and in front of the grand foramen of the fifth pair, and the descending portion of the parietal articulates in front of it to the internal pterygoïd bone. There is no trace of an anterior sphenoïd, or of its wings. Its place is held by a slender membrane, which closes on each side the forepart of the cerebral cavity.

The principal character of the MARINE TORTOISES, or *Chelo-*

nian reptiles, is, that a lamina of the parietal, the anterior frontal, the mastoïd, the temporal, and the jugal, unite together, and with the os tympani by sutures, to cover all the region of the temporal with an osseous penthouse uninterrupted in its continuity. Their muzzle being much shorter and the orbits much larger than in the other tortoises, the nasal cavity is smaller, and of equal width, height, and length. Its hinder paries, belongs entirely to the anterior frontals, and between them is the introduction of the olfactory nerves. The osseous tubes of the back nostrils commence in the lower part of this hinder paries, and as the palatines have a lower lamina, these tubes are a little longer, directed more backwards, and have less resem blance to simple foramina.

From the size of the orbit it also results that the inter-orbital membranous or cartilaginous space is more extended. The piece which we regard as the temporal wing is singularly small in the *chelonia mydas*, altogether at the external face, and simply pasted on the suture of the descending portion of the parietal with the pterygoïd.

The auricular osselet does not pass through a foramen, but by a wide emargination, from the first chamber of the tympanum into the second, and this second chamber is cartilaginous in all its hinder paries. By the same emargination the Eustachian tube descends towards the back part of the mouth. The first chamber of the tympanum is not very concave. There is, properly speaking, no mastoïdean cellules. But the bone of that name completes the roof of this chamber, and thus extends its concavity. The foramen of the fifth pair is oval and very large, between the descending portion of the parietal, the pterygoïd, and the os petrosum.

The most anomalous head of tortoise is that of the MATAMATA *(testudo fimbriata)*. It is so singularly broad and flat, that it has the appearance of having been crushed. The orbits are extremely small, and very near the end of the muzzle. The hinder region of the cranium is elevated, and the two tympanic bones, formed like trumpets, widen on each side of the cranium. The templeis a wide horizontal foss, but not deep. It is uncovered, except on the back part, by the union of the posterior angle of the parietal with the mastoïdean bone. The osseous temporal is reduced to a mere vestige.

The two maxillaries form a transverse arch, at the middle of which, underneath, is a single interparietal bone, and above the external aperture of the nostrils, which in the fresh subject is continued in a small fleshy trunk. The two palatines, and between them the vomer, fill up the concavity of this arch underneath, and have in front of them the two hinder nostrils considerably separated, but the palatines do not surround them below. At the posterior edge of the palatine is a tolerably large pterygo-palatine.

The anterior and posterior frontals form the upper part of the orbits. The principal frontals advance between the anterior as far as the edge of the external nostrils. There are no nasal bones, no more than in the other tortoises.

The jugal bone takes its place from the posterior angle of the orbit, between the maxillary and the posterior frontal, which it does not pass, touching a little on the pterygoïdean behind and underneath, but forming no projection behind to border the temple. The temple is thus separated from the orbit by a postorbital branch of excessive breadth, which takes up the totality of the posterior frontal and jugal bones. The posterior frontal is itself articulated to the pterygoïdean bone by its external hinder angle. The rest of its hinder edge is free, and continues with that of the parietal to cover a wide and flat canal of communication, going from the temple to the orbit, and formed underneath by the pterygoïdean and palatine bones.

The two pterygoïdean bones are enormous. They form the greatest portion of the basis of the cranium and the ground of the temple. Their external edge is curved in its anterior part to continue with the free edge of the posterior frontal. There are neither orbital nor temporal wings. The parietals, which form a large right angle above, unite by their descending parts to the palatines, the pterygoïdeans, the ossa petrosa, and the upper occipitals. They form of themselves almost the entire penthouse of the cranium. At the sequel of the pterygoïdean the temple is bounded behind by the tympanic bone, which partly resembles a trumpet. The frame of the tympanum is complete. A foramen of the hinder paries lets the osselet pass into the second chamber, which, in the skeleton, is only a long groove of the posterior face of the os tympani, which terminates in a cavity, to the formation of which the os petrosum and the external and lateral occipitals concur. It is closed behind only by cartilage and membranes. The fenestræ are in their usual place.

Above this foramen of the first chamber, through which the osselet passes, is another which conducts into the mastoidean cell, which, in consequence of the outward projection of the tympanum, is found within, and not behind. The occipital spine is a short vertebral crest, and the mastoid tubercles are transverse crests belonging entirely to the mastoidean bone.

Underneath, the cranium is smooth and almost plane, and exhibits a sort of regular compartment, formed of the intermaxillaries, the maxillaries, the vomer, the palatines, the pterygoïdeans, the sphenoïd, the ossa petrosa, the ossa tympani, the basilary, and the lateral and external occipital bones. Behind the floor of the temple the os petrosum forms a square compartment, between the pterygoïd, the tympanum, the external occipital, the upper occipital, and the parietal.

In the lower jaw, the space occupied in the crocodile by the two dentary and two opercular bones, is occupied in the marine, fresh-water, and land-tortoises by a single bone analogous to the two dentaries. There is no trace of symphysis, the bone being continued as in birds. But in the *testudo fimbriata* a division is preserved at the anterior part, at every age. The opercular, however, exists at the internal face of the jaw, but is thrown more backward than in the crocodile. Under it is the angular, constituting the lower edge of the jaw, and the external face of this part is occupied by the subangular, which touches not the angular except very far back, being separated from it on the two anterior thirds of its length by the dentary.

Above and towards the back, between the opercular and subangular, is situated the articulary bone. Its dimensions are but small, and it serves only for the articulation and insertion of the depressor muscle, which is analogous to the digastric.

The coronoïd apophysis does not belong to the subangular bone, but to a bone placed between the dentary, opercular, and subangular, and in front of the aperture by which the nerves enter the jaw, which aperture is here on the upper edge, instead of being, as in crocodiles and birds, at the internal face.

The hyoïd bone is very complicated, and singularly varied in conformation, in the genera and species of tortoises, but it is not necessary for our purposes to enter into a detailed description of it.

From what has preceded, it may be observed that there is more difference in the arrangement and mutual relations of the bones of the head in the different tortoises, than probably in the heads of all quadrupeds, and most assuredly than in the entire class of birds. There are proportional though not so considerable, differences in the other parts of the skeleton.

The most general character of the tortoises, as is well known, consists in their having the bones of the thorax outside, enveloping with a cuirass or double buckler what subsists of the muscles, and serving even as a shelter to the bones of the shoulder and the pelvis. This dorsal buckler is formed principally of eight pairs of ribs, united towards the middle by a longitudinal series of angular plates, which either adhere to the annular parts of so many vertebræ, or constitute a part of them. What is most remarkable is, that these annular parts alternate with the bodies of the vertebræ, and do not correspond to them directly.

The ribs catch in by sutures with these plates. They also catch together through all or a portion of their length. There are eight vertebræ in front, which do not share in this arrangement. The first seven, which are the ordinary cervical, are free in their movements. The eighth, which may be considered as the first dorsal, is placed obliquely between the last cervical, and the first of the fixed vertebræ of the dorsal buckler, which position shortens it in front. Behind, its spinous process is elongated, and thickens a little to attach itself by sychondrosis to a tubercle on the lower face of the first of the plates of the middle series of the buckler.

The first of these fixed vertebræ, which is the second dorsal, is rather short, has also its proper annular part, the spinous apophysis of which, shorter than the preceding, attaches itself in like manner to the second plate by a cartilage.

This second plate, more narrow than the first, makes but one bone with an annular part which is underneath; the anterior portion of which is articulated, by two small apophyses, with the articular apophysis of the second dorsal vertebræ. It is, then, properly speaking, the annular part of the third dorsal vertebræ, but the body of this last is only articulated by its anterior half with the posterior half of this third annular part, while its posterior half is articulated to the anterior half of the fourth annular part. This alternation continues, so that the body of the fourth vertebræ corresponds to the annular parts of the third and fourth; the body of the fifth, to the annular parts of the fourth and fifth, and so on as far as the tenth.

The eleventh vertebræ after the cervical is the only one which can be called lumbar. It bears no rib. The twelfth and thirteenth are the sacral. To their sides are attached two lateral pieces, swelled at the end to unite to the hinder and upper angle of the iliac bones. Their annular part is close and complete, and does not make a body with the plates of the buckler which follow that of the eleventh vertebræ.

The vertebræ of the tail are free like those of the neck.

The sea-tortoises have three longitudinal plates after the tenth; thirteen in all. But as the second and ninth are sometimes divided, we may reckon them fifteen. Fourteen have been found in the *emys serrata.* The eleventh and twelfth, however, are very small. There are but eleven in the land-tortoises and chelydes.

The ribs do not always catch in their entire length. Towards their exterior a narrow portion remains, and the intervals between it and those of the anterior and posterior ribs are filled only by a cartilaginous membrane. In the fresh-water tortoises and chelydes, this buckler is entirely filled up in course of time, and the ribs catch, in their whole length, both with each other and the marginal pieces. Ossification goes on still more rapidly in the land-tortoises; and it is only in their early age that vacancies are observable between the external parts of their ribs. This buckler is more or less gibbous, according to the species. There are many other variations which our plan and limits oblige us reluctantly to pass unnoticed.

The sternum, or anterior portion of the buckler, is composed of nine pieces; the atlas of four. The axis and the succeeding vertebræ are composed of a body nearly rectangular, carinated underneath, concave in front, convex behind, and of an annular part which remains distinct from the body during life, by two sutures, is raised above by a crest, instead of a spinous apophysis; and its anterior articular apophyses, placed at first under the posterior apophyses of the preceding vertebræ, rise obliquely to embrace them, as far as the sixth, and resume a little their horizontal position in the two following.

The bone which proceeds from the dorsal buckler to the sternum is suspended by a ligament under the dilatation of the second rib. Sometimes in this ligament are one, or even two particular little bones. One of these bones is at first a little cylindrical. It goes forward, and after having given from its external face a portion of the articulary facet which receives the head of the humerus, it goes, making a greater or less inflexion internally, to attach its other extremity to the internal face of the sternum. The rest of the facet for the articulation of the humerus is furnished by another bone, which goes back more or less obliquely, and towards the central line, widening like a fan. This remains nearly parallel to the sternum.

All the muscles which proceed to the arm are respectively the same as in birds, whatever changes may take place in their position relatively to the horizon, and in their size and figure.

The existence of a clavicle is doubtful. A shoulder with three branches, an omoplate nearly cylindrical, an acromial portion nearly equal in volume to the rest of the omoplate, are characteristic of the tortoises. There is nothing similar in other animals, because in no other is the shoulder within the thorax. Their varied forms present very good characters for distinguishing the subgenera.

In the sea-tortoises, that part of the omoplate which forms the articulary face, is detached in some sort from the bone, and forms a lateral apophysis; and the two branches at the re-entering angle, which they form together, are compressed, broad and flat. The acromium is compressed, but in another direction, and the coracoïd bone is very long and not very wide at its sternal extremity.

In the land-tortoises, in which the dorsal buckler, being more raised, gives more room for the extension of the omoplate and its acromium, the angle is more open and the bone less compressed. The coracoïd is short, and so widened, that its sternal edge is equal to its length.

The shoulder of the fresh-water tortoises is a sort of medium between these two. The coracoïd bone is more long than wide. The acromial branch is compressed. The angle which it makes with the omoplate is marked, but less so than in the marine tortoises.

The chelydes have the coracoïd bone wider and shorter than the fresh-water tortoises, but less so than in the land-tortoises.

In the trionyx, the angle is sufficiently marked, but the coracoïd bone is distinguished by a peculiar form. It is wider than in the other sub-genera. Its external edge is convex, and is continuous with the hinder edge, while the internal is a little concave.

The humerus of the tortoise must turn singularly on its axis, to place the fore-part in the position which the osseous cuirass requires, which leaves it no passage but by a narrow emargination. The head proceeds more out of the axis than in any other animal, and that towards the upper face. It is a segment of a sphere, and very concave. The two tuberosities are very large, very salient, and leave between them a concavity, the same as exists between the condyles of the humerus in the majority of the mammifera. The internal tuberosity is the largest. It has the form of a long obtuse crest, analogous to the deltoïd crest, and which receives the same muscles. The other tuberosity also forms a crest, but much shorter. The body of the bone itself is arched; and its concavity which, in man, would be anterior, is here, in general, lower. The opposite face is convex. In the upper part is a small hollow, opposite the end of the foss, which is between the two tuberosities.

The bottom of the bone is widened, and a little flatted from front to rear. On its external edge may be remarked a furrow, not very distinct in the land-tortoises, deeper in the emydes, the chelydes, and the trionyx, and which in the marine tortoises, separates the lower head of the bone into two unequal parts. This furrow is the best character for distinguishing the lower part of the humerus from that of the femur. This lower head is transversely oblong, and of an uniform convexity, and receives the bones of the fore-arm without presenting to them two distinct facets. There are always two bones in the fore-arm, but of little mobility. They are so placed that the cubitus forms the external edge of the arm, and the radius the internal. The upper head of the radius is semi-circular, and a little concave; the body is slender; the lower head compressed, and cut obliquely, so that it is shorter on the cubital side.

The cubitus is compressed; its upper head is triangular and cut obliquely. The lower head is squared.

The pelvis of the tortoises is always composed of three bones, contributing, as in quadrupeds, to the composition of the cotyloïd foss: these are an ilium, a pubis, and an ischion. At the place where they unite to form the cotyloïd cavity, each bone has three faces, one for each of the two others, and one for the cavity.

The femur might easily be taken for the humerus of a quadruped; its oval head is removed from the body of the bone without being separated precisely by a narrow neck. Instead of a trochanter, there is a transverse crest not much raised, and separated from the head by a semicircular depression. The middle of the bone is slender and round, and the bottom compressed from front to rear, and widening by degrees to form the lower head, which is a transverse portion of a cylinder, a little inflected behind.

The two bones of the leg are nearly straight. The tibia is bulky and semicircular in the top, and again grows rather bulky at bottom. The peroneum is more compressed, and wider in the lower part.

In all the parts now described there are slight variations according to the subgenera, which, if necessary, shall be noticed in our comparison of the fossils.

In the sea-tortoise all the bones of the carpus are flat and cut nearly square. The metacarpian of the thumb is short and broad, the others are long and slender. There are but two phalanges on the little finger, which is not longer than the thumb. The three others, particularly the medius, are elongated. This arrangement produces a pointed hand, in which the thumb and the index alone have the unguical phalanges armed with a claw.

The calcaneum of tortoises, in general, has no prominence behind, so that the tarsus is as flat as a carpus. It is composed of six or seven bones in the sea-tortoises. The bones of the metatarsi of the great and little toe are shorter than the others, and singularly broad and flatted.

We must now speak of the fossil remains of this genus.

The number of existing tortoises is so very considerable, that it is very difficult to decide whether or not a fossil tortoise belongs to an unknown species. It is necessary to compare not only the carapaces, and bucklers provided with their scales, but also the skeletons themselves, to observe the junctures of the ribs and of the other bones which concur to the composition of these cuirasses. The most that can be done for many of them is to assign their subgenus, which, however, is a point of considerable importance, as it tends to throw light on the origin of the stratum which envelopes them, or at least on the existence or non-existence of some dry land in the neighbourhood of the waters where this stratum was formed.

A remarkable abundance of the bones of TRIONYX are found in the same strata with palæotheria, &c. though this subgenus has never been known to exist in Europe at any period of authentic history. There are, in fact, no species nearer to us than in the Nile and the Euphrates. These are the *thirsé* of the Nile and the *rafcht* of the Euphrates, *(Testudo triunguis.)* All the other species whose country is known inhabit the rivers of warm climates, which renders it probable that those whose country is not known have a similar habitat. It seems probable that the tortoise described by Aristotle under the name of *ἐμys*, was the species of the Nile just mentioned. It is the only species of which he could affirm that the head was sufficiently soft to let the humours transpire. It is evident, however, that he was very ignorant of this species, for he employs this character of its organization to explain a very erroneous supposition, that the animal in question had neither reins nor bladder. M. Cuvier has examined numbers of this subgenus, and found them conformed in these particulars, like all other tortoises.

From the remains found in the plaster-quarries of Paris, it appears that at least one trionyx abounded at the period in which the palæotheria, anoplotheria, cheropotami, adapis, sarigues, crocodiles, and all the singular animals which we have already described, existed. But there was nothing in these remains which could determine the characters of the species.

In the plaster stones of Aix some remains were found of trionyx, consisting of a carapace which had lost a great part of its left side, and several of the rib ends of the other, and also a left moiety nearly complete of the sternal portion of the buckler, and a small fragment of the lower part of the right moiety. The portion of carapace was twelve inches long, and eight broad, and differed in its characters from those of all known species. There was, however, some approximation to the trionyx of Java, and of the Ganges, but not sufficient to identify the fossil as belonging to either. The breast-piece showed considerable analogy to those of the Egyptian and Indian species, especially to the former, in the shape of its middle piece, and the small extent of the mutual articulation of its two hinder pieces. But the upper denticulations form a more elongated groupe, and the lower piece has only its middle vermiculated. Its anterior and external contour is smooth, which is the case only with the angles, in the Egyptian species.

The Baron considers this trionyx to be of a species unknown at the present day.

On an estate of the Duke de Caze, in the department of the Gironde, was found a stone analogous to the molasse of Switzerland, which contained fossil remains of many genera;

teeth, fragments of jaws, and other bones of palæotherium were discovered there. It also presented a great number of manifest fragments of tortoises, especially of trionyx. By one of the last, M. Cuvier thinks himself authorized to pronounce these fragments to belong to a different species from the living.

This was a fragment broken in two, and rather mutilated, of the first piece of the carapace, the unequal and transverse, piece which adheres neither to the ribs nor vertebræ. This is easily recognized by the irregular crest which traverses obliquely its lower face, and by the oblique foramina which are pierced there for the vessels. The species of Java most nearly approaches this, but is still very far from resembling it altogether. This trionyx appeared to have equalled that of the Nile in size. In parts of France more fragments have been found referable to this subgenus, but by no means so well characterized as to justify specific distinction.

On the left bank of the Aar, to the north of the town of Soleure, are numerous quarries, in which many discoveries of fossil remains have been made. They are excavated in a small hill which borders the valley, and is situated at the foot of that portion of the lofty chain of Jura, nearest to Switzerland, which ends at that part of the Rhine close to the confluence of the Aar, and partly separates the canon of Soleure from that of Bale and the territory of Porentruy. The stone of which they are composed is a limestone of the recent formations of Jura. It is hard but not brittle, of a whitish colour, approaching to grey, to bluish, and sometimes to yellowish. Its strata are generally horizontal. In many places they rest on the marly banks of intermediate formation, and they constitute no part of complete and regular chains. Many similar strata are found on the other side of the great crest of Jura.

There are eight or nine banks worked in these quarries of Soleure. In the upper one, the stone, left in various directions, serves for no other purpose than making lime. It contains shells of different kinds, and some bones, but the latter occur rarely. The second, of a regular thickness of three feet and a half, furnishes, with terebratulæ, oysters, &c. some debris of the bones of tortoise, and certain portions of the jaws of fishes. In the third, there is the greatest quantity of debris of tortoise, but never in any good state of preservation. They are accompanied by teeth of crocodile, which we have already noticed, and many marine shells of the genera we have just mentioned. The teeth of fish are found in the fourth, with some remains of other bones. In the fifth, shells again appear. The sixth is a small bed of marle. In this are found the bones of tortoise more entire, with shells of different genera. The seventh and eighth are filled with terebratulæ, and contain some vertebræ of fish, but rarely the remains of other bones. In the ninth are crystals of pyrites, but no petrifactions. In the tenth, which is very thick and of a gross and friable grain, are only found terebratulæ. This formation, notwithstanding its peculiarity, is nevertheless marine. It is, therefore, astonishing enough to find there the remains of animals whose genera exist at the present day only in the fresh water, such as the emys and the crocodile. Nevertheless, it is a most certain fact.

A tolerably complete cuirass was found here with its carapace and breast-piece, and the impression of the scales was quite distinguishable. It was twenty-four inches long, and twenty in its greatest breadth, which was towards the lower third part. Its form was a fine oval, rounded at the two ends, and moderately convex. The notches for the paddles are wide. The scales of the middle of the disk appear to have been as broad as those of the sides, but those of the edge were very narrow.

Two other fragments were found, apparently belonging to two other species. One of them, which is a lateral portion, containing the remains of four ribs, and the correspondent marginal pieces, is remarkable for its size. It is very flat, and measures more than eleven inches from back to front. The other is the hinder part of a dorsal buckler, distinguished by three projecting ridges in its anterior and most hollow part.

Many singularities are observable in the arrangement of these bones. The last two ribs join each other in front of a very small dorsal piece, which is followed by another very large, and triangular. This has got another very small one at each of its sides. Then come the last two dorsal pieces, both tolerably broad. It is not without example to see, in the emydes or tortoises of the present day, ribs thus united to one another along the dorsal line, and causing to disappear or contracting much the dorsal plates which should separate them.

A head was found in the same formation, in a fragmentary state. It was broken in such a manner behind, that nothing was left but the anterior paries of the os tympani. The posterior frontal is broader than in the emys of Europe, but not so broad as in the *expansa* or *serpentina*. It does not cover the temple behind until just opposite the anterior edge of the tympanum, as is usual in the common emydes. But the parietal does not unite with it to cover the rest of this foss, which shows that it did not belong to a marine tortoise. The land-tortoises are also equally excluded from any claim to this specimen, because in them the posterior frontal is much more narrow. The size of the orbit, shortness of the nose and muzzle, the marked emargination behind of the maxillary edge, all exist in this head, the same as in the common emydes.

Two plates of the dorsal series found in the same locality, were examined by the Baron. Their form was an almost regular hexagon, arched in the longitudinal direction in the middle, and they were very remarkable for their extreme thickness. The plates of the *emys serrata* approach most nearly to the figure of these, but their hexagon is far from being so regular.

These specimens, and several others, which it would be tedious to dwell upon, perfectly establish, in the opinion of M. Cuvier, the existence of numerous remains of two large and unknown emydes in the quarries of Soleure. Another bone, which seemed to be a fragment of a breast-piece, even appeared to indicate the existence of a particular and unknown genus.

In the forest of Tilgate, where Mr. Mantell collected the bones of crocodile which we have noticed, were also found some remains of emys by the same gentleman. One seemed to have belonged to a part of the carapace, which was a little concave; the other is a portion of the anterior edge. The first was found in a very fine ferruginous sand, strongly agglutinated: the other in an agglomeration of divers little rolled stones, or gravel, partly agglutinated by the sand and partly by spathic infiltrations.

The immense beds of soft sandstone, called by geologists molasse, which fill all the lower parts of Switzerland, and again appear over great spaces in the south of France and in Hungary, are considered as well as the lignites and the other subordinate beds which they contain, to be superior to the chalk, and inferior, or perhaps, in some places, contemporaneous, to the coarse coquillaceous limestone, and some other more recent tertiary strata. These strata are rich in fossil remains, which belonged to the land, and to the fresh water, in crocodiles, in trionyx, and in palæotheria. It is, therefore, not surprising to find among them, in the same strata, the bones of emydes. Fragments have been found in the quarries of La Grave, which appeared to have belonged to very large species. They corresponded in form to the analogous portions of the buckler of the *emys serrata*, but many of them were three or four times as thick. One of these fragments M. Cuvier supposes to have come from an individual of more than three feet in length, which is an uncommon size in the existing species of emys. There was even found the head of a coracoïd bone, which indicated a still greater size in the individual to which it had belonged. Similar fragments were found in the molasse of Switzerland, near the town of *Aarberg*, but which seemed to have rather more analogy with the emys of Europe.

In the argillaceous formation of the Isle of Sheppey, which is a continuation of the plastic argilla of the neighbourhood of

Paris, very evident remains of this sub-genus have been discovered. A portion of carapace was sent to the Baron by Mr. Crow of Feversham, which, though a little compressed and deformed, still clearly exhibited all the characters of emys. Five pairs of ribs were distinguishable, and the remains of a sixth, with six vertebral plates. The fifth of these plates is separated from the sixth by a point formed by the ribs of the fifth pair, which unite together in front of the sixth plate, which is very small. This arrangement somewhat resembles what has been seen in some remains of emys from Mount Jura. There were also the entire impressions of two scales of the middle series. The ribs which remain are of an equal breadth throughout; a constant character of the emydes. The vertebral plates are more narrow than in the existing species; and, from the impressions remaining, it appears that the scales of the middle range were more long than broad. M. Cuvier is of opinion that the *emys expansa* most resembles this fossil. The marine tortoises have, it is true, like this specimen, their ribs of equal length; but their middle scales are rhomboidal, and of greater breadth than length.

Mr. Parkinson, in his "Organic Remains," has given a figure of a breast-piece from the same locality. The parts which compose it do not appear to have been completely joined by sutures, which might give rise to the opinion that they belonged to a marine tortoise, or a trionyx; but M. Cuvier prefers attributing them to a young emys, whose ossification had not been completely terminated. The similarity of the forms of the bones renders this notion extremely probable.

Some remains were found in the neighbourhood of Brussels, which were at first attributed to the sub-genus of the marine tortoises. Even M. Cuvier himself was inclined to this idea, though he clearly considered them as specifically distinguished from any existing sea-tortoises. A close and more extensive examination, however, of the carapaces of the different subgenera, and of the sutures which unite their bones, fully convinced him that the remains in question belonged to the subgenus on which we are now writing. The description of a carapice found in the quarries of the village of Melsbroek will prove this. Its contour is oval, a little narrowed behind, but not more so than in the *emys centrata*. The ribs unite uninterruptedly with the marginal pieces, as they do in all the emydes and land-tortoises. The curve of these ribs is pretty nearly the same as in the *emys centrata*. The vertebral plates are singularly narrow, more so than in any living emys, and even than in the fossil emys of Sheppey, which we have just spoken of. This peculiarity may be remarked in this specimen, that the seventh and eighth ribs are each of them united to their opposites, between the eighth and ninth vertebral plate. This is a circumstance which also takes place in relation to the seventh pair in the emys of Sheppey, and which in the eighth pair, but only in the internal face, is found in the *emys centrata*, but is just the same in the *emys expansa*, as in the specimen of which we now speak.

Comparing this carapace with that of any sea-tortoise of the same size, a specific character, very strongly marked, is instantly discovered. The fossil tortoise has the interval of its ribs completely ossified, and no vacancy remains between them and the pieces of the edge, which are also much more broad in proportion than those of the sea-tortoise. In the *testudo mydas*, for instance, at the age when its carapace is no more than thirteen or fourteen inches in length, a vacancy remains between the ribs, not ossified, which almost equals one-half of the length of the rib. A part of this vacancy has been found remaining, even in an individual whose carapace was nearly four feet in length; and it has been verified upon many of a size intermediate between the last two mentioned. All this serves to prove that the fossil tortoises of Melsbroek cannot belong to *testudo mydas*: nor in fact to any marine tortoise, for the ossification does not take place more rapidly in any of

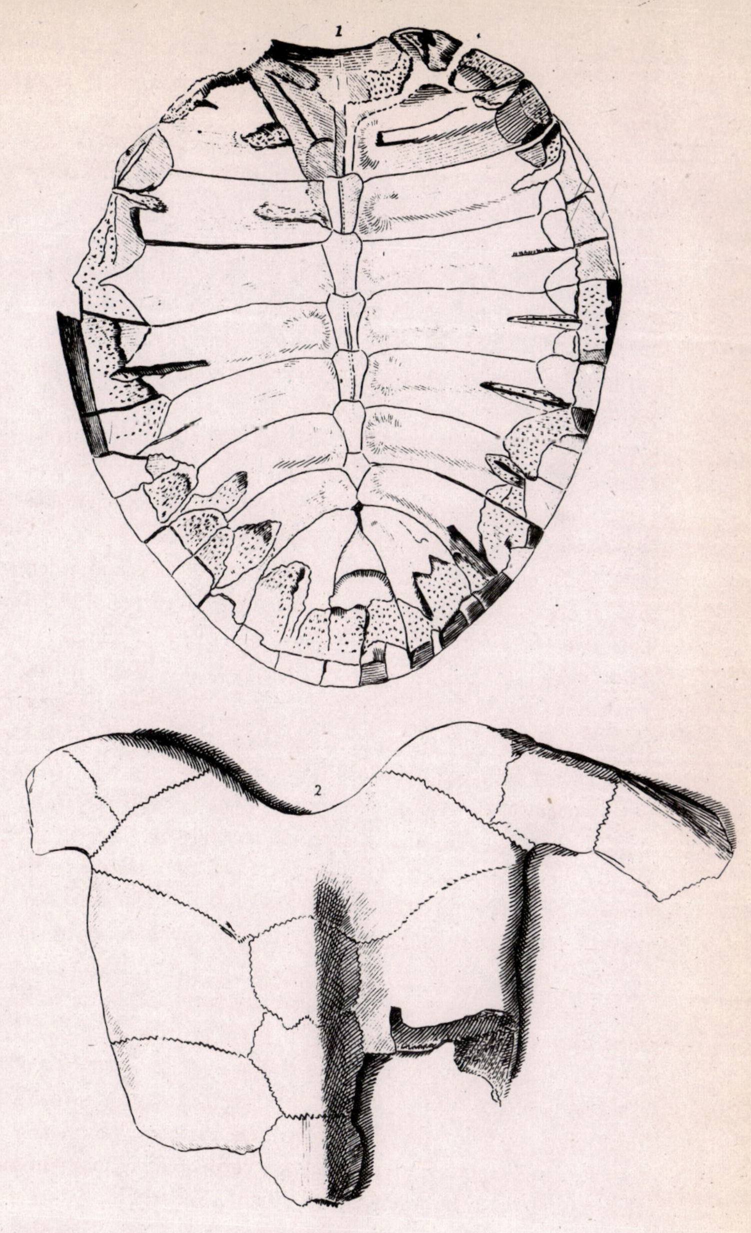

1. *Fossil Tortoise from Melsbroeck near Brussels.* —

2. *Fossil marine Tortoise from Maestricht.*

them than in the species between which and the fossil emys in question we have instituted a comparison.

It is said that many other emydes, or fresh-water tortoises, are found in diluvial formations with the bones of elephants, &c.; but nothing satisfactory is ascertained concerning them.

The remains of sea-tortoises, turtles, or Chelonian reptiles, are found in the neighbourhood of Maestricht, in celebrated quarries of a sort of coarse and sandy chalk in the mountain of St. Pierre. They are mixed with marine productions of many kinds, and with bones of a gigantic reptile of the Saurian order. Some incomplete portions of the upper *testa*, or carapace, were found here, which M. St. Fond, in his "History of the Mountain of St. Pierre," thus speaks of:—

"The upper part, towards the top, bears a sufficient resemblance to a military cuirass, provided with a fore-arm, and indicates that the fore-paws were partly covered with scales adhering to the buckler. This constitutes a marked character, from which a distinct genus might be formed, as nothing of the kind occurs in any of the living species of tortoise."

This opinion, however, of M. St. Fond is totally without foundation. There is nothing extraordinary in these pretended fore-arms, nor anything which is not found in all the sea-tortoises, and those of the land and fresh-water, the trionyx alone excepted. This is easily proved by a comparison of these fossil carapaces with such as have been deprived of their scales, and reduced to merely their osseous frame-work, and not by a comparison with those which are still covered by their exterior envelope. What M. St. Fond calls the fore-arm, is only the commencement of the edge which surrounds the carapace, and which is usually formed by twenty-four osseous pieces: only two or three of these pieces remained in the specimen of which we have been speaking, the others having fallen. The emargination which separates this commencement of the edge from the disk of the carapace, is produced by the unossified space which remains in the tortoises in general, but more especially in the sea-tortoises, to a greater or less-advanced age.

Accordingly, it is impossible to believe that the *testæ*, or carapaces, represented in M. St. Fond's "History of the Mountain of St. Pierre," indicate a new genus. They exhibit no part which does not exist in the *testæ* of all tortoises, or nothing which does not resemble the sea-tortoises in general.

M. St. Fond was also desirous to establish another genus, or at all events a new and unknown species, on some other remains, found in the same mountain, but apparently with as little foundation as in the instance just cited. This, however, was entirely established on mutilated specimens.

At all events, it is quite certain that the tortoises of Maestricht bear the generic character of the sea-tortoises, or Chelonian reptiles; and it is equally certain that they appertain to a species very different from all existing sea-tortoises. They have the ribs ossified scarcely for one-third of their length; whereas any of the existing species of similar size would have them ossified almost to the end. It is, however, observable, that in these tortoises, as in others, the progress of ossification is in proportion to age.

Near Glaris, in the mountain called Plattenberg, is a slate-quarry, with strata inclining to the south, which is very abundant in impressions of various kinds of fish, and in which some remains of tortoise appear to have been found. Those who endeavoured to determine the species took these for the remains of our common emys, or fresh-water tortoise *(testudo Europæa)*. Thus was it named, by Andreæ, who observes, that these animals formerly existed in the lakes of Switzerland; just as if the formation of the slate-mountains could have anything in common with the present lakes of that country.

There can be no doubt, as M. Cuvier observes, that the tortoise in question belongs to the marine sub-genus. A decisive proof of this is the elongation, and more especially the unequal elongation, of the toes. In the fresh-water tortoises the toes

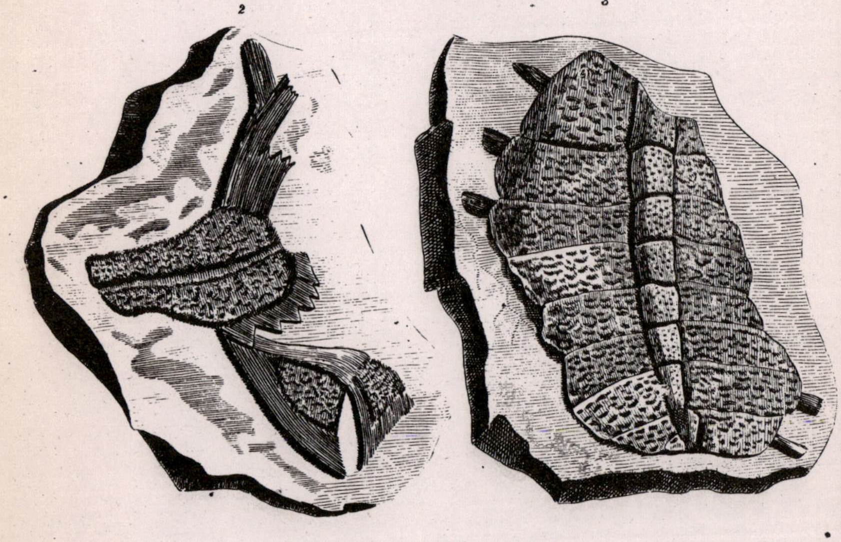

1. *Emydis from the Isle of Sheppey Kent.*

2 & 3. *Fragments of Fossil Trionyx from Aix.*

are of moderate length, and nearly equal. In the land-tortoises they are nearly equal, and all remarkably short. But in the sea-tortoises they are strongly elongated, and those of the fore feet form a pointed fin, as they go on increasing from the thumb to the medius, and then decreasing. This is exactly to be observed in the tortoise of Glaris; but it is too badly preserved to determine the species, or even to determine whether or not it belong to any species now existing.

In the neighbourhood of Aix, in Provence, some remains were found in 1779, which the Baron, with good reason, refers to the sub-genus of the land-tortoises. The figures given of them by M. Lamanon, in the "Journal de Physique," are, it is true, very imperfect; but there can be no doubt of their having belonged to the genus of tortoises, and the very convex figure of the carapace leads infallibly to the conclusion that they must be referred to the sub-genus we have mentioned. They were at first taken for human heads. Guettard imagined them to be nautili. Lamanon was the first who recognized them to be what they really are. All the laminæ and sutures did not appear in the petrified tortoise, until what remained of the shell had been removed. The substance of the stone, while it was yet soft, had taken the place of the animal, and formed a mould, over which all the parts of the shell were clearly to be distinguished. Eight ribs remained on each side very much curved, and came to an end at the vetrebral plates, which are arranged longitudinally, and separated by a tolerably deep furrow. This is the description of M. Lamanon; but the furrow was caused by the projection of the body being imprinted concavely on the mould. The writer in question adds a character, which, united to the great convexity of the carapace, proves that those remains must have belonged to the land-tortoise. This is, that the laminæ are not of equal breadth throughout their whole length; they go on growing more narrow, and are emboxed into each other, so that after a base comes a summit, and so on. This is an exclusive character in the skeleton of the shell, or carapace, of the land-tortoises.

The height of the carapaces of which we have been speaking, was seven inches on a breadth of six, which is as great a convexity as was ever seen in any existing land-tortoise.

These remains were found in a calcareo-gypseous rock, at no great distance from Aix, mingled with grains of rolled quartz, situated at the foot of a little mountain, in which are excavated the plaster-quarries of the same city. It is very probable that the bed containing them is of the same formation as those plaster-quarries, in which numerous impressions of fish and palm-leaves are found.

This rock, we are informed by M. Lamanon, contained fossil bones of all kinds, such as tibiæ, femora, ribs, jaws, teeth, &c. Some of the femora, adds M. L., were too long and too bulky to have appertained to *man*. There were also, in the same formation, bones smaller than those of a mouse.

The last fossil bones of tortoise which we shall mention have nothing more interesting about them than the locality in which they were found. This was certain volcanic formations in the Isle of France, and they are the first fossil remains which geologists have been able to produce from the warm climates of the old continent. They were found with other bones of the same genus, in a place called "Quatre Cocos," in that island, in digging a cistern, in a very thick chalky bed, situated under the lava, which forms a grand plain along the entire eastern coast of the island. This lava, whose surface decomposes, and becomes singularly fertile, is the most recent formation in the island, and has in all probability been produced since the æra in which authentic history has commenced. The same would appear to be the case with the bones of which we speak, and M. Cuvier is inclined to refer them to the enormous land-tortoise of India *(T. Indica)*. It is true that he found a tibia, in the collection of M. St. Fond, from the same island and the same strata, a little longer and less bulky than that of the living species; but, in our humble opinion, no conclusion of any great weight could be drawn from this fact.

It will easily be seen from the details which we have now given, that far less precision is attached to the results drawn from the remains of tortoise, than was afforded by those of the crocodiles. The difference, however, of certainty belongs less to the relations of this genus with the strata in which its remains are found, than to the difficulty of determining the species from the simple osteology of the shell, or carapace. The equal antiquity of the tortoises with the crocodiles is clearly established, for they generally accompany them in the same formations. As the greater number of their debris belong to subgenera, whose species are proper to fresh water, or to terra firma, they confirm the conjectures which the bones of the crocodiles gave rise to, concerning the existence of islands or continents, which nourished reptile races before viviparous quadrupeds began to exist, or, at all events, before they were sufficiently numerous to leave remains at all comparable in quantity to those of the said reptiles. This is a great geological fact, of which we shall soon find further and more astonishing confirmation.

We must now, as briefly as we can, follow the Baron in his observations on the osteology of existing LIZARDS, which will be necessary for fixing the comparison with the extraordinary remains of other reptiles of the SAURIAN order, which have been found in various localities. We shall confine ourselves to the most important points.

It must be observed, that M. Cuvier, in his "Fossil Osteology," takes the word lizard in its most general acceptation, as embracing all that remains of the old genus *lacerta* of Linnæus, after the crocodiles have been separated from it on the one side, and the salamanders on the other. All the animals thus included have an osteological system pretty nearly alike in the most important points, with very trifling differences of composition and proportion, but they differ materally from crocodiles and tortoises, and still more from salamanders. M. Cuvier would also join with them the serpents of the family of *anguis*, for their osteology, especially that of the head, greatly resembles that of many lizards.

As to the head, the principal characters of this family are as follow:—

The four common occipital bones form the ring which surrounds the encephalon behind. The lateral occipital is not divided into two portions, as in the tortoise. In front of the occipitals are placed the sphenoïd, underneath, and the os petrosum laterally. The parietal covers the whole as a roof or penthouse.

The sphenoïd is visible throughout its entire lower face. The pterygoïdean bones, forming a simple continuation of the palatines, are prolonged as far as the internal edge of the ossa tympani, not touching the sphenoïd, except on a lateral tuberosity of that bone, and not uniting together. The sphenoïd is prolonged in front, by a cartilaginous process, on which is raised the interorbital partition, and in this last are seen many points of ossification which belong to the ethmoïd. The bone which is analogous to the petrosum, and which is not concealed by the tympanic box, extends externally, and forms between the sphenoïd, and the occipitals, all the hinder lateral paries of the cranium. The lateral and anterior paries of the cranium, from the os petrosum to the interorbital partition, is membranous, and contains only, on each side, a bone differently configurated according to the species, and which represents the temporal and orbital wings.

An osseous stem arises from the upper edge of the pterygoïd, where it is articulated into a fosset, as far as the lateral edge of the parietal, where it attaches by a ligament. Some anatomists have imagined that this bone was analogous to the temporal wing, but it does not perform the functions of that bone. Others have named it the *tympanic bone*, for which there does not appear to be the slightest foundation. It cannot even be averred that it is comprised in the paries of the cranium; and this paries has also sometimes in the thickness of its membranes a point of ossification which truly represents the temporal wing. M. Cuvier, who regards it altogether as a new and peculiar bone, has given it the name of *columella*. Its object is to support the vault of the cranium, which is not supported in front, because the orbital and temporal wings and the ethmoïd are in a great measure membranous.

The lateral occipital gives out a projecting part externally, to which are united by their extremity the mastoïdean, which is much reduced, and the temporal bones. To this common junction of these three bones is suspended the tympanic bone, which descends vertically to serve as a pedicle to the lower jaw. This bone, for the most part, gives attachment only to the anterior edge of the tympanum; and the rest of the contour of that membrane, as well as the hinder paries of the cavity, is cartilaginous, or simply membranous.

The Eustachian tube is merely a wide communication from the cavity into the back part of the mouth, between the extremity of the pterygoïdean and the sphenoïd. In the fresh animal it corresponds to the inside of the mouth, near the articulation of the jaws; and the communication is sometimes so open, that one can scarcely tell whether the auricular osselet is in the mouth, or in the pharynx.

The cavity of the vestibulum is formed in common by the *os petrosum*, the lateral, and upper occipitals. The fenestra ovalis, where the auricular osselet is attached, is common to the *os petrosum* and the external occipital. Under it is a wider opening, pierced in the lateral occipital only, and at the bottom of which are two foramina; one, anterior, which goes into the cranium; the other, posterior, which is the fenestra rotunda, and opens into a fosset of the vestibulary cavity which represents the cochlea.

A transverse bone unites the pterygoïd to the jugal, and to the maxillary, as in the crocodile.

The palatines have no palatine laminæ, or, at least, these laminæ are not sufficiently extended to unite; and the hinder osseous nostrils are large holes in the anterior part of the vault of the palate, between the maxillaries, the vomer, and the palatines.

Such are the principal peculiarities of the head of lacerta, in general, and all that are necessary to be noticed here.

This family is sub-divided into two tribes, in relation to the composition of the muzzle: that of the monitors of the ancient continent, which have but a single nasal and two principal frontal bones, and those American lizards, such as the *bicarinata*, *teguixin*, *ameiva*, &c. of Linnæus, and which are termed *sauvegardes*, by M. Cuvier, and other French naturalists. This sub-division embraces most of the other genera of Saurian reptiles, and the animals comprised therein have two nasal and one principal frontal bone.

The first family comprises only the monitors of the old continent, with small scales under the belly, and on the tail. We shall consider, as an example of this, the great *Monitor of the Nile*, *(Lacerta Nilotica,)* called *Ouaran* by the Arabs.

The general confirmation of the head is that of an elongated cone, depressed, blunt at the point, with the anterior frontal and parietal region plane. The orbits are round, and pretty nearly in the middle of each side. The external osseous nostrils rise almost to the elevation of the orbits.

There is but one inter-maxillary widened in front, where it has four teeth on each side, and mounting by a long compressed apophysis as far as towards the middle of the nostrils, where it unites to what represents the nasal bone. This is also equally uneven, widening in the top, and there bifurcating to unite itself to the two frontals. These occupy their usual space between the orbits, and have, each of them, underneath, an orbital lamina, which approaches and unites to its corresponding one, to complete the canal of the olfactory nerves.

The maxillaries embrace in front, by a depressed part, the

widened part of the inter-maxillary, under which, and behind the teeth, is a projecting apophysis; and the bone then proceeds to unite itself by a short process, which is furcated and marked with a groove, to the vomerian bones, which occupy the middle, of the palate.

The maxillaries, as usual, form the edges of the palate leaving on each side between themselves, the vomer, and palatines, a wide hinder nostril, which, consequently, opens into the palate. The maxillaries, also, form the sides of the muzzle or the cheeks, and terminate by widening towards the orbit, from which they are separated by the anterior frontal, the lachrymal, and the jugal bones.

The anterior frontal has, as usual, a frontal and orbital part, which serves as a hinder partition to the nasal cavity.

The lachrymal is partly on the cheek, and partly in the orbit. It has a projecting point at the edge of the orbit, a lachrymal foramen within, and leaves another hole tolerably large between it and the anterior frontal.

The jugal touches on the lachrymal, the palatine, and the transverse bones. It is an arched and pointed stilett which does not reach the posterior frontal, nor the temporal bones, so that the orbit remains incomplete. No other example of this is found among the Saurians except in the genus of Gecko.

There is a particular bone which has nothing analogous in other genera, and which M. Cuvier terms the *superciliary* bone. It is articulated by a wide portion to the orbital edge of the anterior frontal, and directs behind a pointed apophysis, which protects the upper part of the eye. This is also found in birds. The line of the union of the frontals with the parietals is nearly straight. On the two extremities of this line the posterior frontals are articulated, one half on the principal frontal, and one half on the parietal. Each of them presents an orbital apophysis, and one behind, which unites obliquely to the temporal bone to form the zygomatic arch.

This last is narrow, and a little crooked towards the summit. It is chiefly formed by a temporal of similar configuration, which is closely united by its posterior extremity to the mastoïd. The mastoïd is equally narrow and crooked, and is similarly fastened on the lateral point of the parietal.

The parietal is in the form of a buckler, widened in front, hollowed on the sides of the two temporal fossæ, furcated behind, and giving out there two long points, which, with the temporal and jugal, and a salient apophysis of the lateral occipital, proceed to give a point of suspension to the tympanic bone.

A foramen should be remarked which is naturally pierced in the parietal bone, very nearly towards the centre, and which is again found in many of the Saurian reptiles, and even in the *ichthyosaurus*.

In the bifurcation of the parietal behind is placed the upper occipital, which attaches to the emargination of the parietal only by a round ligament, and not by a suture. The ossa petrosa are tolerably extended, and cover, both at top and in front, the vacancy which remains on each side between the occipitals and the sphenoïd.

Besides the vacant space which descends into the temporal fossa, between the parietal, the posterior frontal, and the temporal, there is another which penetrates behind, between the point of the parietal, and the occipitals. These are great spaces which correspond to the foramina which exist in the crocodiles, but which are much smaller, because there the bones are less dilated.

The *fenestra ovalis* is, as usual, common to the os petrosum and to the lateral occipital.

The *fenestra rotunda* is pierced in a fossa of the lateral occipital. This last bone proceeds laterally, having the os petrosum before it to unite by its external extremity to the lower extremity of the mastoïd, on the outside of which also terminates that of the temporal.

There is found in this place, between the occipital and the mastoïd, and above the tympanic bone, a very small osseous piece distinct from all the others, and which is a sort of epiphysis, or inter-articulary bone for the os tympani. This last, suspended to a pedicle, to which, as we have already seen, five bones contribute, is prismatic, almost straight, and slightly hollowed into a semi-canal at its external face. It only supplies the anterior paries of the cavity. The tympanum behind is extended only on membranous parts, and when the throat is opened and the pterygoïdean muscles a little removed, the tympanic cavity appears as a simple hollowing in the roof of the pharynx.

The floor of the cranium on the sphenoïd and basilary bones is concave; the foss of the pituitary gland is very large, and separated almost horizontally from that of the cerebrum by a projecting lamina of the sphenoïd.

The palatines are short, concave in front to conduct to the back nostrils, uniting to the vomer, the anterior frontals, the maxillaries, the transverse, and pterygoïdean bones; not leaving there a large empty space, but forming, as usual, a part of the floor of the orbit. They are, each of them, pierced with a small hole analogous to the pterygo-palatine.

The palatines are continued by the pterygoïdean bones. These remaining considerably separated from each other, and becoming vertical, are supported in passing, on the lateral apophysis of the sphenoïd, and proceed to their termination in a point near the internal lower edge of the tympanic bone. They present on their external side an apophysis for their articulation with the transverse bone, which is short and broad, and unites the pterygoid to the palatine, the maxillary and the jugal on each side, leaving between itself the pterygoïd, and the palatine, an oval foramen tolerably large, though much less so in proportion than in the crocodile.

That particular, straight, and narrow bone already mentioned, called the *columella*, is articulated on the pterygoïdean bone, in a fosset for the purpose. Its other extremity unites to the anterior extremity of the junction of the parietal and the os petrosum. It is nearly parallel to its correspondent bone, and it is between them both that the membranous partitions, which close the Cranium in front, commence to approach each other and be confounded in the partition, equally membranous, which separates the two orbits. The bottom of this partition is supported by the prolongation of the anterior and middle apophysis of the sphenoïd, which diminishing in thickness and consistence in front, finishes by attaching itself between the two vomeres.

In the anterior membranous partitions of the cranium is an osseous branch, at first, crescent-formed, to surround the posterior or external edge of the optic foramen, and then giving out a point in front and one above, which extend themselves in the membrane and assist to support it. This is the sole representative of the orbital and temporal wings.

The vomeres form the middle of the under part of the palate, going from the intermaxillary bone to the palatines, and hollowed each of them in front into a small canal.

All the anterior and lower part of each great osseous nostril is occupied by a bone formed like a spoon, which seen from above is concave behind, and convex in front, and which manifestly corresponds to the lower cornet of the nose. It proceeds in all this part from the vomer to the maxillary, leaving underneath, in front, between the maxillary and the vomer, a hole which penetrates into its convex part.

A little more in front, on each side, is an incisive foramen between the maxillary and intermaxillary bones.

Besides its eight intermaxillary teeth, this monitor has usually eleven teeth in each maxillary bone, and as many on each side of the lower jaw. The anterior are conical and pointed, the hinder blunt.

With the exception of some trivial differences of proportion, such is the structure of all the monitors. The principal of such differences consist in the number and form of the teeth, which in many species are trenchant, and the number in the intermaxillary unequal.

The *Lacerta teguixin*, or American safeguard, is taken by M. Cuvier as the type of the second family.

Compared with the monitor of the Nile, the head is shorter and less depressed. The muzzle is a little more raised. The intermaxillary is also unequal; but its nasal apophysis is much shorter, and instead of a single nasal, there are two large peculiar bones of the nose which cover the greatest part of the nasal cavity, so that the external osseous nostrils are small and altogether towards the fore part of the muzzle. On the contrary, the principal frontal is single. The point of the edge of the orbit appertains to the anterior frontal, and not to the lachrymal, which is very narrow, and even not pierced. The single lachrymal foramen is between the two bones, and below it is seen a pterygopalatine foramen, or a posterior suborbital one, formed between the anterior frontal, the palatine, the maxillary, and the lachrymal.

There is no foramen in the parietal.

The jugal rejoins the posterior frontal and closes the framework of the orbit. The descending laminæ of the principal frontal project very little. There is no suborbital; but what is very remarkable, the posterior frontal is divided by an oblique suture into two bones, one of which attaches only to the frontal and parietal, the other to the jugal and temporal.

Underneath, the intermaxillary, instead of forming a process behind, suffers an emargination there, and into this emargination enter the points of the maxillaries and the vomeres. The incisive foramina are extremely small.

The lower cornets of the nose are ossified as in the *Ouaran*, but they are not seen so easily in the head when entire, because they are covered by the proper bones of the nose. But the vomeres are shorter, wider, and not hollowed. The palatines advance more, which renders the back nostrils more narrow.

They are, however, continued under the palatines in a concavity of their surface. These two bones are less separated from each other.

In the pterygoïdean bones, that part which is between the apophysis of the sphenoïd and the tympanic bone is hollowed into a canal, deeper at its lower or internal face. The tympanic bone is widened at the summit, and slightly concave externally.

The basilary has on each side a descending tubercle, which is wanting in the ouaran. The lamina which separates the pituitary foss from that of the cerebrum is less prominent.

The osselets, representing the alæ temporales, are in the form of a Y, the two upper branches of which end at the frontal and parietal, and the inferior at the sphenoïd, at the place where it comes forward in the form of a crest, or ridge, to serve as a basis to the interorbital partition.

In this same partition are also certain ossified parts, representing the orbital wing, and distributed so as to leave a foramen common to this interorbital partition, and to that of the cerebrum. Into this foramen pass the optic nerves, before the two temporal wings of which we have spoken. Behind them, but in front of the point opposite to which is the *columella*, pass the nerves of the third, fourth, and sixth pair, and the nerve of Willis; and behind this *columella*, into an emargination of the os petrosum, passes the rest of the fifth pair. Thus this emaginated division corresponds with the fenestra rotunda and the fenestra ovalis.

The *Dracœna* of Lacépède, not Linnæus, resembles almost in every thing the last-mentioned species. The resemblance also is still stronger in the *bicarinata* and the *ameira*.

The lizards, properly so called, such as the *lacerta agilis*, independently of some details of forms and proportions, have all the characters of the last, except what follow:—The principal frontal is longitudinally divided into two bones. The anterior frontal descends but little into the orbit, where the lachrymal occupies much more space. The posterior frontal is united to the parietal; a wide suborbital, divided into many pieces, is united to the anterior frontal, the principal, and the posterior frontals. The pterygoïdean bones have each a range of small teeth near their internal edge, almost midway of their length. There is a small foramen in the middle of the parietal.

In the *stellio uromastix* (Merrem.), of which there are two species, the head is depressed and widened externally, so as to produce a swelled appearance of the cheeks from the size of the jugal bones The frontal is very narrow; the nasal bones small and short; the external nostrils and the orbits very large; the hinder branches of the parietal very long. When this bone joins the frontal, it is emarginated by a wide foramen, closed by a simple membrane. The anterior and posterior frontals are very small; the palatines broad and short. There are no teeth in the intermaxillary bone, the edge of which projects between the maxillary teeth.

In the common Stelliones the frontal is shorter; orbits and nostrils not so large. There are two teeth in the intermaxillary, and the second of the maxillary teeth is a sort of canine.

In the *iguanas* the muzzle is swelled and convex; the frontal is flat; the anterior frontal is broad upon the cheek, and has a tubercle in front of the orbit. The posterior frontal is divided into two parts. The internal nostrils are very long, and the palatines very wide.

The *geckos* differ much from the other lizards, by the extreme smallness of the osseous parts of the jugal and temporal, and the longitudinal division of the parietal into two bones. The muzzle varies in elongation and depression, according to the species. The principal frontal is broad, and slightly concave. The orbit is large, round, and incomplete on the side of the temple. A great part of the hinder edge of the orbit is visible only by a ligament, in consequence of the smallness of the jugal. The pterygoïdean bones, greatly separated from each other, have no teeth.

The head of the *cameleon* is very singularly formed, still its composition presents sufficient analogy to that of other lizards. The casque of the occiput is supported by three ridges, one of which appertains to the parietal, and the two others to the temporal bones. In fact, it is that the parietal is very narrow, and instead of sending branches to the temporal, it rises into a point like a sabre, and the temporals also send similarly pointed ridges or crests, which unite their points to that of the parietal.

There is but one principal frontal, bordered on each side by the anterior and posterior frontals, above the orbit, the frame of which cavity they unite together to form, and also to form the sort of denticulated crest which the cameleon has in this part. The rest of the bony orbit is formed by the lachrymal and jugal, which last unites to the posterior frontal and temporal bones. The muzzle is formed by the upper maxillaries, between which is a very small intermaxillary.

The most extraordinary part of this arrangement is, that the external nostrils are pierced in the maxillary bone, one on each side, whose edge is somewhat completed above by the anterior frontal. Still, on the muzzle, in the skeleton, are two foramina, covered externally by the skin, and between which are two very small nasal bones. The tympanic bone is cylindrical, straight, and without concavity. Though the cameleon has no external tympanum, there is, however, a tympanic cavity tolerably large, closed on every side by muscles or bones, and on the side of the mouth by a membrane which doubles that of the palate, and is extended between the basilary and the hinder point of the pterygoid. In this place, on each side, is a narrow hole, which holds the place of the Eustachian tube.

The lower jaw of lizards in general is composed of six bones, like that of the crocodile and tortoise, but somewhat differently disposed, and producing a form a little different. This is principally occasioned by the coronoïd apophysis being very projecting and more forward, the lower angle of the jaw being more forward, and the dentary part shorter in proportion. It is not necessary here to enter into a detailed description of these six bones, nor of their variations, and those of the jaw in the different subgenera

But the teeth cannot be passed over without notice. They are not in alveoli, like those of the crocodiles, and those which should replace them are not produced in their cavities. The gelatinous nuts of the teeth adhere to the internal face of the dentary bone, without having any osseous partitions between them, and sometimes without being protected on the internal side by a lamina of this bone. Their bases are therefore separated from the cavity of the mouth only by the gum. This basis is not divided into roots; but when the tooth has arrived to its full growth, the same phenomenon occurs as in the fish. The gelatinous nut becomes ossified. It unites intimately, on one side, to the bone of the jaw, contracting, on the other, a close adherence with the tooth which it has exuded. The tooth then appears as a prominence, an apophysis, in fact, of the jaw, only that it is covered with enamel; but its base is naked and purely osseous, and around this base are seen striæ, and little pores, through which the vessels have penetrated, or are still penetrating, into its interior cavity, and which also mark the place where the rupture will be made when this tooth shall yield its place.

The new teeth spring, not in the cavity of the old, and passing through them, as in the crocodiles, but near the internal face of their basis, or, in certain species, in the thickness of the bone, above or below this basis, according to the jaw. In this last case a cavity is formed in the bone, which lodges, for a certain time, the pulpy nut and the cap which springs above it. This cavity opens by degrees to the internal face of the dentary bone. In the other way the pulpy nut is simply developed under the gum; but in proportion as the dentary cap grows, it often forms for itself a notch in the base of the nearest tooth in place, where it is partly enclosed. The new tooth is then certainly in the old, but not entirely enveloped by it.

In whichever way the new tooth comes, the time arrives when its growth completely displaces the old one. It produces on its ossified base, a sort of necrosis, or dry gangrene, which breaks its adherence to the jaw, and makes it fall. This is not a spontaneous rupture like that of the old antlers of stags, which fall before the new ones have pushed forth. The agency of the new tooth always goes for something. There can, therefore, be no difficulty in distinguishing the teeth of lizards from those of crocodiles, nor even in distinguishing, to a certain degree, the teeth of one genus of lizards from those of another.

In the monitors there is no internal alveolar edge, and the new teeth grow in the thick part of the gum, between the bases of the teeth in place, or at the internal face of their bases. They are easily discovered by detaching the gum. These teeth are conical. In the majority of species they are moreover pointed, compressed laterally, and a little hooked. The aquatic monitor of Egypt, and one or two other African species, have only the hinder teeth in straight obtuse cones, or even entirely rounded and blunted at the top. The species with trenchant teeth have the edge very finely crenulated, but the crenulations are sometimes visible only with a convex lens. These teeth are not very numerous, not more than a dozen or fifteen on each side, and none in the palate. Neither have the *teguixin*, *ameira*, *dracæna*, and others of this subgenus peculiar to America, any teeth in the palate. Many others also want these palatine teeth.

The subgenera which have these teeth are the lizards proper, the iguanas, the polychrus, the anolis, and many skinks. There are other variations respecting the form of the teeth in lizards, into which it is impossible, in a sketch of this kind, to enter minutely.

We have already seen that the hyoïd bone is simple in the

crocodiles. In different tortoises it varies in formation. In the lizards it presents some analogies with that of birds, but its composition is more complex. To enter, however, into all its variations in this great family is not necessary here. It will be sufficient to observe, that it has, in general, a simple body, and two pair of cornua, and sometimes even a third. From the body in front is a process like a thin stem, which is more or less prolonged in a cartilage, which penetrates into the tongue. The anterior cornua are variously folded, and the posterior variously directed, according to the species. Those of the third pair are not very often found, and when they do exist may be considered rather as posterior processes of the body of the bone, than particular cornua.

We must now give some attention to the vertebræ and ribs of the lizards, as they are of such importance in the determination of the fossil remains of this immense family.

The atlas of the monitor is a ring composed of three pieces; two upper ones emarginated, in front and back for the nerves, and united to each other at the dorsal part, and one lower piece.

The piece of the axis analogous to the odontoïs penetrates into the ring of the atlas, and fills nearly half its breadth, leaving, however, in front, a concavity for the condyle of the head. A triangular piece, underneath, on the junction of the atlas, the odontoïd, and the axis, forms a sort of pointed crotchet directed backwards.

The axis is compressed, and its annular part above is formed like a sharp longitudinal crest. Its anterior articulary facets are turned outwards, the posterior downwards. The body is terminated in a transverse convexity, of the form of a kidney. At each of its lateral faces is a small crest, not projecting much, with a small point towards the third part of its front. Underneath, is a crest, under the lower part of the bone, which widens behind.

The succeeding five vertebræ resemble the axis, but have no

odontoïd process; their anterior face has a concavity proportioned to the convexity of the preceding vertebræ, with some other small variations.

The dorsal, after the second, have always a square dorsal crest, an anterior face concave, and a posterior convex, both kidney-formed, and horizontal articular apophyses, the lower facing downwards, the anterior one upwards. On each side, under the anterior, the whole transverse apophysis consists in a tubercle oval-formed, and vertically directed to support the rib.

Of these vertebræ there are twenty-two, and no lumbar, though there are twenty-seven pairs of ribs, including the five cervical, but the first and the last are very small. The total absence of lumbar vertebræ seems to be a general rule in this family. There are but two sacral vertebræ in the monitor. The caudal vertebræ after the eighth are very numerous, being sixty-six, eighty, and upwards. They are easily recognized by their spinous and transverse apophyses being long and narrow, and their articulary apophyses almost vertical.

The differences which characterize the vertebræ of the various subgenera consist chiefly in the respective length and bulk of their bodies, and the respective length and breadth of their apophyses.

The ribs of lizards are slender and round, and the anterior ones alone have the costal head a little bulky and compressed.

A remark worthy of being made here is, that a great number of the caudal vertebræ of the lizards proper are divided vertically, in the middle, into two portions, which are separated very easily, even more easily than any two distinct vertebræ at the place of their articulation, because this articulation is complicated, formed by many apophyses, and strengthened by ligaments, while in the other case a separation is prevented only by the periosteum and the surrounding tendons. This is, probably, the cause that the tails of lizards are so easily broken. The tail will grow again after it has been broken, but neither its skeleton nor teguments continue of the same quality. The scales of the skin are generally small, without crests, and without spines, though they possessed them in the former tail; and internally, instead of the numerous vertebræ, with all their apparatus of apophyses and ligaments, there is nothing but a long cartilaginous cord, all of a piece, and exhibiting only numerous annular wrinkles of no great projection.

The sternum of lizards forms with the shoulder a sort of cuirass for the protection of the heart and large vessels. It is more complicated than in the crocodiles, and on a different plan from that of the tortoises. It consists essentially in a long, narrow, depressed bone, which gives out in front two branches directed on each side, and between which its point sometimes passes, and proceeds more in front under the neck. This bone, from its hinder part, penetrates into a cartilaginous lamina, of a rhomboidal form, which has two sides in front, and two behind, and often exhibits traces of a longitudinal division into two halves.

Its anterior sides are continued with the edges of the anterior part of the bone, but departing a little to the right and left. They are sometimes ossified, particularly their edge, which has a groove, supporting, like a mortise, the sternal edge of the clavicular bone. The hinder sides of the rhomboïdal cartilage serve for the insertion of the false ribs. Hitherto there is little difference between the conformation of this sternum and that of the crocodiles, except in the anterior branches of the oblong bone, which give it the figure of a T, of an arrow, or a cross, according to the species.

A more considerable difference, however, consists in the development of the coracoïd bone, and in the constant presence of clavicle of a greater or less size. The coracoïd furnishes nearly one half of the glenoïd foss. But its principal peculiarity is giving out one or two apophyses to support a large cartilaginous arch, which passes over the narrow bone in front of the sternum, and crosses with that of the coracoïd on the other side. This singular crossing, which is found even down to the lowest of the batracian reptiles, is generally the cartilage of the right side, passing over that of the left.

There is always a small hole for the vessels pierced in the neck of the bone, between its apophyses and the glenoïd facet.

The cartilaginous semicircle just mentioned acquires hardness and consistence with age, but not so much as the other bones. It is hardened by the accumulation of small calcareous grains, like bones of chondropterygian fishes.

The omoplate forms the other portion of the glenoïd facet. In the midst, or at about one-third of its length, the osseous part suddenly terminates, and is continued by a cartilaginous portion. This frequently becomes hardened, and then the omoplate is divided constantly into two bones.

The clavicle rests on one side against the slender bone of the sternum, or its lateral branch, and often it touches the opposite clavicle. On the other side it rests against the anterior edge of the omoplate, either on its osseous or its cartilaginous portion, which often puts forth a tubercle, or little crest, to receive it. Sometimes an apophysis proceeds from the osseous omoplate which sustains the body of the clavicle, and has some slight resemblance to an acromion, which, however, is better represented by the tubercle of the cartilaginous part.

This is the general structure of these parts in all lizards. The differences in the various sub-genera are of small importance.

The pelvis of the lizards is composed of three bones, which, as in viviparous quadrupeds, concur to the formation of the cotyloïd cavity. The ilium takes the upper part; the pubis and ischium unite each to its opposite in the lower middle line, but the pubis does not join the ischium, and the two ovalary foramina are separated only by a ligament. There is a foramen in the neck of the pubis, of tolerable size, and a pointed process from its anterior edge, which curves below and externally.

The humerus has great analogies of form with that of the

birds, but is easily distinguished from it by not being hollow, or pierced with holes for the admission of air.

The cubitus is compressed, and trenchant on its radial edge. Its olecranon projects but little.

The radius is slender; the femur much resembles that of the crocodile. The rotula is very small, and often scarcely visible.

The leg is always composed of two bones, of which the tibia is the most bulky, and the peroneum or fibula is flatted and widened in the lower part, and united to the tarsus by a narrow line.

The carpus is composed of nine bones, like that of tortoises, and its composition will also bear a comparison with that of the simiæ. The tarsus has but four bones, like that of the crocodile.

The first four metatarsians are slender, and nearly straight. There are two phalanges to the thumb, three to the second toe, four to the third, and five to the fourth. The latter is the longest toe, and gives the peculiarly elongated and unequal form characteristic of this family. The unguical phalanges of all the toes are trenchant, hooked, and pointed.

In the cameleon, the toes are grouped in an inverse manner. The thumb and first toe are together, and directed inwards; the three others are together, and directed outwards. Aristotle has remarked this singular conformation.

Having thus prepared ourselves for the investigation, we now proceed to the remains of

Fossil Saurians.

The Saurians, indeed, constitute in the "Animal Kingdom" an entire order of the class reptilia, and include the crocodiles and gavials. The rest of the order consists of the lizards which we have been now describing, and which, in the *Règne animal*, are divided into different families. But it is more convenient to consider their fossil remains under this general title; and we have, therefore, adopted M. Cuvier's arrangement in the

"Ossemens Fossiles" here, as well as in the greater part of this compilation.

The first we shall treat of are the **Monitors**, which are found in the pyritous schists of Thuringia and other countries of Germany.

In almost all the parts of Thuringia and Voigtland, and in the bordering portions of Hesse, as far even as into Franconia and Bavaria, a bed of marly and bituminous schist predominates, which M. Werner regards as the lowest of what he names the *first formation of the secondary limestone*, and which is found for the most part strown with grains of coppery pyrites, containing silver. It is worked in many places for these two metals, though it does not produce them in anything like abundance. It is of no great depth, rarely more than two feet, and sometimes not above an inch or two in thickness. It rests upon a red sandstone, which contains pit-coal in divers places. Above the coppery slate are calcareous strata, belonging to what geologists term the Alpine limestone, containing the most ancient shells and zoophytes, such as belemnites, &c. Above this is gypsum, accompanied with mineral salt, which, in its turn, is surmounted by sandstone, which is covered by a second sort of gypsum without salt, and over it is another limestone analogous to that of Jura. In some strata of the latter are those famous caverns, containing the bones of bears and other carnivora, mentioned in the earlier part of this account of fossil remains. Thus we find that this formation of bituminous schist is among the most ancient of those which contain the debris of organized bodies.

It is from these slates that an immense number of fossil fish have been derived, which has rendered the districts of Mansfield, of Eisleben, of Ilmenau, and other places of Thuringia and Voigtland, so celebrated among the describers and collectors of petrifactions. The general opinion is, that these fish belonged to the fresh water, an opinion further corroborated by the remains of oviparous quadrupeds of which we are about to speak, and which, though not crocodiles, as was at first supposed, were animals whose genus invariably frequents marshes and the banks of rivers. Thus, again, we find productions of the fresh water, covered by immense marine productions of the most ancient date; an added proof, if any more were wanting, that the sea has repeatedly covered and again left dry the continents of our part of the globe, during an astounding series of innumerable ages.

From the remains in question, and their representations, sufficient ground has been afforded to determine the genera, and to characterize, to a certain point, even the species which they exhibit.

Four specimens, found in these strata and engraved in different works, not necessary to mention here, enabled the Baron to determine that they had belonged to animals of the same species, judging from the resemblance of size and conformation in all the common parts, particularly the spine, the tail, and part of the limbs. They may be all employed to reconstruct a complete individual, by attaching to the common trunk the isolated parts in each specimen. In one are the head, the fore-feet, and almost the entire of the tail. The latter is also found in another, with one hind and two complete fore extremities, and a good part of the trunk. The ribs, almost the entire tail, the two hinder extremities very complete, and many parts of the fore ones, were engraved in a treatise of the celebrated Count Swedenborg; and in the last specimen from which an engraving was taken, is an impression of a portion of the pelvis. These different parts are more than sufficient to throw light upon the nature of the animal.

The form of its head, its teeth all sharp, the size of the vertebræ of the tail, sufficiently prove that it was an oviparous quadruped, without the assistance of the hinder limbs, which confirm it still more.

The head is not without some resemblance to that of a

crocodile, though not to that of the subgenus gavial, for its muzzle is very short. But its teeth prove that it could not have belonged to the genus at all. Were it a crocodile, it would have at least fifteen teeth in each side in the lower jaw, and seventeen or eighteen in the upper, which would continue as far as under the middle of the orbit. It has but eleven, which conclude under the anterior angle of the orbit. This is the character of one of those numerous species which have been confounded together by Linnæus, under the name of *lacerta monitor*.

Every other character confirms this. The hind-feet, which are in admirable preservation in the impression of Swedenborg, have five very unequal toes, of which the fourth is the longest; and the respective numbers of their articulations, beginning with the thumb, and reckoning the bones of the metacarpus, will stand thus: 3, 4, 5, 6, 4. The same proportions and the same number of articulations were in another specimen.

This number, and this proportion of the toes, and this number of the articulations of each toe, are exactly the same as in the monitors, the common lizards, and the iguanas, but by no means the same as in the crocodiles, which have only four toes on the hind-feet, little differing in length, and whose articulations will stand thus: 3, 4, 5, 4. The opinion of Swedenborg, that this animal was an *ape*, or a seal, needs no refutation in the present state of geology and comparative anatomy.

In the fore-feet are to be distinguished four toes, nearly equal. The crocodiles have five toes, but their little toe is sensibly less in proportion.

The length of this animal appears to have been about three feet, which is the usual size, pretty nearly, of the monitors of the most common species.

The comparison holds good in the bones of the thighs, arms, legs, and fore-arms. The vertebræ of the tail, with the high and narrow spinous apophyses, are also exceedingly like those of the monitors. In fact, the Baron observed but one or two specific differences. The first was, that the spinous apophyses of the dorsal vertebræ are much more raised than in the existing monitors, whose skeletons he examined, being almost equal to those of the tail; the other, that the leg was a little longer in proportion to the thigh and foot. But these differences can in nowise affect the just and rigorous determination of the genus.

We have now to speak of the great and most celebrated fossil Saurian, discovered in the quarries of Maestricht, and which has given rise to many controversies, having been sometimes taken for a crocodile, sometimes for a Saurian of some other genus, and even sometimes for one of the cetacea, or of the fish.

It would appear that its bones have as yet been discovered only in a confined district, in the hills, by which the western side of the valley of the Meuse is bounded, in the environs of Maestricht, and principally in that on which Fort St. Pierre stands, and which forms a sort of cape between the Meuse and the Jaar.

The formation in which they were found is a soft, crumbly, calcareous stone, many parts of which are easily reduced to powder. Other portions are sufficiently hard for the purposes of building; and the quarries are now very much extended. Those of Fort St. Pierre are above twenty-five feet in height. The massive limestone above them has been found to be two hundred and eleven feet, and two hundred and thirteen feet has been dug down without any other stone being found. All is of the same character, with the exception of about sixteen feet of vegetable mould, which covers the summit of the hill.

This massive limestone is then at least four hundred and forty-nine feet in thickness. Lumps of silex have been found in many parts of it. What proves that it belongs to the chalky formation is, that the stone changes by degrees into a true chalk, as one retraces a few leagues the valley of the

Meuse. It contains the same fossils as are found in the chalk of the environs of Paris, such as teeth of squali, gryphites, echinites, belemnites, and ammonites. These shells are found with the bones in the lower parts of the mass, which are also the most crumbly. The upper parts are harder, and also contain more madrepores, many of which are changed into silex.

The multiplied productions of the sea with which this stone is filled are generally in good preservation, although seldom petrified: most of them have lost only a part of their animal substance. The most voluminous of all these objects, and which by their most extraordinary form, must have chiefly attracted the attention both of the workmen and the curious, are assuredly the bones of the animal which we have now to notice. The quarries excavated under Fort St. Pierre furnished the greater number of these interesting objects; but they have also been found in all the other hills of the chain we have mentioned.

These remains appear to have excited no attention before the year 1776, when an officer named Drouin began to make a collection of them, which afterwards passed to the museum at Haarlem. He was followed by Hoffman, a surgeon of the garrison, and afterwards by Peter Camper, who transferred some of his specimens to the British Museum.

It was the opinion of Peter Camper, that these were the bones of some cetaceous animal, which opinion was followed by M. Van Marum, who described the specimens in the Haarlem Museum.

M. Faujas St. Fond, in his "Natural History of the Mountain of St. Pierre," will have it that the bones in question belonged to a crocodile.

M. Adrien Camper, however, son of the illustrious anatomist just mentioned, was convinced, on examination of the pieces left by his father, that they neither belonged to a cetaceous animal, nor a fish, nor a crocodile, but to a peculiar genus of saurian reptiles, exhibiting some relations to the monitors, and some to the iguanas. To this opinion M. Cuvier accedes, — first refuting the arguments opposed to himself and M. Adrien Camper. We must follow him a little in his reasonings here, as they are absolutely necessary to the understanding of the osteological peculiarities of the animal in question.

Peter Camper's arguments in favour of these remains being cetaceous are briefly these. All the objects which accompany the bones of Maestricht are marine, and not fluviatile. The bones are polished, and not rough. The lower jaw has externally many foramina for the issue of the nerves, like that of the dolphins and cachalots. The root of the teeth is solid, and not hollow. There are teeth in the palate, which are seen in many fishes, but not in the crocodile. The vertebræ have no suture which separates their annular part from their body, as there always is in the crocodile. There are certain differences between the fossil ribs and phalanges, and those of the crocodile.

These reasons, except the first, which is of no great value, prove most assuredly in a demonstrative manner that the animal was no crocodile, but none of them prove it to belong to the cetacea rather than to the reptiles. Many reptiles, and notoriously the monitors and iguanas, have smooth bones, numerous foramina in the lower jaw, the root of the teeth osseous and solid, and vertebræ without suture.

Moreover, the presence of teeth in the palate would of itself prove that this animal was neither cetaceous, nor a crocodile, for neither one nor the other have teeth in the palate; but Camper, in common with all of his day, confounded the cetacea and fishes together, many of which last have certainly this character. The genus Hyperoodon, of M. de Lacépède, cannot be opposed to this opinion, for the Baron has clearly proved that the supposed teeth in the palate of this animal are nothing but cartilaginous or corneous points, adhering to the skin of the palate, as in the echidna, and not teeth implanted in the palatine bones. Accordingly, it was with the echidna that M. de Lacépède compared his hyperoodon.

Contrary to the opinion of M. St. Fond, this animal has nothing in its dentition which is peculiar to the crocodile. All that it has in common with it in this respect, it has in common with an infinitude of fishes and reptiles. On the contrary, it has many characters which the crocodile has not, and which, of themselves, would be sufficient to distinguish it, were there not a crowd of others.

We have already, in treating of the osteology of the crocodile, observed that, in this animal, the tooth in place always remains hollow—that it is never fixed to the bone of the jaw, but always remains merely emboxed there—that the succeeding tooth springs in the same alveolus, and that it often penetrates into the hollow of the tooth in place, and causes it to start and fall out.

The animal of Maestricht, on the contrary, had the teeth hollow only while growing, like all other animals. They were filled at last, for the most of them have been found entirely solid. They ended by being fixed to the jaw by means of a truly osseous and fibrous body, quite different from their proper substance, though intimately united to it. The tooth of succession grew in a peculiar alveolus, formed at the same time as itself. It pierced sometimes at the side, sometimes through the osseous body, which supported the tooth in place. As it increased in size, it finally detached this body from the jaw, with which it was intimately connected by vessels and by nerves. That body then fell by a sort of necrosis, like the antlers of a stag, and brought with it the tooth which it carried. By little and little the tooth of succession, and its body, improperly called its osseous root, occupied the place which the old tooth had quitted.

The dentition of osseous fishes, that of monitors and many other Saurian and Ophidian reptiles, is exactly of the same character as this. In fact, this cellulous and osseous part, which unites itself to the maxillary bone, is simply the nut of the tooth, which, instead of remaining pulpy as in quadrupeds, until it is destroyed, ossifies, and makes one body with its alveolus. The tooth has no true root, but it adheres strongly to this nut which has secreted it, and is still retained there by the remainder of the capsule, which furnished the enamel, and which, also ossifying and uniting itself to the maxillary bone and to the nut, become osseous, enchases or sets the tooth with new force. It is easy to conceive that this nut, identified with the maxillary bone, must suffer the same changes as it—that the alveolus of the succeeding tooth must penetrate its solidity—that compression must detach it, either by breaking it, or obliterating the vessels by which it is nourished—and that, in fact, as was said before, it must be exposed to revolutions analogous to those of the antlers of the stag.

The cetacea exhibit nothing of this, neither do the crocodiles. The teeth of the cetacea, it is true, fill with age, and become solid; but far from adhering to the alveolus by an intermediate osseous piece, they are only feebly retained there by the fibrous substance of the gum, when they are once filled with the substance of the ivory, and their pulpy nut is obliterated.

The only hesitation, then, that can remain respecting the place of this animal is between the osseous fishes, and the iguanas and monitors. An examination of the jaws will put an end to this doubt, and confirm the exclusion of the cetacea and of the crocodiles.

The lower jaw, from a specimen in the French Museum, exhibits fourteen teeth on each side, all conformed, as we have just described, after the fashion of the monitors. But the monitors have only eleven or twelve. The crocodiles have fifteen, but very unequal; these are all equal, or nearly so. In the iguanas the number is more considerable.

There are in this jaw ten or twelve large and tolerably regular foramina. There are five or six in the iguanas, six or seven in the monitors. The crocodiles have an infinite number of them, small and irregular. A dolphin would have but two or three towards the end.

There is a coronoïd apophysis, raised, obtuse, whose anterior edge is widened as in the monitors. No crocodile has anything of the kind. That of the dolphin is much smaller and farther back. In the iguana it is more pointed.

The articulary facet is concave, and very near the posterior end of the jaw, as in all the saurians, but it is lower than the dentary edge, as in the monitors. In the crocodiles and the iguanas it is higher, or at least, upon a level. The dolphins have it convex, and placed altogether at the end.

The apophysis for the muscle, which is analogous to the digastric, is short, as in the iguana. The crocodile has it longer, and the monitor still more so. In fine, the composition of this jaw shows greater relations with the monitor, than with any other existing saurian, and entirely excludes the cetacea; these last having, like all the mammifera, each side of the lower jaw of one piece.

With respect to the further composition of this jaw, there is no great oval foramen at its external face; the coronoïd apophysis is a bone apart, analogous to that which, after M. Cuvier, we have called *supplementary*. The articulary bone makes by itself the hinder apophysis, and pushes out the angular bone very forward. The sub-angular is united squarely with the dentary, and there is a small opening in the opercular.

In all these respects, the animal of which we speak approaches most to the monitor. It approaches it even more than it does the iguana, as well in the lower jaw, as in the structure of the teeth, their figure, and insertion. Though, in this respect, there is something particular.

In the monitor, as in the iguana, the teeth simply adhere to the internal face of the two jaws, without the maxillary bones being raised to envelope them in their alveoli. But here the pediments or osseous nuts which support the teeth adhere in hollows or true alveoli, formed in the thickness of the edge of the jaw.

The upper jaw of the fossil head has eleven teeth, but the intermaxillary bone was wanting in the specimen ; and if, as might be supposed, it had three teeth, as in the monitors, that would make the number above and below equal. The river monitor of Egypt has fourteen above, but only twelve below.

In the fossil animal, all the teeth are pyramidal, and a little hooked. Their external face is plane, and distinguished, by two sharp crests, from their internal face, which is round, or rather a demicone.

Some of the monitors have the teeth conical, others compressed and trenchant. The *lacerta teguixin*, the *ameiva*, the iguanas, and other subgenera, have them with a denticulated edge. There are only some fine denticulations, and nearly microscopic, in the monitors with trenchant teeth. In the fossil the crest is entire, and without notches.

In all the characters now stated there is a greater approximation in the fossil to the monitors, than to the other saurians. But in the pterygoïdean bones there is a character which removes it from them to approach it to the lizards proper, and the iguanas. This is the teeth by which these bones are armed.

The crocodiles, monitors, teguixin, dracæna of M. Lacépède, ameiva, dragons, stelliones, cordyles, agames, basilisks, geckos, cameleons, many skinks, and the chalcides, all have the palate deprived of teeth. The iguanas, the anolis, the common lizards, the marbres *(polychrus)*, and a certain number of skinks, alone among the saurians, partake with many serpents, batracians, and fish, of this singular character.

But the iguanas and other saurians have these teeth in the pterygoïdean bones only; the serpents in the palatine, as well as pterygoïdean bones. The frogs, *hylæ*, and salamanders, have them in the vomerian bones, the first on a transverse, the others on a longitudinal line. Many fishes have them also on a longitudinal line; and this, probably, helped to cause the mistake of Camper about this fossil reptile; but the comparison of the bones which have these teeth, will prove that they belong to reptilia, and not pisces.

The pterygoïd bone in the monitor and iguana is not united to its consimilar as in the crocodile, nor widened into a large triangular plate. It is a bone with four branches: one proceeds forward, and unites itself to the anterior palatine; the second goes sideways to join the transverse bone, which unites itself to the upper maxillary; the third rests by a facet provided with a cartilage, on an apophysis of the basis of the cranium; the fourth goes back, and gives an attachment to the muscles, but is not articulated to any bone. On the edge of the anterior branch is implanted the series of teeth which characterizes the iguanas. The anolis have this bone wider in all its parts, and the lower branch shorter. The monitors, on the contrary, have all parts of the bone more slender, and have no teeth there.

There is a great resemblance to the iguanas in the pterygoïd bones of the fossil. The posterior apophysis of the right pterygoïd, though seen from a mutilated specimen, being broken at the end, still appeared to be equally long in proportion as that of the iguana. The four apophyses of the other were very distinct. The principal specific difference was, that the internal one was longer in proportion than in the monitor and iguana. But there did not appear the slightest relation of form with the palatine bone of fishes, and still less with their pterygoïdean bone.

This bone, in the fossil, appeared to have had eight teeth, which grew, were fixed, and replaced like those of the jaws, though considerably smaller.

From a large fragment preserved at Haarlem, M. Cuvier observes that the upper jaw was elongated and not much raised, and that its edges along the external aperture of the nostrils were entire over a long space, nearly like the monitors and the horned iguana, which would induce a conjecture that these nostrils were large, and the bones of the nose not much extended, a circumstance which absolutely excludes the crocodiles and the teguixin.

The principal foramen for the issue of the suborbital nerve is nearer the edge of the nostrils than in any known species.

Whatever doubts, says M. Cuvier, may subsist respecting these scattered pieces, they do not in the slightest degree affect the determination of the place of this animal. The head fixes that irrevocably between the monitors and the iguanas. But the size of the animal was enormous, in comparison of all the known species of these two subgenera. None of these, perhaps, has the head longer than five inches, while the fossil head approaches to four feet.

In zoology, when the head, and particularly the teeth and jaws, are given, almost all the rest may be concluded as far as the essential characters are concerned. Thus, there was little difficulty, in this case, of recognizing and classing the vertebræ.

All these vertebræ, like those of the living crocodiles, the monitors, the iguanas, and, in general, most of the saurian and ophidian reptiles, have their bodies concave in front and convex behind, which distinguishes them remarkably from those of the cetacea, which have it nearly *plane*, and still more from that of fishes, where it is hollowed on both sides into a concave cone.

The anterior vertebræ have the concavity and convexity much more strongly marked than the others.

Of these vertebræ there are five sorts established on the number of the apophyses. The first have an upper spinous apophysis, long and compressed; a lower, terminated by a concavity; four articulary, the hinder ones of which are shorter, and face outwards; and two transverse apophyses, bulky and short. These are the last vertebræ of the neck, and the first of the back. Their body is more long than broad, and more broad than high. The faces are of a transverse oval form, or kidney-shaped. Others are minus of the lower apophysis, but in all the rest resemble the preceding. These are the middle dorsal. Some follow which have no articular apophyses: these are the last dorsal, the lumbar, and the first caudal. Their peculiar place is recognized by their transverse apophyses, which are elongated and flattened more and more. The articulary faces of their body are nearly triangular in the first caudal. Those which follow have, beside their upper spinous apophysis and the two transverse, two little facets at their lower face, to support the chevron-formed bone. The articular faces of their bodies are pentagonal. Then come some more which do not differ from the preceding, but by the want of transverse apophyses. They form a large portion of the tail, and the faces of their body are ellipses, at first transverse, and then more and more compressed at the sides. The chevron-formed bone is not articulated, but soldered, and forms a body with them.

Finally, come the last vertebræ of the tail, which have no apophyses whatever. In proportion as they approach the end of the tail the bodies of the vertebræ are shortened, and almost from its commencement they have less length than breadth and elevation. The length ends by being one half less than the height.

This series of vertebræ gives rise to manyimportant remarks. The first is relative to the chevron-formed or rafter bone, and the position of its articulation. Its length, and that of the spinous apophysis opposed to it, prove that the tail was very much raised vertically. The absence of transverse apophyses on a great portion of the length of the tail, proves at the same time that it was very much flatted at the sides. The animal was therefore aquatic, and swam after the manner of the crocodiles, causing the oar of its tail to act right and left, and not up and down like the cetacea. The monitors have the tail more round, and the transverse apophyses prevail much farther in it.

In the crocodiles, iguanas, &c., and, in general, in all the saurians, except the monitors, and even in the cetacea, and in all quadrupeds with large tails, the chevron-formed bone is articulated under the juncture, and is common to the two vertebræ. The monitors alone have, under the body of their vertebræ, two facets to receive it, like the fossil animal; only, the body of their vertebræ being more elongated, these facets are at the hinder third part. In the fossil, where the vertebræ are very short from front to rear, the facets are almost in the middle. But there is no reptile known in which this bone is soldered and makes body with the vertebræ, as in this one. It is a character belonging to fish, and must have greatly augmented the solidity of the tail.

Another character which distinguishes the fossil from the monitors and all the saurians, is the prompt cessation of the articular apophyses of the vertebræ, which are wanting from the middle of the back, while in the majority of animals they predominate as far as very near the end of the tail. The dolphins exhibit this character, which, united to the shortness of the bodies of the vertebræ, may have contributed to the mistake of Camper above mentioned.

The first dorsal vertebræ have their transverse apophyses short, and terminated by an oblong, gibbous facet, the direction of which is oblique in relation to the axis of the vertebra. This facet, which supports the rib, is single; consequently, the rib is attached to it by a single head. This is a character of the monitors and most of the saurians, the crocodiles alone excepted, in which it does not take place; for in the neck there are two tubercles on each side for each rib. In the back the transverse apophyses are long, depressed, and trenchant, and the anterior ones have two facets for each rib, one at the anterior edge, the other at the extremity. The last three ribs are the only ones which have but a single head. This, again, puts the crocodile completely out of question.

A part of these anterior vertebræ, which have one tubercle or lower spinous apophysis, doubtless belonged to the neck; but as the two tubercles, which, in the crocodile, bear the little false rib on each side, are not found in any of them, it proves that the animal was not a crocodile, and that it had more liberty of turning its head aside. The lower spinous apophyses do certainly exist in the crocodiles, but they also exist in other saurians and in many serpents; nay, some are found in the ruminantia and solipedes. In the cetacea there is not the slightest appearance of these tubercles, which would be utterly incompatible with their cervical conformation. This lower apophysis in most genera of saurians is compressed, and at the posterior edge of the bone. It is round in the crocodiles, and at the anterior edge. In the fossil it is also round, but truncated, and at the middle of the vertebra.

It would appear, from some other drawings of remains, that the extraordinary breadth of the axis was a distinguishing feature in this animal from all other reptiles. The number of vertebræ in this animal is thus given by M. Cuvier:—

The atlas; the axis; eleven vertebræ with the lower apophysis, the articular and transverse apophyses; five, without the lower apophysis; eighteen, without articular apophyses, in which number the sacral are perhaps comprised; twenty of the tail; twenty-six more with the two lower facets; forty-four without transverse apophyses, and seven without any apophyses: in all, one hundred and thirty-three.

This is more than double the number of the vertebræ of the crocodile, but accords very well with the number in the monitors, which is from a hundred and seventeen to a hundred and forty-seven.

The great number of vertebræ in the base of the tail, not bearing the chevron-formed bone, M. Cuvier considers as a distinctive, and indeed a generic, character of this animal.

As the jaw measured three feet nine inches, the entire animal is calculated to have been upwards of four-and-twenty feet in length. The head approached one-sixth of the total length. This proportion resembles that in the crocodile, but is very different from that of the monitors. But again, the tail being ten feet, is but nearly one-half of the total length, while in the crocodile it exceeds the length of the body by one-seventh. Its shortness is owing to the shortness of the body of the vertebræ. But it must have been very robust, and the breadth of its extremity must have made it a very powerful oar, and enabled the animal to contend with the most agitated waters. There is no doubt, then, from the nature of all the other debris found with its remains, in the same quarries, that it was a marine animal.

On the articulation of the ribs, either to each other or the sternum, the Baron hazards no opinion, in consequence of the imperfection of their remains. All, however, which have been discovered were round, as in the lizards, not flat, as in the crocodiles.

The bones of the extremities which have been collected are small in number, and nothing very definite is concluded concerning them. The bones of the hands and feet, as far as known, seemed to have belonged to a contracted sort of fins, not unlike those of the dolphin or the plesiosaurus.

It seems, upon the whole, quite certain that this great animal of Maestricht must have formed an intermediate link between the tribe of saurians without palatine teeth, the monitors, &c., and those which have palatine, or rather pterygoïdean teeth, as the common lizards, iguanas, &c., but that it had no relation to the crocodiles, except in some partial characters, and in those general bonds of connexion which re-unite the immense family of oviparous quadrupeds.

It may doubtless appear strange to some naturalists to see an animal exceed so much in dimensions those genera to which it approaches most in the natural order, and to find its debris with marine productions, while no saurian of the present day appears to live in the salt water. But these singularities are nothing in comparison of a multitude of others presented to our consideration by the numerous monuments of the natural history of the ancient world. We have already seen a tapir of the size of the elephant, and in the megalonyx a sloth as large as the rhinoceros. Why, then, should we be surprised to find in the animal of Maestricht a lizard as large as a crocodile? We shall presently meet with more of equal and even of greater size.

But the reflection which strikes us here as of the greatest importance is, on the admirable constancy of those zoological laws, which are never falsified in any class or in any family. "When occupied with the teeth and jaws," says the Baron, "previously to the examination of limbs or vertebræ, a single tooth has been sufficient to put me in possession of all the generic characters. The genus once determined by it, all the rest of the skeleton, in some measure, was arranged of itself, without trouble or hesitation on my part. I cannot too much insist upon these general laws, the basis and principle of the methods which in this science, as in all others, have an interest far superior to that of all particular discoveries, however curious."

The animal which we have now described has been called by Mr. Conybeare the *mosasaurus*.

The remains of another large saurian were discovered by M. de Sœmmering in the environs of Manheim, which M. Cuvier considers as a new sub-genus, intermediate between the monitors and crocodiles, and to which he has given the name of *geosaurus*.

The remains of this remarkable animal were discovered in a district called Meulenhardt, at ten feet deep, and a few paces from the remains of the *crocodilus priscus*, which we formerly noticed. They were enveloped in a marly bed, softer than that in which the crocodile was incrusted. They were not so well preserved as the latter, and it was with difficulty that certain parts could be sufficiently disengaged for the recognition of the

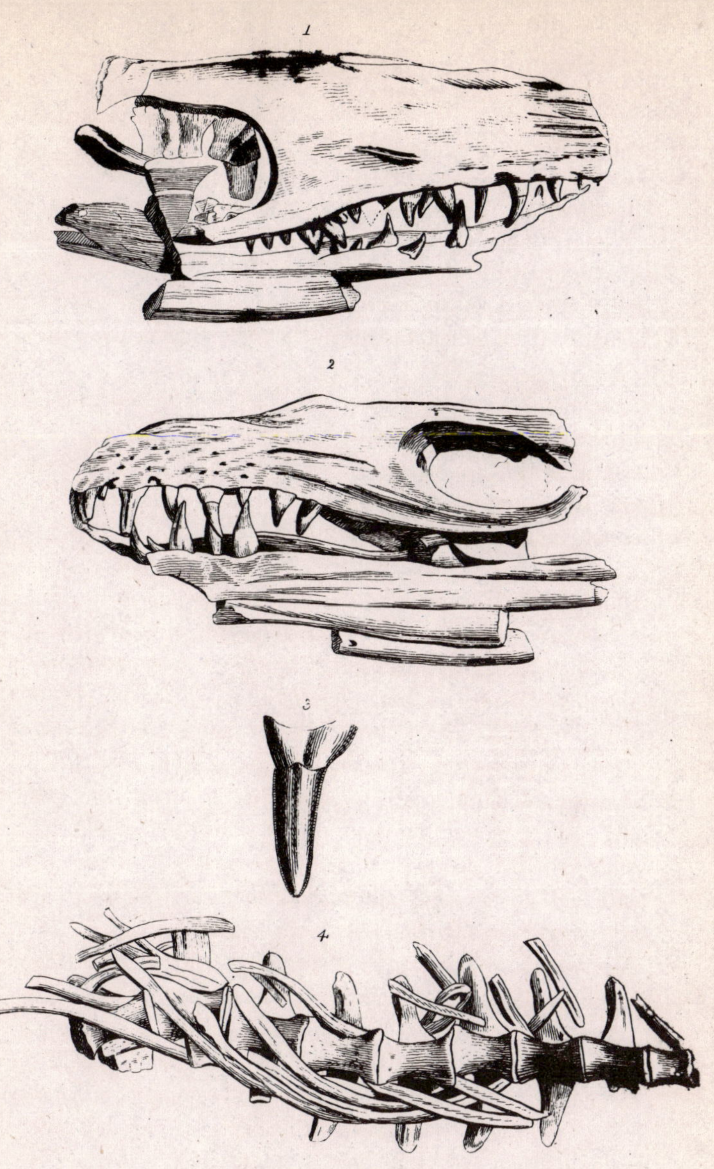

1. Left side of Head and Jaw.
2. Right side of same.
3. Single tooth.
4. Vertebræ, &c.

Geosaurus of Manheim.

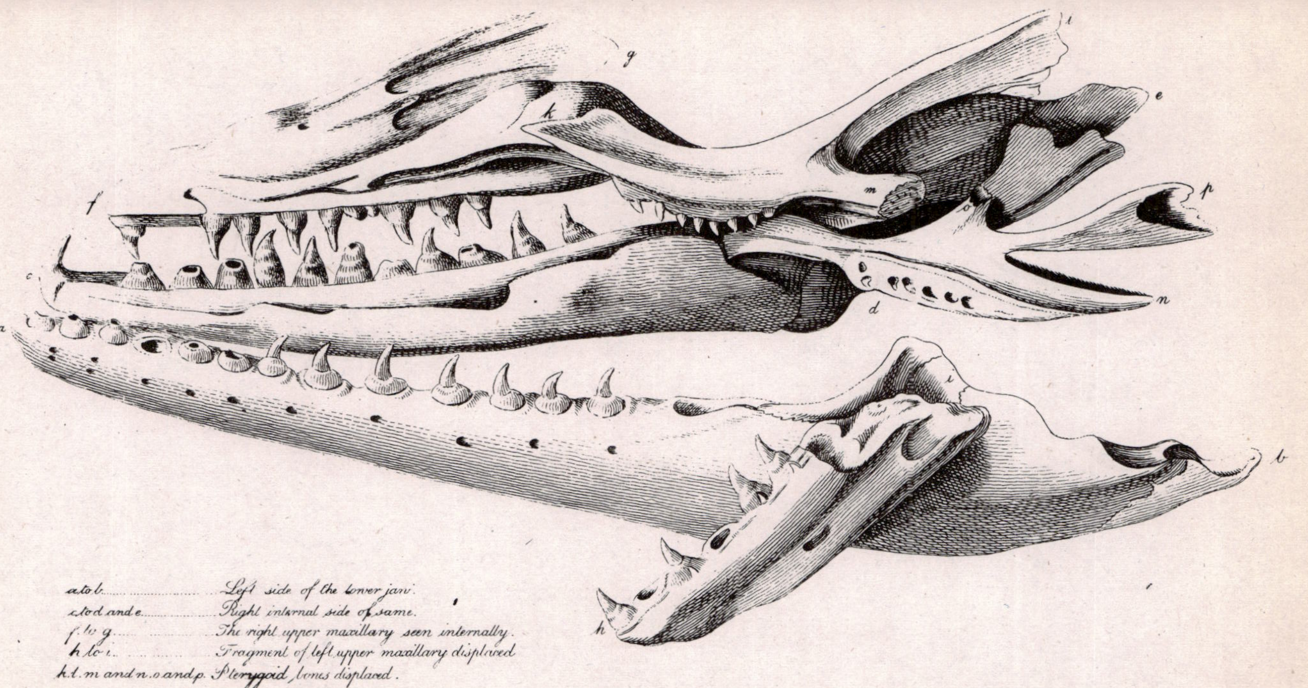

The Mosasaurus of Maestricht.

characters. M. de Sœmmering, in the "Memoirs of Munich," has given a very exact description, and a fine lithographic engraving of these remains.

The bones were almost calcined. Near them was seen a flat ammonite, four inches broad, a fragment of a bluish shell, and a great quantity of small scales, which M. Sœmmering believed to have belonged to some fishes, or perhaps to the animal itself, if it be true that it was a monitor, or some other lizard with small scales.

The teeth have preserved their enamel, which is hard, brown, and shining. The head was compressed, and its two sides brought so near as almost to touch, and to render it impossible to see the disposition of the palatine bones, or whether there were teeth in the pterygoïdean. Neither was it easy to distinguish the bones of the face, nor the sutures which separated them. All that could be perceived was, that the orbit was large, and that the muzzle could not have been much prolonged. This would render the whole configuration of the head pretty similar to that of the monitors. The form of the teeth appeared pretty well to confirm this result. They are a little compressed, trenchant in front and behind, pointed, a little hooked, and their edge exhibits a fine and close denticulation, altogether similar to what is observable in the land monitor of Egypt and in many Indian species of this genus when their teeth were not much worn. It was also seen in the fossil teeth of the crocodile of Argenton.

Fourteen or fifteen of these teeth were on the left side above, but some were wanting in front. A fragment was found which might have belonged to the anterior end of the muzzle, and which still contained three teeth. The last teeth were smaller than the others, and came even under the orbit, as in the crocodiles, and *teguixin*. On the right side but seven teeth remained. The hinder had fallen. Nor were they better preserved in the lower jaw, which had only five on each side, but in different positions.

The jugal bone appeared to have been prolonged behind and under the orbits more than in the common monitors, where it finishes in a point. This prolongation in the fossil is such, that it might lead to a belief that it rejoined the temporal bone behind, and by it the back part of the head, as in the crocodiles and many saurians different from the monitors. A particular character in this head, which is very remarkable, is a circle or ring of osseous laminæ, which occupies the internal part of the left orbit, and is evidently composed, as Mr. Conybeare has remarked, of osseous scales which invested the sclerotica of the eye of the animal, like that in birds, tortoises, a great number of reptiles, especially the monitors, and which we shall see still more strongly marked in the enormous eyes of the ichthyosaurus. In fine, without entering into further details, it is sufficient to repeat that M. Cuvier considers this animal as constituting a new subgenus of the saurian order. It might have been from twelve to thirteen feet in length, and is not deserving of the name of *lacerta gigantea* given it by M. Sœmmering, the last-mentioned lizard, and the one we are about to describe, far surpassing it.

This is the *megalosaurus*, discovered at Stonesfield, near Oxford, by our respected countryman, Dr. Buckland, whose eminent services to geological science are not less appreciated abroad than at home. The remains were discovered in a stratum of calcareous slate, which in some parts becomes sandy, and which the Doctor names the Stonesfield slate. This stone is situated a little below the middle region of oolitic strata, and above the lias which contains the ichthyosaurus. This formation, which is never above six feet in thickness, is very much extended in this country. It is a formation equally regular and ancient, and there is no possibility that the fossil bones which it contains could have got there through any cleft or any other accidental aperture.

The pieces collected there consist of the fragment of a jaw, containing a tooth developed, and many germs; a femur, a

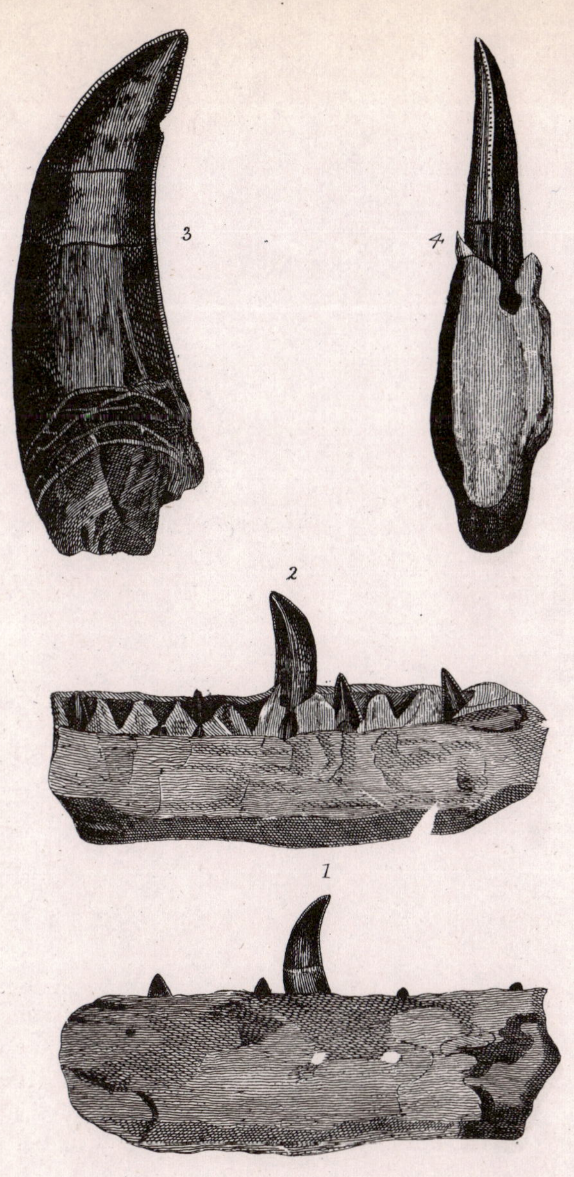

1.2. Fragment of Jaw of Megalosaurus.

3.4. Single teeth of same.

series of five vertebræ, a large flat bone which appeared to be a coracoïd, and some other bones less determinable, a part of which appear to have been rolled and worn by rubbing. Unfortunately, they were not found together, nor, with the exception of the vertebræ, even by two and two, or three and three, so as to render it probable that they proceeded from the same individual; and, in fact, it is only from their zoological relations, and their existence in the same locality, that it can be concluded that they came from one species. It must be added that these zoological relations are by no means very unequivocal.

The teeth much resembled those of the animal last described, being compressed, sharp, and hooked backwards, with two edges finely denticulated. Their anterior edge is not so thick, and the denticulations there seem to have been worn more quickly. It appears that they sprung from alveoli well cleared, and that the germs which should replace them pierced the jaw at the internal side of the teeth in place, and in distinct alveoli. These teeth had no adherence to the jaw, which would approximate them to those of the crocodile, but the external edge of the jaw rises nearly an inch higher than the internal, and thus forms a sort of parapet for the teeth on the external side. This is a character of the monitor.

The external face has some rugose foramina for the issue of the threads of the lower maxillary nerve. The portion preserved of this jaw exhibits no curve in its length, which seems to indicate that the muzzle was straight and elongated.

The most remarkable of these bones was that supposed to be a coracoïd. It is flat; a little concave at one face, a little convex at the other. More slender towards its arched edge, and especially at its thickest apophysis. One of its edges is, as we said, in an arc a little curved. The opposite edge is strongly emarginated in its middle. On one side the bone is terminated in a point, and on the other it is, as it were, truncated, and divided into two apophyses by a narrow emargination. This bone, M. Cuvier thinks, is probably a coracoïd of a saurian. Dr. Buckland was inclined to think it an ilium.

Another flat bone, widened on one side, and slender towards its edge, seems to have been an ischion.

The femur, like the teeth, exhibits a sort of mixture of the characters of the monitor and the crocodile. It is arched in two directions, being, at first, concave in front, and then behind. Its articular head, directed forwards, has behind it a compressed and rather silent trochanter. It thickens towards the bottom, and is terminated there by two unequal articular condyles. Within a little of the third of its height, it has, on its two faces, a swelling like the one which is seen on the internal face in the crocodile. The femur of a monitor would be less arched. The medullary cavity of this is wide and filled with spath.

The three long bones were taken from the same slate quarries, the first of which Dr. Buckland regards as a metatarsian, but M. Cuvier thinks that it more resembles a humerus, but one very different from that of reptiles in general. Another may be either a radius or a fibula. The third, Dr. Buckland thinks, is probably a clavicle. If so, from its proportions, it must have belonged to an animal fifty-five feet and upwards in length. But it is very difficult to decide respecting these bones.

The vertebræ that remain resemble those of no living crocodiles, monitors, or other lizards; and they can only be compared to the first of the Honfleur fossil crocodiles, or some other fossil species of that genus. They are a third more in length than in breadth. The annular part is united to them by a very marked articulation, which approximates them more to the crocodiles than the monitors. It is raised and hollowed into a cavity as in the Honfleur species. Their body is a little narrowed in the middle, but less so than in the crocodile of

Honfleur. Its two faces are plane. The spinous apophysis is but little raised, and is cut squarely. The transverse apophyses, long and depressed, rise a little obliquely.

Although these different bones should not all come from the same animal, it is not less certain that the most of them cannot come from any known animal, and on the demonstrations given by the femur and the teeth, and even on the characters derived from the femur alone, it may be affirmed that there are in the Stonesfield slate the remains of a very large reptile, allied to the geosaurus already described, and approximating in many points to the crocodiles and monitors. If the coracoïd bone we described be referred to it, we cannot hesitate to pronounce it a lizard. It appears, assuredly, to have exceeded in size the largest crocodiles known, and approached the size of a small whale. From the trenchant form of the teeth, its disposition was excessively voracious. All that accompanies its remains in these quarries in which it was buried, announces that it was a marine animal—immense numbers of nautili, ammonites, trigoniæ, belemnites, some teeth of squali and other fishes, and one or two species of crabs. Among these innumerable marine fossils, however, have been found some long bones, which appear to have come from birds of the order of grallæ, and even two fragments of jaw which we formerly noticed, which appeared to belong to didelphis. Dr. Buckland even adds that the elytræ of more than one species of coleopterous insect were found there.

Judging from the femur, the dimensions of the animal to which it belonged must have been, had it the proportions of the crocodile, more than thirty feet in length; allowing it those of the monitor, about forty-five. But from some of the other bones, were they properly determined, a much greater length might be calculated for this animal, even up to seventy feet.

Mr. Mantell, of Lewes, also discovered some remains of megalosaurus, in the forest of Tilgate. Their dimensions were enormous. One fragment of a femur was twenty-two inches in circumference; which caused Mr. Mantell to conclude that its length must have been fifty-four. Teeth were found, altogether of the same form as those found by Dr. Buckland, and unquestionably belonging to the same species. Some fragments of bones of the metacarpus and metatarsus were so thick, that M. Cuvier says that at first sight he might have taken them for those of a large hippopotamus.

With these bones of megalosaurus, Mr. Mantell found many others, such as those of crocodile, tortoise, plesiosaurus, cetacea, and birds, and he also collected some of which it is not possible to assign the genus. The most singular of these were certain teeth, respecting which it is not determined whether they come from a fish or reptile. M. Cuvier says, that it is not impossible but that they may have belonged to a saurian, and if so, to one still more extraordinary than any of which we have had hitherto any knowledge.

What gives them their singular character is that their point and shaft are worn transversely, like those of the herbivorous quadrupeds. The first of these teeth which was presented to M. Cuvier he decidedly thought to be the tooth of a mammiferous animal in a state of detrition. It seemed especially to resemble the cheek-tooth of a rhinoceros, which, were it the case, would go far to overturn the now received ideas of the relations of bones and strata. But having subsequently received a series of entire teeth, and of some more or less worn, the Baron was convinced, on inspection, that his first notion had been erroneous.

The largest of these teeth has a root a little curved, which grows slender towards its deep extremity. The crown is prismatic, and broader at its external face. This face alone is covered with enamel, or, at least, it has an enamel more thick and hard than the rest of the circumference, just as it is in the incisor teeth of rodentia. It widens at first proceeding from the root, and then its edges approximate to form the trenchant point which terminates the tooth. The two edges which, from the wide part, proceed to unite in this point, are strongly serrated. The external and enamelled face of the tooth has two longitudinal ridges, very obtuse and not projecting, which divide it into three parts also longitudinal and very slightly concave.

In the business of mastication, this tooth at first is worn in the point, and gradually all the part whose edges are denticulated disappears by detrition. A truncation at the same time is produced on the tooth, which grows broader and broader, but which is always oblique, because the external and enamelled face is less worn than the rest. It is only when all the denticulated part is removed that one might be tempted to take these teeth for those of herbivorous mammalia, worn as far as the root, for there are no lineaments of enamel on the crown, and even the want of them would oblige us to suppose these teeth to be incisors, were we to attribute them to mammalia. But this would be a supposition exceedingly difficult to admit, for there are no incisors of mammalia which in the slightest degree resemble the teeth in question.

Some of these teeth are smaller than the others, and the smallest have usually on their external face but one obtuse longitudinal ridge, but on the sides many smaller and sharper ridges are observable which form striæ there. Some are also found with a simple trenchant edge, without denticulations, slightly convex at their two faces, and terminated by an obtuse point, and which tolerably resemble the canines or lateral incisors of tapirs, or other animals with short canines. These differences are probably attributable to the different positions of the teeth in the mouth of the animal. These teeth do not afford the only indications of the existence of gigantic species of saurians, equal, or nearly so, to the megalosaurus, the animal of Maestricht, and the crocodiles, in those remote ages.

Among the bones collected at Honfleur, in the possession of M. Cuvier, are vertebræ of several sorts, which he could not refer to any of the species hitherto described, and which may serve as bases for the future re-erection of additional monuments of the ancient world.

Some very large ones, from Havre and Honfleur, have a cylindrical body, nearly as long as wide, marked on each side with a small fosset, with plane, circular faces, a medullary canal very narrow, and an annular part not articulated. The spinous process is high and straight. The transverse processes, at the level of the medullary canal, are gross, cylindrical, and vertically dilated at the end; and, what is very remarkable, the posterior articular processes are small, pointed, approximating, and go into two small fossets, between the anterior processes, and in front of the base of the spinous.

These would appear to belong to a species of saurian, much resembling the plesiosaurus, yet to be described. Their only differences from the vertebræ of this last genus relate to a greater proportional breadth of body, and that the little fossets, or dimples, are hollowed at the sides of the bone, instead of underneath. Some from Newcastle were smaller, but their proportions of body were the same, and the little fossets only were wanting. Some large bones of the extremities were found along with the vertebræ now described, which appeared to have belonged to the same species, whatever it was.

In the environs of Luneville the remains of a saurian have been discovered, in many respects approximating to the crocodiles. This species, as new to geology as to zoology, was found by M. Gaillardeau, a physician of that city, in certain quarries in the neighbourhood, used for the purposes of building. The stone composing them is compact, in horizontal strata of moderate thickness, separated by thin strata formed of debris of shells, or entire shells accumulated together. Terebratulæ and mytulites are especially in great abundance there. This formation, in which many other fossil remains were found, belongs to the inferior strata of the order of Jura, or to those above what is called Alpine limestone.

Here were found bones which manifestly had appertained

to a saurian. They consisted of a vertebra, one side of the lower jaw, some ribs, and some bones of the shoulder, and pelvis. They preserved no gelatine, and did not become black in the fire.

The vertebra resembled a caudal of some species of crocodile, in which both faces are plane or slightly concave. The body alone was preserved.

The jaw has some characters of crocodiles, and others of lizards. It is more long and slender than that of the common crocodile. The alveoli of the teeth are well separated; they are ranged on a single line. Twenty-seven were observed in the fragment of which we speak, though it was mutilated at the front. The teeth appeared to have been alternately more bulky and slender, and to have been hollowed internally.

All this agrees with the crocodile, but the composition of the jaw was different. The coronoïd process, which is short and obtuse, appertains to the supplementary bone, which, instead of being small and crescented as in the crocodile, proceeds forward between the dentary and opercular bones, and along the internal edges of the teeth over a length of more than twenty-eight alveoli. The opercular is also carried very forward; and, instead of a simple foramen intercepted between it and the angular, and independent of the grand opening behind the opercular and angular which is observed in the crocodile, there is only a single very long aperture, which predominates from this posterior point of the opercular as far as the articulation, and which has the complementary and subangular bone above, and the angular below. At the external face, the subangular exhibits a sharp longitudinal ridge. The articulation resembles that of the crocodile, but the post-articular apophysis seems to have been a little shorter in proportion.

From these two pieces alone, it may be pronounced, without hesitation, that they proceed from an unknown reptile; very probably from some genus intermediate between the crocodiles and saurians, like those already described. Some other bones were found serving to confirm this result, and indicate an animal not unlike the plesiosaurus or ichthyosaurus.

It was not merely in size that the class of reptiles stood pre-eminent in those ancient days. They were also distinguished by forms more varied and singular than any which we now behold existing. We are about to examine a genus, the *pterodactylus*, or wing-toed reptile, which had the power of flying, not by means of its ribs, like the *draco volans*, nor by a wing without distinct fingers, like that of birds, nor by a wing in which the thumb alone is free, like that of bats, but by a wing sustained principally on one very elongated toe, while the others preserved their usual shortness and their claws. At the same time, these flying reptiles (almost a contradiction in terms) had a long neck and the beak of a bird, which must have given them a most remarkable and anomalous aspect. There were two species.

The first is termed by M. Cuvier, *Pterodactylus longirostris*, from the length of the muzzle. The first knowledge of it was owing to M. Collini, Director of the Museum of the Elector Palatine at Manheim. The skeleton was found in one of those marly, foliated, grey, and sometimes yellowish rocks of Aichstedt, which abound in dendrites and animal petrifactions. The skeleton was wonderfully entire.

M. Blumenbach thought it was a bird of the order palmipedes; but, as M. Cuvier observes, a bird would have larger ribs, and each provided with a recurrent apophysis. Its metatarsus would have formed but a single bone, and would not have been composed of as many bones as there are toes. Its wing would have had but three divisions after the fore-arm, and not five like the pterodactylus. Its pelvis would have had quite a different extent, and its osseous tail a totally different form; it would be widened, not narrow and conical. There would have been no teeth in the beak, for anything of that kind in birds appertains to the corneous envelope, not to the bony skeleton. The vertebræ of the neck would have been more numerous. No bird has less than nine. The palmipedes, in particular, have from twelve to twenty-three, and in this animal there are but six or seven. The vertebræ of the back, on the other hand, would have been less numerous. Here there are more than twenty. In birds they are from seven to ten, or at most eleven.

The teeth present the reptile character without any equivocation. They are all simple, conical, and nearly alike among themselves, as in the crocodiles, as in monitors, and other lizards. The dolphins alone, among the mammifera, show anything like this; but that genus is out of the question here.

M. de Soemmering considered this animal as belonging to the class Mammalia, and would range it in the neighbourhood of the bats. He rested his opinion much on the variation of numbers in the teeth of the bats. But it is certain that the bats have never more than two forms of cheek teeth, neither of which are at all applicable to the dentition of the pterodactylus.

In the side of the head which was preserved, there were nineteen teeth below and eleven above, sixty in all: but some had probably been lost from the upper jaw. What completes the proof that these are reptile teeth is, that in the jaws, along their bases, there are foramina, from which the teeth to replace them must have issued. Similar are to be seen in the safeguards, and especially in the dragon (dracona of M. de Lacépède).

The lower jaw is equally that of a reptile, having no salient condyloid apophysis, nor coronoïd prominence. The pangolins alone show some relations with this jaw; but they have no teeth. In the bats there is not the least approximation.

It is the same with the enormous elongation of the muzzle. The vespertilionidæ have all short muzzles. The roussettes, indeed, have it a little longer, but not beyond the proportions of a dog or fox.

No mammiferous animal has the cranium so small in proportion to the muzzle as this fossil reptile.

The length of the neck is proportioned to that of the head, muzzle included. There are five vertebræ, large and prismatic, like those of long-necked birds, and a smaller one at each extremity; perhaps there may have been two towards the head, so that the total number would have been seven, as in mammifera and crocodiles, or eight, as in tortoises.

What is most astonishing is, that this long head and neck are supported on so small a body. Birds alone exhibit similar proportions, which, doubtless, added to the length of the great toe, determined some naturalists to refer this animal to that class from which it is removed by so many other characters.

The neck is so much recurved behind that the occiput touches the pelvis in the specimen. There are nineteen or twenty dorsal and lumbar vertebræ. M. Oken admits twenty-two, including the sacral.

It is difficult to say how many of these vertebræ supported ribs, but there remain at least a dozen in their places on the right side. The bodies and the spinous apophyses of the vertebræ are visible, but the left side of the annular part is removed from almost all of them, so that the medullary canal is to be seen. The anterior spinous apophyses are a little the longest. The posterior are short and cut square. Neither birds nor bats have any so formed.

There are transverse apophyses, at least to the first seven vertebræ, to which apophyses the ribs are attached. Beyond the ninth, says M. Oken, there are no more, and the rib is attached immediately to the vertebra.

All the ribs are singularly slender and filiform, which entirely excludes this animal from the birds, where the ribs are broad, and each provided with an oblique and recurrent apophysis very peculiar.

The tail is very short and very slender, and but twelve or

thirteen vertebræ are reckoned in it. The faculty which the animal, without doubt, possessed of flying, and the difficulty under which it laboured of creeping or walking, by reason of the disproportioned length of the head and neck, are probably the reasons why a long tail was not necessary for it.

Though the shoulders and sternum were not very well preserved, M. Cuvier could ascertain in them all the true characters of the reptile. The same may be said of the bones of the carpus and metacarpus.

There are, at first, three small toes, one with two phalanges, and one with three. The last phalanx in both is unguical, compressed, hooked, and pointed. The third has three phalanges; and as it was mutilated, it probably had a fourth, which was unguical. These numbers are exactly those of the first three toes in crocodiles and lizards.

Finally, comes the toe, enormously elongated into a thin stem, which eminently characterizes the animal. It has four articulations without a claw. The fourth toe of lizards would have five articulations with a claw. In the crocodiles there are four and no claw, but not this extraordinary elongation.

The crocodiles and lizards have a fifth toe besides, which in the lizards has four articulations, and in the crocodile but three without a claw. It appears that, in the fossil animal, there remains but a vestige of the fifth toe very obscure. The great toe is probably the fourth, for it is the fourth which is longest in the lizards. The three others precede it in the inverse order of the number of their articulations.

To complete the resemblance, the penultimate phalanx is the longest. That which precedes it in the third toe is the shortest, absolutely the same as in the lizards.

The form of the unguical phalanges is also the same, that of a semi-crescent, compressed, trenchant, and pointed.

It is scarcely possible to doubt that this long toe served to support a membrane over the entire length of the fore-leg, which constituted for the animal a wing, much more powerful than that of the dragon; and, at least, equal in force to that of the bat. The animal could fly as long as the vigour of its muscles permitted, and would then use its three short toes, armed with hooked claws, to suspend itself to trees.

It is unnecessary to pursue this investigation farther. We find an animal which, in its osteology from the teeth to the end of the claws, presents us with all the classic characters of the saurians. We cannot doubt, therefore, that it had the same characters in its teguments and soft parts; that it had their scales, circulation, organs of generation, &c. But, at the same time, it was an animal provided with the means of flying, and one which, in a stationary position, would make but little use of its anterior extremities, if, indeed, it did not keep them folded up as birds do their wings. It could yet employ its little fore-toes to suspend itself to the branches of trees, but its tranquil position would be ordinarily on its hind feet like birds. Like them, too, it would keep its neck straightened and curved back, to prevent its enormous head from destroying its equilibrium.

After these data, the figure of this animal, as it was when living, might be drawn. But it would be one of the most extraordinary kind, and to those who had never examined into the subject, would appear to be a monster, the product of a distempered imagination, rather than, by any possibility, the work of nature. Something approaching to it, or analogous, has been seen in the fantastic paintings of the Chinese. There is an engraving in the German Journal, entitled *Naturforscher*, taken from a Chinese book of "Natural History," and which represents a bat, with the beak of a hawk, and the long tail of a pheasant; but this would be no representation of the animal of which we have been treating.

The second species is called by M. Cuvier, *Pterodactylus brevirostris*, from having a shorter muzzle than the last. This, too, was discovered in the same strata as the last. The stone

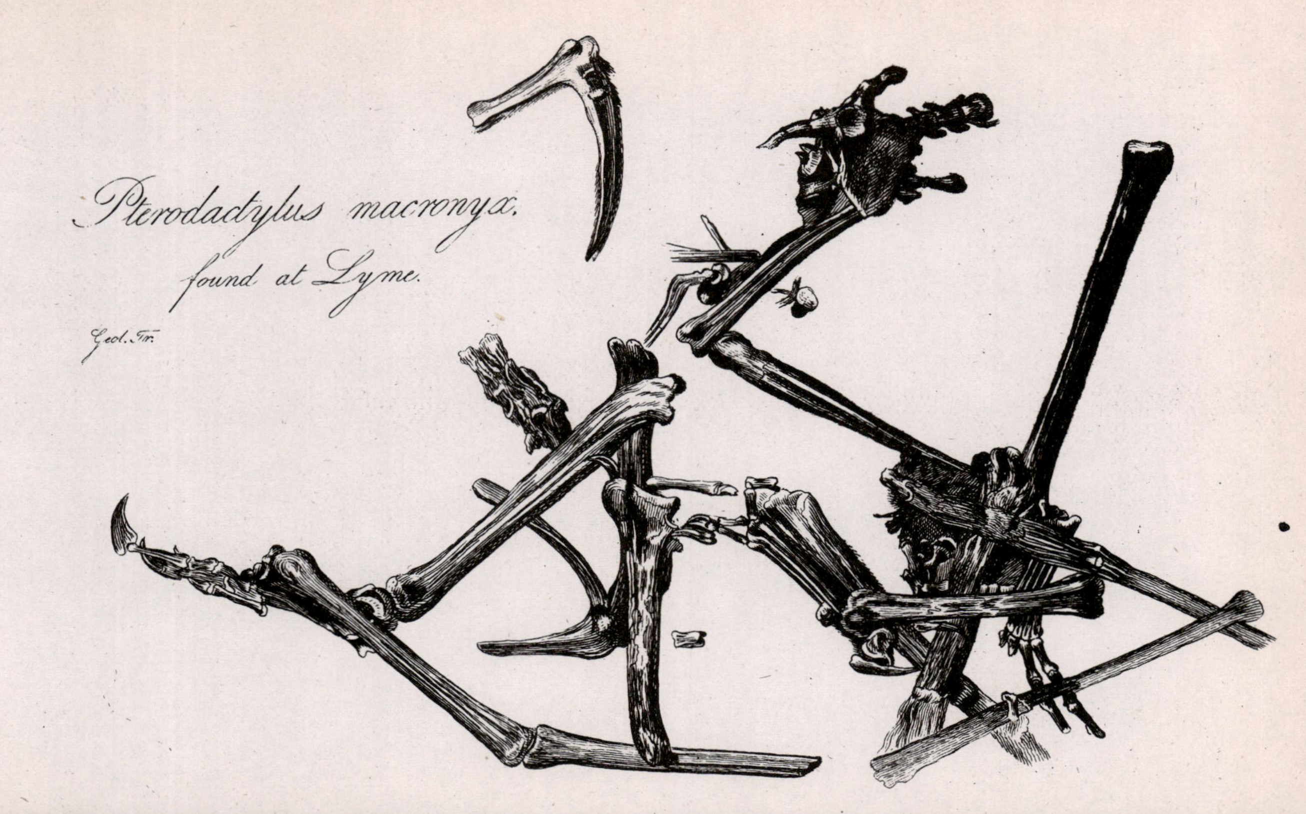

Pterodactylus macronyx.
found at Lyme.

Geol. Tr.

in which it was, presented also the remains of a very small fish, and some small asteriæ. Like all those schists it is of a greyish red. The bones are distinguished by a browner tint. Their cavities are filled with a whitish spath.

The individual is one third smaller in the trunk than the preceding, and its head and neck much less elongated in proportion. Its head is by no means so well preserved, and, seen in an isolated state, it might more readily be taken for that of a bird than a reptile.

There are no teeth marked in the figures of it. But M. de Soemmering says, that there were some small ones in the two jaws, some of which resembled the front teeth of the bats, others, their molars, and that, on the opposite stone, there were eight pointed ones in the lower jaw, and in the upper five. M. Oken does not mention the teeth, but thinks that he could distinguish a tympanic bone.

There were seven vertebræ in the neck, to the last three or four of which M. de Soemmering has marked spinous apophyses. There are ten or eleven vertebræ between the cervical and the sacral, all provided with spinous apophyses, not much raised, cut square, and which all appear to have carried ribs. These last are slender and simple, as in the other species.

The tail is also short, slender, and pointed, but it is not easy to reckon the vertebræ. It would seem that the last are divided into two parts. The pelvis is exactly that of a lizard. The arm is the same as in the large species.

The three small unguiculated toes show distinctly the same number of phalanges as in the large species, and in the lizards. The long toe has the same proportion as in the large species, and is composed in the same manner of four articulations, the last of which has no claw. The hind feet were composed each of four toes, with a metatarsus with four bones, like the large species.

These details, and more that we have omitted through fear of tediousness, prove that in this district, at the epoch when these strata were formed, and when crocodiles, monoculi, and so many other beings existed there, which are now found only in the torrid zone, there were two species of saurians which could fly by means of a membrane sustained by a single one of the toes of the anterior extremity—which suspended themselves, and, perhaps, crept by means of the three other toes of this extremity—which sat upright on their hind-feet only, and whose large head was cleft by an enormous mouth, armed with small pointed teeth, proper for seizing insects, and other small animals.

Of all the wonderful beings that the researches into fossil osteology has brought to light, these are incontestably the most extraordinary, and the most entirely different from any beheld in existing nature.

In the same strata some more bones were found, which might have belonged to some large species of this genus. Two of them appeared to be the second and third phalanges of the great toe of the pterodactylus. The second, which had but eighteen lines in length in the preceding species, was more than seven inches long in this.

Fossil Batracian Reptiles.

The animals analogous to the frog, the toad, the salamander,—these naked reptiles subject to metamorphoses, which constitute a small family so isolated in the Animal Kingdom by its whole organization, did yet exist in those lands destroyed by the revolutions of the globe, and, though differing in species, were subjected to the same laws of co-existence and conformation of organs. They must, however, have been rather few in number, if we may judge by the small quantity of debris which remain of them, and the small number of places in which they have been discovered. Even where they have been found, there is a portion of doubt or obscurity thrown around them.

As this is the case, it is not necessary here to follow the

Baron in his admirable and most interesting account of the osteology of the living species. Nothing, indeed, can be in itself more curious and important, but it is not necessary, as such researches were in former instances, to elucidate our account in the way of comparison. The details of the living batracians will be given in their proper place in the present work ; and we shall now proceed at once to what can be said respecting their fossil remains.

We shall first speak of the pretended *fossil man* of the quarries of Œningen, described by Scheuchzer, which other naturalists considered a *silurus*, but which, in reality, is nothing but an aquatic salamander of gigantic size, and unknown species.

It was natural that those who attributed all petrifactions to the deluge, should be astonished at never encountering amongst so many remains of animals of all classes, any human bones that could at all be distinguished.

Scheuchzer, who supported this theory, with more detail and continuity than any other writer, was accordingly more interested in discovering some remains of our species. He, therefore, received, with a sort of transport, a schistous rock from Œningen, which appeared to present to him the impression of a skeleton of a man. He described this specimen, in brief, in "The Philosophical Transactions for 1726," (vol. xxxiv. p. 38). He then made it the subject of a particular dissertation, entitled "*Homo Diluvii testis.*" He reproduced it in his "Sacred Physics," assuring us, "that it is indubitable, and that it contains a moiety, or nearly so, of the skeleton of a man—that the substance even of the bones, nay more, of the flesh, and of parts still softer than the flesh, are there incorporated in the stone. In a word, that it is one of the rarest relics which we possess of that cursed race which was overwhelmed by the waters."

It required all the blindness which the spirit of system can produce, for such a man as Scheuchzer, who was a physician, and must have seen human skeletons, to deceive himself so grossly—for this fancy, which he reproduced so pertinaciously, and which has so often been repeated on his authority, does not rest on a particle of foundation, nor will bear the slightest examination.

John Gesner, again, quotes this specimen for an anthropolite in his treatise of Petrifactions, printed at Leyden, in 1758. It appears, however, that this naturalist, having become proprietor of a similar piece, was afterwards the first to raise doubts on the species which he had furnished, and to conjecture that it might be nothing but the *silurus glanis* of Linneus, an opinion which naturalists then adopted with a confidence equal to that which they had accorded to Scheuchzer.

This last specimen was not engraved, no more than another which was said to be in the Convent of the Augustine friars at Œningen. But another more complete one than Scheuchzer's was discovered, which belonged to Dr. Ammann, of Zurich, and has since passed to the British Museum. An engraving of it was published by M. Karg, in the Memoirs of the Society of Suabia.

The simple comparison of the first of these specimens with the skeleton of a man, ought to have been at once sufficient to disabuse any one of the notion that it was an anthropolite.

The proportions of the parts present of themselves the most sensible differences. The size of the head is nearly that of a man of middle size, but the length of the sixteen vertebræ is some inches more considerable than it ought to be; and we find, accordingly, that each vertebra, taken separately, is longer in proportion to its breadth than in man.

The other differences derived from the form of the parts are not less striking. The roundness of the head, which must have been the principal cause of illusion, presents, however, a very remote relation with that of man. What is become of all the upper part, all that there should be of forehead? And if it be supposed that the front has been removed, then the total roundness can only be the effect of chance, and will prove nothing.

How have the orbits become so large? Let us suppose that the head has been compressed from front to back, or that there is nothing of it but a vertical section; on either supposition this size of the orbits is equally inexplicable. The deeper such a section was sunk, the smaller the orbits would become.

The interval of the orbits is furnished with entire bones, distinguished by a longitudinal suture. There is nothing analogous to this structure in man. Why are not the bones or cavity of the nose to be seen, and if there are remains of the posterior part only, how was this suture formed there?

How does it happen that in a head, whether compressed or cut, no trace of teeth should remain? We know that the teeth are the parts which are always best preserved in the fossils. Scheuchzer supposes that the bones placed at the two sides of the first vertebra are the remains of the lower jaw. But, in fact, there is no resemblance, and the total want of teeth is decisive.

These reasons, and many more, caused naturalists to seek for some other type for this fossil, beside man. But instead of having recourse to direct comparison, they began to argue on the subject. The quarries of Œningen, they averred, abounded with remains of fresh water fishes, which all appeared to be fish peculiar to Europe. Among these fish, therefore, they reasoned, among fresh-water fish, and among the fish of Europe, we must find our animal. Now what species is sufficiently large to have furnished this skeleton? It was then remembered that the *silurus glanis* often attains a very large size, and that its head, externally, presents a rounded contour. This was deemed sufficient to solve the problem without any further trouble of examination or comparison.

It was singular enough that M. Karg should have adopted this opinion, after having observed, and caused to be drawn, the specimen of Dr. Ammann, the resemblance of which to a salamander is so striking; and yet, he says, that he does not doubt but that the fossil is a silurus, and that the head and fins are distinguishable in the clearest manner. His editor, M. Jæger, has taken a very simple method of refuting this opinion. At the side of the fossil, he has given a drawing of the skeleton of a *silurus glanis.* M. Cuvier has done the same, and placed the figure of that fish beside the fossil skeletons of Scheuchzer and Dr. Ammann.

On the first glance of the eye it may be remarked, that, with an equal size of head, the *silurus* would not be more than two thirds of the length of the fossil skeleton of M. Ammann—which, nevertheless, is not complete; and that, within the same space, where the spine of the silurus contains fifteen vertebræ, that of the two fossil skeletons show but five or six. There is no relation of form between the yet shorter vertebræ of the rest of the spine of the silurus, and the vertebræ of greater length than breadth in the fossils. The entire, too, of the spine in the silurus consists of seventy vertebræ, while there are but thirty or thirty-two in the much longer spine of the fossil.

The fossils present no vestige of the long spinous apophyses of the tail of the silurus. It is by mere chance that in the fossil there are bones of the extremity opposite to the place where the ventral fins of the silurus are attached. But the correspondence is illusory; for, in the fossil, it is the anterior extremity, and in the silurus, the posterior. The posterior extremity of the fossil is very far behind, and opposite to the very point where it is attached, the tail of the silurus is about to terminate. The two extremities of the fossil present solid, cylindrical bones, like those of the legs of quadrupeds and reptiles, and not articulated or spiny radii like those of the fins of fishes. The silurus exhibits nothing like the little ribs spread on the two sides of the spine in the individual of M. Ammann.

Finally, if the head be compared, which, probably, has altogether given rise to the supposition here combated, no resemblance will be found either in the general contours, or in the details. The contour of the silurus is much less rounded, and yet this is owing to the lower jaw, while in the fossil the lateral branches appear to appertain entirely to the zygomatic arch. The parts placed behind the orbit have not nearly the breadth which they would have in the silurus.

This rounded figure of the head for a long time struck M. Cuvier as bearing a singular resemblance to the head of a frog, or salamander; and he had no sooner seen the figure of Dr. Ammann's specimen, than he observed, in the vestiges of the hind feet and of the tail, strong confirmation in favour of the last-mentioned genus. This would be rendered very obvious by placing the skeleton of a salamander beside the fossil, without suffering one's self to be prejudiced by the difference of size. Everything would then be explained in the clearest manner.

The rounded form of the head, the size of the orbits, the suture in the middle of their interval, the lateral angle for the articulation of the lower jaw, the length of the vertebræ in relation to their breadth, the little ribs attached to their two sides, the remains of anterior extremities very sensible in the fossils, those of the posterior extremities still more so in one of them (that of Dr. Ammann), in which are seen the femora, a part of the tibiæ, and some fragments of the pelvis; all, in short, strongly demonstrate for the family of the salamanders to the exclusion of all others.

M. Cuvier, in the first edition of the "Ossemens Fossiles," says, "I am even persuaded, that if one could have the disposal of these fossils, and examine them with a little more minuteness of detail, still more numerous proofs would be found in the articular faces of the vertebræ, in those of the jaw, in the vestiges of the very small teeth, and even in the parts of the labyrinth of the ear; and I invited the proprietors, or the keepers of these fine specimens, to proceed to this examination."

He had, subsequently, the advantage of making this examination himself. Being at Harlem in 1811, he was permitted by M. Van Marum, Director of the Teylerian Museum, to dig into the stone which contains the pretended anthropolite of Scheuchzer, in order to discover the bones which still might lie concealed there. The operation was performed in his presence, and in that of M. Van den Ende, Inspector-general of Studies. These gentlemen placed before them a drawing of the skeleton of the salamander, and it was not without pleasure, that in proportion as the chisel removed portions of the stone, they beheld bones appear which fully confirmed their anticipations.

First of all, around the rotund part, on right and left, they found a double range of small teeth, which proved that the rotund appearance was produced by the jaws, and not by the cranium. They then discovered little ribs at the end of each of the transverse apophyses, as in the specimen of Dr. Ammann, and as in the salamanders. They were assured that these ribs were very short, and could by no means embrace the trunk. They proved that the head was articulated on the first vertebra by a double condyle, as in all batracian reptiles.

Passing, then, to the anterior extremities, which had only been indicated by a small face of the left shoulder, they discovered them both. On each side was found an omoplate, very much dilated at its spinal edge, the contour of which is semicircular. It is altogether like that of the aquatic salamander. But it appears that the clavicle and coracoid were last. Near the omoplates are two humeri double the length of the omoplates, a little widened in the top and bottom, with a furrow for the separation of the condyles, absolutely again the same as in the aquatic salamanders.

At the end of the humerus are some bones of the fore-arm, one half shorter, and one a little thicker than the other. Finally, come the bones of the toes, incomplete on the right side, but complete, though a little in disorder, on the left. They exhibit exactly the same number of parts as in the aquatic salamanders, namely, four toes, with one metacarpian and two phalanges for each, except the third, which has three phalanges.

Some years after, the Baron, being in this metropolis, had an opportunity of personally examining the specimen of Dr. Ammann, in our Museum.

The omoplates and humerus are the same as in that of Scheuchzer, but the fore-arms and hands are wanting. The thighs and legs are to be seen opposite the nineteenth vertebra. In the living aquatic salamanders the position of the pelvis varies much; but in one species, *triton cristatus*, the Baron has seen it suspended at the eighteenth vertebra, which accords with the position of the posterior extremities in this fossil.

The bones of the legs are one half shorter than the femora, and the tibia is very broad. Some remains of the pelvis and of the toes are visible.

Behind the pelvis are still sixteen vertebræ, and by the size of their transverse processes, one may judge that there were many more to come after them in the entire tail.

In the head of this specimen, in the British Museum, are some teeth. Its form is exactly the same as that of the specimen at Harlem, and broader in proportion than in our living salamanders. The great species of the Allegany Mountains approaches it most in this character, and also resembles it in the breadth of the pterygoïd bones, and the prominence of its occiput behind the lateral processes which bear the lower jaw.

There are two bones placed on each side of the occiput, and which are found in both specimens. M. Cuvier at first imagined that they announced a considerable and permanent branchial apparatus, and was, therefore, inclined to refer these animals to the genus Proteus. But his present opinion, founded on a more extended study of the hyoïd bone, and its pieces in the aquatic salamander, is, that the two bones in question are the two pieces of the posterior cornet. The first is yet, in part, concealed under the cranium. The second is very entire, and exactly of the same form as in our living aquatic salamanders. Thus, there can be no doubt but that the pretended anthropolite of Œningen was an aquatic salamander, of a size gigantic in its genus.

We shall conclude this account by saying a few words repecting those celebrated quarries where the above described fossil was found, and an immense number of others of various kinds.

The Rhine, after having formed the lake of Constance, and narrowing near the town of the same name, widens again to form the lake called Zellersee, and does not resume the ordinary width of its bed until it approaches the little town of Stein.

On the right bank, a little above Stein, is the village of Œningen, formerly belonging to the Bishop of Constance, and at present, like the rest of the bishopric, to the Grand Duke of Baden.

The quarry, which abounds in impressions of fish, &c., is three-quarters of a league from this, on the southern declivity of a mountain termed Schiener-berg, and at least five hundred feet above the level of the lake. A small stream runs along its eastern side. The elevated part of the mountain is of a soft micaceous sandstone, and rolled blocks of red and green granite are found in the fields.

The quarry is extended over about two hundred and seventy feet in length, and is about thirty deep, but its bottom is often full of water. Beneath the vegetable soil is found, at first, a friable bluish marl, two feet in thickness, which is employed in making tiles and bricks, for want of better clay. Under this marl are many feet of a primitive schistus, yellowish grey, soft, with very thin laminæ, and filled with vegetable impressions. Then comes a second bluish marl, like the first, about half a foot in thickness, and without organized bodies. All the succeeding strata are calcareous, and when they are disturbed, give out an odour more or less strong of petroleum. They are distinguished into several beds. The first is named by the workmen, the *gross bed,* or *sulphureous stone.* It is from two to six feet thick, and does not divide into foliations. The second is called *white slate.* It is four inches thick, very argillaceous and soft, and divides into very slender laminæ. Plants, insects, and the first fishes are to be seen there. Another schistus follows it, called *small pieces,* two feet thick, divisible into thin leaves, composed, in a great measure, of the debris of vegetables, and containing many bivalve shells, excessively small, round, and pearly.

The next bank is named *thick pieces.* It is a foliated limestone, two feet thick, and exhibiting scarcely any traces of vegetable debris. Then come two beds, scarcely two inches high, named by the workmen *black plates,* which appear tinctured by vegetable debris.

The *first white plate* follows them. Flagstones for apartments are made of this ; some large fishes are visible in it, and fine dendrites. It is three inches high, and divisible into thick leaves or slabs.

At last comes the *fishy plate,* so named from the immense quantity of fish which it contains, with small limneæ. It is a white limestone of a fine grain, with slender leaves, and of moderate hardness.

Under it is what is called the *little skin* very thin, and of a blackish grey. Then comes the *third black plate,* two inches and a half high, which is followed by the *knotted* or *Indian stone.* This is a grey, coarse-grained schistus, dotted and radiated with white and yellow, and filled with fishes and other animal and vegetable impressions. It is in much estimation, and its thickness is about four inches.

The *mussel-stone* is a blackish, micaceous limestone, full of debris of vegetables, of small limneæ, and of fragments of muscles still pearly. It is about a foot in thickness.

The *dill strecken* is a calcareous schistus, a little micaceous, with thick foliations, of a whitish grey, ten inches thick, and without fossils.

The *little white skin* is a soft calcareous schistus, with slender leaves, and about an inch thick.

The *small muscle-stone* is a coarse-grained calcareous schist, dry and yellowish. It contains an innumerable quantity of small limneæ, and various other shells of the fresh water, or their nodules, and vegetable impressions.

The *gross* or *thick plate* is a grey schistus with thick leaves, about half a foot thick, and contains only some vegetable fibres.

The *white plate* (No. 2) is a calcareous schist, with coarse grains, very rich in petrifactions and impressions of every kind; and all that exists separately in the other strata is found assembled here.

The last bank is a grey or reddish schist with thin leaves, containing innumerable limneæ, and very fine impressions of leaves of various colours. It reposes on a bluish coarse sandstone, which forms, in general, the banks of the Rhine in this country, and where some veins of pit-coal are found, and sometimes numerous muscles, which we are assured belong to the fresh water.

At about a quarter of a league above the quarry of Œningen, on the same side and nearer to the lake, is another quarry belonging to the village of Wangen, where similar petrifactions were found, and as it would appear, in analogous strata.

M. Karg, who has described the quarries of Œningen, proceeding on the supposition that all the animals of these strata are the same as in the neighbouring waters, has endeavoured to prove that the strata must have been recently formed in a pond, which became empty in consequence of some accident happening to its dykes. This hypothesis doubtless hindered him from giving us more information regarding their relative position to the neighbouring strata, and thus affording a criterion by which to judge of their antiquity. But his opinion is not that of the profoundest geologists. M. de Humboldt and M. de Reuss agree to regard the schists of Œningen as belonging to an ancient and regular formation: the last considers them as belonging to his third calcareous formation. M. Brogniart regards them as subordinate to the molasse of Switzerland, and as contemporaneous and, perhaps, posterior to the gypsum of Paris. What is certain is, that they constitute a formation of the fresh water, which contains animals utterly unknown at present.

The Ichthyosaurus and Plesiosaurus.

We are at length arrived at beings which, of all the reptiles, and perhaps of all the fossil animals whose remains have been found, bear the least resemblance to any of the living inhabitants of our present world; which exhibit combinations of structure well calculated to astonish the naturalist; and which would doubtless appear incredible to any one who either had not the opportunity of observing them himself, or was not fully satisfied of the authenticity of those authorities on which the relation of their existence rested. In the first genus we behold the muzzle of a dolphin, the teeth of a crocodile, the head and sternum of a lizard, the paddles of the cetacea, but four in number, and the vertebræ of a fish. In the other, with the same cetaceous paddles, we find the head of a lizard, and a long neck resembling the body of a serpent.

Such are the astonishing characters which the ichthyosaurus and plesiosaurus present to our inspection, after having been buried for so many myriads of years under enormous masses of stone and marble. It is to the most ancient secondary strata that they belong. They are found only in those strata of marly stone, or of greyish marble filled with pyrites and ammonites, or in the oolitic beds; all formations of the same order as the chain of Jura. It is in England that their remains are most peculiarly abundant; and we are proud to add, that it is to the zealous assiduity of our scientific countrymen that the discovery, description, and determination of them are principally owing. They have spared no pains in the collection of remains, or in putting them together as well as the state of the fragments would permit. This is a tribute to their merit which the Baron Cuvier himself, with all the characteristic liberality of genius, most willingly pays, and we have made but little alteration in his language in repeating it.

We shall soon see, notwithstanding the anomalies of structure in these extraordinary animals, that they approximate more to the lizards than to any other genus. We shall begin with the

Ichthyosaurus.

It is to Sir Everard Home that the scientific world owes its first knowledge of a characteristic specimen of this singular genus. He published, in the Philosophical Transactions of 1814, a description of a very well preserved head, and some other bones, deposited in Bullock's Egyptian Museum, in Pall-Mall. They came from the coast of Dorset, between Lyme and Charmouth. They were taken from a rock twenty or thirty feet above the level of the sea.

Sir Everard quickly observed that the shoulder exhibited some relations with that of the crocodile; but the position of the nostrils, the circle of osseous pieces, surrounding the sclerotic tunic of the eye, appeared to him, as also did the vertebræ, to present some approximations to the class of fishes. On this account, M. Kœnig, of the British Museum, conceived for it the name of ichthyosaurus, literally fish-lizard.

Two years after, in the Transactions for 1816, Sir Everard added many details to his first description. Mr. Johnson, a native of Bristol, who for many years had been in the habit of collecting fossils from the cliffs of Lyme, procured for him some specimens from which he deduced the form of the articulation of the ribs, the shoulder-blade and the entire anterior fin, which he then compared to that of sharks, and was more and more inclined to conclude that it was a fish.

But after two years more, some pieces were collected by different individuals, and the attention of Sir Everard was drawn to them by Dr. Buckland. These acquainted him with the nature of the sternum, the clavicle, and the coracoïd bone, as well as with the relations of these parts to those of the ornithorhynchus, which bear no indistinct resemblance to the same parts in the lizard tribe. Sir Everard then abandoned the notion of the ichthyosaurus having been a fish. In the same article he states the probable existence of more than one species.

In 1819, some magnificent specimens—and, among others, an entire skeleton, discovered by Mr. de Labêche, and Colonel Birch, of Lyme—enabled Sir Everard to perfect his description, and particularly to convince himself that the ichthyosaurus had four feet. But a head, in which the nostrils were closed up, led him into an error, and made him think, wrongly, that what he had hitherto taken for these apertures was simply the effect of accident. In the Transactions of the same year, the author, in consequence of the resemblance of the concave faces of the fossil vertebræ with those of the proteus, the siren, and the axolotl, proposed for his animal the name of *proteosaurus*.

Lastly, in 1820, the indefatigable researches of Colonel Birch furnished Sir Everard with materials, from which he determined the composition of the vertebræ, and the mode in which the annular part is articulated with the body, and likewise the singular structure of the fins.

This series of articles and notices renders it impossible to avoid acknowledging that the entire honour of having completely acquainted naturalists with this extraordinary genus is owing to Sir Everard Home. Messrs. Conybeare and Labêche, however, must come in for their share of acknowledgment, having added many interesting particulars and extensive details to what this scientific anatomist had already advanced upon the subject.

In a memoir inserted, in 1821, in the Transactions of the Geological Society, they published their grand discovery of a new genus of the same tribe, but more approaching to the common lizards, which they named *plesiosaurus*. They have there described the composition of the lower jaw of the ichthyosaurus, that of the muzzle, and a considerable part of that of the posterior and inferior faces of the cranium. They showed that the ring of osseous pieces in the sclerotic is a character of lizard, and not of fish, and entered into new details on the vertebræ and articulation of the ribs.

A second memoir, by the same gentlemen, in the Geological Transactions for 1823, while it extended the description of the plesiosaurus, served to determine the notions concerning the teeth of the ichthyosaurus, clearly expressed the characters of its species, re-established the truth relative to the position of its nostrils, and marked the relations and differences of structure between its head and that of the lizards.

With materials so abundant, and given with so much care and accuracy, it was possible to compose an osteological description of the ichthyosaurus at least as complete as that of any lost animal. But M. Cuvier came to the task with additional advantages. Many drawings were sent to him by different lovers of science. From Mr. Cumberland, of Bristol, he received a sketch of an entire skeleton, four feet ten inches in length, found in 1818, near the sea, at Watchet, in Somersetshire, and belonging to Mr. Morgan, of Bristol: that of a head, and teeth of several species, from the collection of Mr. Johnson, of Bristol: that of many pieces, and among others, of an eye, found at Weston, near Bath, and preserved by Mr P. Hawker. He also had the good fortune to acquire certain valuable pieces which supplied him with the means of furnishing some additional characters to those which had been recognized by his predecessors. It remained for him to show the forms of the frontal and its accessory bones, the foramen of the parietal, like that of lizards, and the sphenoïd also much more like that of lizards than it had appeared in the remains previously analyzed.

The ichthyosaurus is, as we have said, peculiarly abundant in its remains in England. Its debris are deposited from the new red sandstone up to the green sand, which is immediately under the chalk. Thus it belongs to almost every epocha of the secondary strata, which continental geologists call the formation of Jura. Fragments of it, in fact, are found in a marl, associated with green sand, at Bensington, and in the calcareous sandstone under the oolite at Shotover Hill—all places in Bedfordshire. There are some also under the oolite at Kimmeridge, in Dorsetshire.

But it is especially in the *lias*, a bluish, grey, marly, and pyritous marble, that its sepulchre is most usually found. This has furnished innumerable remains of it in the counties of Dorset, Somerset, Gloucester, and Leicester, and principally in the valley of Avon, in Somersetshire between Bath and Bristol, and on the coast of Dorsetshire, where the cliffs between Lyme and Charmouth have been inexhaustible deposits of it. The ichthyosauri appear to be as abundant there as the remains of ancient mammifera are in the gypsum of Paris, and their bones are generally surrounded by quantities of small ammonites.

In the same formation, further to the North, many remains of ichthyosaurus are to be found, and M. Cuvier has received many specimens from Newcastle-upon-Tyne.

The bones of the ichthyosaurus are much more rare on the continent of Europe. Some vertebræ, however, were evidently to be distinguished among the groups of crocodiles' bones from Honfleur. From the coast of Calvados M. Cuvier received some fragments; and even from the interior of France not a few very strongly characterized; from Condat, in Agenois, and from Reugny, near Corbigny, in the department of the Nievre. These last were manifestly in the oolite, but the gangue of most of the rest was very similar to the lias.

There was even one very celebrated fragment, the history of which proves with what levity of judgment naturalists, otherwise of ability, have attributed to the human species fossil or petrified bones.

Scheuchzer, walking one day in the environs of Altorf, a town and university in the territory of Nuremberg, with his friend Langhans, went to make researches at the foot of the Gibet. Langhans, who had penetrated into the inclosure, found among the stones a piece of ash-coloured marble, which contained eight dorsal vertebræ, tinted black and of a brilliant appearance. *Seized*, says Scheuchzer *with a panic terror*, Langhans threw this stone over the wall, and Scheuchzer having taken it up, kept two vertebræ, which he considered to be human, and caused to be engraved in his *Piscium Querelæ*. He also tells this story to Bayer, on the occasion of two similar vertebræ, probably from the same place, which he had caused to be represented in his *Oryctographia Norica*, pl. vi. fig. 32, and Bayer had Scheuchzer's letter printed in the supplements to this Oryctography, which contain the sequel of his description of his collections.

These vertebræ, copied by Dargenville in his *Oryctologie*, and quoted by Walch and many other describers of petrifactions, have, until within a comparatively short time, passed, without contradiction, for human bones.

The slightest knowledge of osteology, however, or rather the mere view of a human skeleton, would be sufficient to show that these vertebræ could never have proceeded from man. They might have been taken from those of crocodiles or fish. But since the bones of the ichthyosaurus have been made known, there can be no hesitation in referring them to that genus. M. Cuvier has seen similar ones, from the same place, in the collection of the Grand Duke of Tuscany.

Lastly, there was a skeleton nearly entire, and many debris of ichthyosaurus found at Boll, in Wirtemberg, the same place where crocodiles and other fossils belonging to the secondary formation have been found in great abundance. They are in a calcareous schistus analogous to that of Solenhoffen; and M. George Frederic Jæger, director of the Royal Museum of Stuttgard, has figured and described them in an especial dissertation. This able naturalist has even recognized many specimens taken from the same place during a number of years, and scattered through various collections, where nobody took the trouble of attempting their determination.

The pieces which M. Cuvier has employed in his description of the ichthyosaurus he has figured, and thus describes:—

A skeleton about three feet and a-half in length. There are wanting to the spine only some vertebræ of the end of the tail, which have even left their impression. But there remains very little of the ribs. The head is crushed, but tolerably complete, as are also the two anterior extremities and the left posterior. There are but few fragments of the pelvis. The omoplates, clavicles, and anterior part of the sternum, have disappeared.

Another skeleton, from a larger individual, with the teeth less narrow. The tail and a part of the loins are wanting, likewise the sternum, the omoplates, and the clavicles; but the rest of the anterior extremities are complete, various bones of the head in a good state, many ribs in their entire length, a considerable remnant of the pelvis, and a posterior extremity almost entire. The Baron had also numerous isolated vertebræ, or joined together in series of eight, ten, or more.

For the description of the head his materials were also very complete. He had a head to which nothing was wanting but the anterior end of the muzzle and a part of the occipital and basilary region. The teeth are the same as in the preceding skeleton. This head was described by Sir Everard Home, in the Philosophical Transactions for 1819, pl. xiii. But M. Cuvier disengaged it better from the stone which covered it, and discovered some new peculiarities, especially the nostrils and the foramen of the parietal. A head whose muzzle was still more truncated, but which was valuable from possessing the entire basilary and palatine region, also belonged to the

Baron. There is but a small number of teeth preserved, which are narrow and straight.

Another head, truncated from the front as far as the parietal, and which preserves no teeth ; but the temporal region is very entire, and it also has the hyoïd bone.

Two other heads, flattened vertically, but nearly entire. The teeth are the same as in the first and second skeleton. These heads furnished very satisfactory details concerning the sutures and foramina, and confirmed every thing respecting the hyoïd bone.

An enormous lower jaw, which, though very much truncated at both ends, was yet nearly two feet in length. This was a sufficient indication of the size to which the genus might arrive.

Finally, there were also some isolated bones, especially of the occiput, which proved of no small utility in illustration of this important part.

Nothing was wanting for the shoulder and the entire anterior extremity. There was a fragment of a very large individual, where, with many vertebræ and ribs, were found the sternum, the clavicles, the coracoïds, an omoplate, two humeri, and two fore-arms. From an individual of smaller size, came the sternum in its proper place, the clavicles, the coracoïds, an omoplate, a humerus, a fore-arm, and a bone of the carpus.

The same bones in another specimen, with the fin almost entire.

A more complete piece for all these parts from rather a small individual. The left fin was entire, and attached to its shoulder, which was also complete, as was the sternum and the coracoïd of the opposite shoulder.

The Baron declares that he was not so fortunate in regard to the posterior extremity. The pelvis was mutilated in his two skeletons, and he did not meet with one separately. His opinion was that these parts were more weak, less adherent, and more easily detached after death.

We must now enter on a description of the ichthyosaurus, beginning with the teeth.

The teeth of all the ichthyosauri are conical, and their crown is enamelled, and striated longitudinally as in the crocodiles. It is more or less sharp, more or less swelled, more or less compressed, according to the species. The root is more bulky, not enamelled, but it is striated like the crown.

These teeth remain a long time hollow internally. They are not enchased in alveoli as deep and close as those of the crocodile, nor are they naked on the internal side like those of the lizards. And it appears that they are simply ranged in a deep furrow of the maxillary bone, the bottom of which alone is hollowed with fosses correspondiug to each tooth.

Their mode of replacement is analogous to that of the crocodile, with this difference, that in the crocodile whose teeth are always hollow, the new tooth penetrates into the interior of the old ; while in the ichthyosaurus, the root being ossified, the new tooth penetrates only into the cavity formed by the caries, a cavity which augments in proportion as the new tooth increases in size, and which, finally causing the root to disappear, determines the fall of the crown of the old tooth.

This crown of the tooth preserves still in its anterior a cavity, usually filled with spath, long after the root is ossified. The new root begins to ossify even before the old tooth has fallen.

The number of teeth is considerable. Mr. Conybeare does not reckon less than thirty on each side in each jaw. Sir Everard Home has pointed out forty-five on each side in each jaw, in the individual figured in the " Philosophical Transactions of 1820."

Messrs Conybeare and De Labêche have found sufficient differences between these teeth, to deduce from them the characters of four distinct species.

The *Ichthyosaurus communis*, whose teeth have a conical crown, moderately sharp, slightly hooked, and deeply striated. This species is, in general, large, and to it the most gigantic individuals belonged.

The *I. platyodon*, in which this crown is compressed, and presents on each side a trenchant ridge. The individuals of this species vary in length from five to fifteen feet.

The *I. tenuirostris* has the teeth more narrow, and the muzzle longer and more slender.

Lastly, the *I. intermedius* has the teeth more sharp, and less profoundly striated than those of *I. communis*, less narrow than in *tenuirostris*.

The last two species do not attain to more than half the size to which *I. communis* may arrive.

As these species do not differ in the rest of their conformation but by slight variations in the proportions of the bones, but not in the composition of the parts, we shall first give a generic description of the head as if it belonged but to one species, noticing the differences after.

The elongated and pointed muzzle of the ichthyosaurus is formed principally by two bones furnished with teeth, and which, in all its anterior moiety, are united to each other, above and below, by a suture, but are separated in the upper moiety by two other bones, which advance between them in a point.

On each side of this same upper half is seen a narrow bone, in which the series of the teeth is continued, and which is prolonged behind as far as under the anterior angle of the orbit.

There are no nostrils at the point of the muzzle; their apertures are two oblong holes, in the summit of the intermaxillaries; from their anterior edge they form an emargination in the upper edge of the intermaxillaries.

The upper and internal edge of the nostrils is made by the nasal bone, which widens to arrive there, forming a notched suture, with the base of the intermaxillary.

To the upper edge, that is, nearest the orbit, two bones concur, or, at least, approach, which from their lower part also concur to the formation of the circle of the orbit; and even the upper one extends over a good part of the superciliary arcade. This appeared to M. Cuvier to be the anterior frontal. The other, which is smaller, he thought might be the lachrymal, but could discern no lachrymal foramen.

The two nasals re-ascend between the anterior frontals as far as the principal frontals, to which they are articulated by a notched suture, which varies in direction according to the species.

The principal frontals are placed, as usual, on the middle of the interval of the orbits, but it does not appear that they came as far as their upper edge.

The posterior frontals proceed along the superciliary arch, and the external edge of the principal frontals, to join the anterior. They also form all this posterior edge of the orbit in descending to join the jugal bone. This last is slender. It is placed obliquely on the maxillary to form all the lower edge of the orbit, and remounts a little behind to join the posterior frontal, with which it closes the frame of the orbit. But this ascending part is not considerable, so that the emargination exhibited by the zygoma underneath, in the lizards, is less in the ichthyosaurus.

A broad bone behind the orbit is peculiar to the ichthyosaurus, and distinguishes it from the lizards. This bone articulates with the posterior edge of the posterior frontal, and of the jugal, and proceeds from its other extremity to take part in the articular face which bears the lower jaw. The bone which joins the rest of this articular face is placed more within than the preceding, and suspended to the mastoïd and the lateral occipital. These two bones, M. Cuvier thinks, are the temporal and tympanic.

The temporal, in its form, much resembles that of the lizards, only that it is articulated by a higher line to the posterior frontal and the jugal bones. But its peculiar character is its descending like that of the crocodile, as far as the articulation; but, though articulating with the posterior frontal, it does not leave, as in the crocodile, a second temporal foss behind the orbit.

The temporal of the sea-turtle has much analogy with this, both in form and connexions. But in that animal, the mastoïd and posterior frontal unite to the parietal above the temporal, to form a vault at the temple, while here, on the contrary, a great vacancy remains, as in the lizards. There is one, also, of variable dimensions in the crocodiles.

The mastoïd is articulated on one part to the posterior frontal and temporal, and, on the other, to the posterior lateral apophysis of the parietal.

In the lizards, in which it is very small, it is articulated only to the parietal and temporal bones, because the last is interposed between it and the anterior frontal. In the crocodiles it is articulated only to the anterior frontal and parietal, because the tympanic is interposed between it and the temporal. But in all those genera it contributes to bear the tympanic bone, and it does the same in the ichthyosaurus.

The parietal bone of the ichthyosaurus perfectly resembles that of an iguana. The temporal crests approach it like the two branches of an *x*. Behind it bifurcates into two apophyses, each elevated by a crest, which proceed to the mastoïd bones, and attach themselves there to form by their union the posterior angle of the temporal foss. On the suture of its junction with the principal frontals, it is emarginated by a large foramen, which, in certain species, is prolonged into a fissure, over almost its entire length.

The upper occipital is very like that of an iguana, in the general form, in the large emargination which it has in the lower part for the occipital foramen, in the small one, which is sometimes in the top, for the ligament which united it to the parietal, and in the rough faces which it presents to the lateral occipitals, and to the ossa petrosa. Its external face only is of a more equal convexity. There are two foramina for the vessels, differently situated according to the species.

The lower occipital is very thick, and terminated behind by a very gross tubercle, which serves nearly alone for the articulation with the atlas, for the lateral occipitals scarcely have any sensible share in this. Its inferior face is convex, not concave as in the lizards, nor has it the lateral apophyses which theirs possesses, so that it takes no part in the parietes of the external ear, or of the cavity. This is a marked relation with the tortoises.

Another still more sensible relation is the division of the lateral occipitals. These bones, articulating with the upper and lower occipitals, have, externally and above them, a void, and present conjointly with the upper occipital an indented edge, announcing a suture, which cannot be filled but by an external occipital analogous to that of the tortoise, and which occupies the space which the os petrosum leaves behind it.

The sphenoïd is as thick as the lower occipital. A transverse crest of its upper face distinguishes the cerebral region from that in which the pituitary gland reposes. The last is pierced at the bottom by a canal which goes obliquely back, and goes out at the lower face of the bone by one or two holes, according to the species. In front, the sphenoïd gives out a very long point, to support, as in the lizards, the vertical and membranous partition between the orbits. Laterally, it presents on each side a truncated apophysis, to touch the pterygoïd bone, and a little further back, a rough face for its juncture with the os petrosum. In all these relations it resembles exactly the sphenoïd of a lizard. The os petrosum is articulated, relatively to the sphenoïd, in such a manner that the pterygoïd must proceed nearly parallel to the external face of the former, and that if they do not actually touch, as in the tortoises, at least very little space must be left between them. The external face of the petrosum is simple, and a little convex, like that of the lower occipital, and has not the crest, which, in the iguana, protects the concavity, at the bottom of which is the fenestra ovalis—an additional reason for believing that the external ear was more simple in the ichthyosaurus than in the lizards.

At the internal face of the petrosum is a deep cavity which concealed the vestibulum, and which was closed on the side of the cranium by the upper and external occipitals. In its parietes some remains of semi-circular canals were visible.

From the mutilation of the lateral occipital and the os petrosum in the Baron's specimens, he was unable precisely to judge whether there had been two fenestræ, or only one.

He inclines, however, to the latter opinion; and also thinks that the auricular osselet was reduced to a simple platine, corresponding to that of the stapes, as it is in the salamander, the siren, and the proteus.

The pterygoïds form two bands, long, broad and flat, separated behind by the entire breadth of the sphenoïd, and approaching each other in front, a little under the anterior edge of the orbit. They sharpen there into a long point, which is inserted between the palatines, to which it is united by a very oblique suture. They grow wide, laterally towards their middle, probably to give an attachment to the transverse bone, which would appear to join the external posterior extremity of the palatine, and the external edge of which should, in that case, unite, as usual, to the jugal. The posterior extremity of the pterygoïds is terminated by a slight enlargement, somewhat in a concave arch behind. Neither they nor the palatines appear to have had any teeth.

Such is the general composition of the head of the ichthyosaurus—a muzzle formed almost entirely by the intermaxillaries; the maxillaries thrown to the sides of its base, the nasals to the upper face of this same base, the nostrils pierced between the nasals, the intermaxillaries, and the anterior frontals; the frontal, the parietal, the occipital, the ossa petrosa, the sphenoïd, the pterygoïds, pretty nearly the same as in the lizards, especially as in the iguanas.

But in the region of the ear and temple are some characters

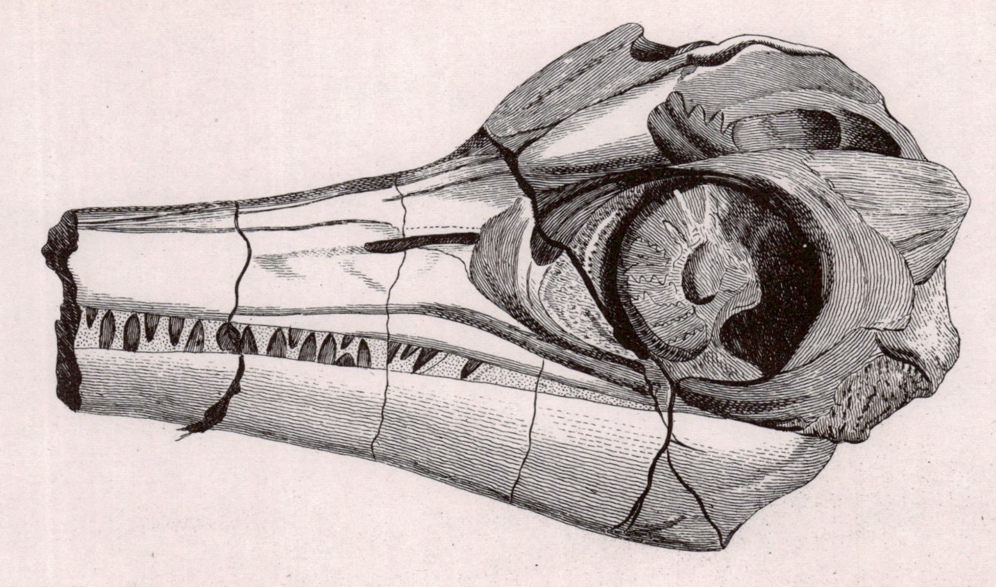

Part of Head of Ichthyosaurus in the British Museum from Lyme.

of greater peculiarity, namely, an orbit surrounded by the anterior frontal, the posterior, and the jugal—the temporal united to the tympanic bone, placed at its internal face to furnish the articulation of the lower jaw; the region of the cranium where the concavities for the auricular chamber should be, smooth, and even a little convex, and probably no other auricular osselet than the platina of the stapes. The osteology of the head of any existing animal could scarcely be better known.

The most striking character in this head is the enormous size of the eye, and the circle of osseous pieces which strengthen the sclerotic in front. These pieces form a character common to birds, tortoises, and lizards, to the exclusion of crocodiles and fish. In fact, in the crocodiles, the sclerotic is simply cartilaginous. In the fish it is often osseous, either in the whole or in part, but never furnished in front with a ring of osseous pieces, as in the birds. This simple character alone is sufficient to approximate this animal to the lizards.

The total form of the head, and that of many bones, taken separately, announced, as well as the teeth, specific differences in the ichthyosaurus, but the limits of which are not so easy to determine.

In the head, for instance, which the Baron refers to *I. communis*, are seen at the root of the nose between the orbits, two angular prominences in front, and between which there is an angular concavity behind. The parietal is more long than broad, and pierced with two oblong holes, one in front, and the other behind, and its temporal crests unite in a single line.

In another and a smaller head, the parietal is nearly alike, but pierced with only a single hole. Nevertheless, as the teeth are similar, this can only be an accidental difference.

In another head, on the contrary, the parietal is of equal breadth and length, and altogether flatted behind. There is but a single round hole in front. There are no teeth in this head; but from the resemblance of the parietal and the sphenoïd to those of a little skeleton which the Baron had already referred to *I. tenuirostris,* he is inclined to refer this head to the same species.

In a fourth head the parietal was flat and short, but the round hole in front is continued by a fissure which is widened behind into a second hole. The Baron refers this to *I. intermedius.*

Lastly, in the hinder portion of a head, as large as the first-mentioned, the parietal is equally elongated, and without a hole. This M. Cuvier refers to *I. platyodon.*

There are other differences observable in the sphenoïd, but it is needless to insist upon them here.

The lower jaw of the ichthyosaurus, equally elongated and pointed at the muzzle, is formed of two branches, which approach each other without curving much, and which are symphysized on a little more than half of their length. Each branch is composed of six bones, as in all the lizards and crocodiles, but somewhat differently disposed than in either of these families.

Neither at the external nor internal face of the jaw are observable the two large holes which are seen in the crocodiles.

The dentary bone forms its external face from the point as far as under the middle of the orbit.

The opercular occupies the lower edge and the internal face of the jaw almost on an equal space, penetrating into the symphysis almost to its point.

The angular and subangular share the external face at the posterior part.

To the sub-angular belongs the coronoïd apophysis, contrary to the arrangement in the lizards, and conformably to what is seen in the crocodiles. This coronoïd apophysis is small, and very obtuse.

The complementary bone is very small, and thrown to the external face of the jaw, as in the crocodile.

The articular is not considerable, and its greatest extent is at the external face, as in the crocodile.

M. Cuvier has seen in three individuals the two anterior horns of the hyoïd bone, otherwise the styloïd bones, in their place, large, prismatic, and as osseous as any of the other bones. He has even observed between them an osseous disk of greater breadth than length, emarginated behind, which he suspected to be the body of the hyoïd. Having seen nothing to announce the existence of bronchial arches, he considers that this animal respired the elastic air, and had neither gills like a fish, nor any bronchial apparatus like the siren or the axolotl. He could, however, discover nothing which appeared a remain of larynx, or windpipe.

The number of vertebræ in the ichthyosaurus is considerable. Mr. Conybeare rates it at between eighty and ninety. M. Cuvier possesses an individual which could not have had less than ninety-five. In the fine specimen of Sir Everard Home, the vertebræ amount to seventy-two, at least.

As much as the ichthyosaurus resembles the lizards in the forms of its osseous head, so much does it differ from them in the conformation of its vertebræ, and in this respect decidedly approaches the fishes and cetacea, as Sir Everard has well remarked.

It has not the atlas and axis differently formed, but all the vertebræ are nearly alike, as in fish. Their bodies are shaped like the pieces of a draught-board; that is, the diameter is greater than the axis, sometimes even double or treble. Both faces of their bodies are concave, just as in fishes.

The annular portion is attached to them on one part and the other, by a face somewhat rough, which takes the whole length of each side of the medullary canal. The adherence must have been weak, for in most cases this annular portion is gone. It was raised above in a compressed spinous apophysis, which, in the commencement of the spine, is pretty nearly the height of the body. These apophyses, placed obliquely, and almost as broad as the bodies, formed, on this part of the spine, almost a continuous crest. That of one vertebra rests its base behind, on that of the vertebra which follows, and for this purpose each of these apophyses has an horizontal prominence in front which passes under that which precedes it. This arrangement holds the place of articular apophyses.

The annular parts grow narrower towards the tail; their spinous apophyses diminish in all directions, and also their articular laminæ.

There are no true transverse apophyses, but in a certain number of these vertebræ, the body has on each side two tubercles, nearer its posterior than its anterior edge. The most elevated is contiguous to the annular part, and convex. It is the only vestige of transverse apophysis which is visible. It serves for the articulation of the tubercle of the rib. The other is a little lower and slightly concave. It receives the head of the rib. According to the observation of Mr. Conybeare, verified by M. Cuvier, this disposition of the lateral tubercles continues, from the first, as far as the seventeenth or eighteenth. Afterwards, the upper tubercle ceases to be contiguous to the annular part, and approaches, by degrees, to the lower one. It has been found still convex, though very much lowered, as far as the thirty-fourth vertebra.

These tubercles have been found distinct as far as the forty-third vertebra, and very near the pelvis. They are then small, and both concave. But here specific and even individual variety must be taken into consideration; for Mr. Conybeare found them reduced to a single one at the fortieth vertebra in one of his specimens.

After the pelvis, the caudal vertebræ have, on each side, but a single small and concave tubercle, nearly approached to the suture of the annular part. They narrow, by degrees, to the end of the tail, which terminates in a point.

In entire individuals we may be assured that the tail is shorter than the trunk by almost one-fourth of the length of the trunk, and that the head is nearly one-fourth of the total length. These proportions are taken on individuals of small dimensions.

The forms of the vertebræ, like those of the head, announce different species.

In individuals of the middle or small size, the length of the body from front to back, in the vertebræ of the trunk, is nearly one-half of its transverse diameter, but there are many much thinner.

Mr. Conybeare has represented some, the length of which does not make one-third, and scarcely a fourth, of the transverse diameter. Their absolute size also differs much. One, in the Baron's possession, was five inches and a-half in transverse diameter. Comparing them with those of an individual four feet long, they indicate one of at least six-and-twenty.

The ribs of the ichthyosaurus are very slender, for so large an animal, not compressed, but rather triangular. Almost all of them are bifurcated in the top, and attached to their vertebræ by a head and a tuberosity, which is rather a second pedicle or stem, than a second head. These existed, as in the lizards, without exception, to all the vertebræ, from the head to the pelvis, for the costal tubercles to the vertebræ are visible the whole length of the trunk.

It is possible that the cervical and lumbar ribs were short, but those of the greatest part of the trunk were large enough to take in nearly its semicircumference. Their mode of union underneath, whether to the sternum, to each other, or their correspondents, has not been ascertained.

The shoulder and sternum of the ichthyosaurus are arranged, in all essential points, like those of the lizards. The osseous sternum is composed of an odd stem, which in front has a lateral cross-piece, like a large T, and which, consequently, essentially resembles its analogue in the monitor and the ornithorhyncus. To the branches of this T are attached, by a suture, two clavicles, arched and tolerably strong.

Behind this T, and partly above its odd stem, is the line of meeting of the two great coracoïds, cut a little fan-like, very broad at the middle line, and a little narrowed towards their external part, where they proceed to unite to the omoplates.

The omoplate is also a little dilated, like a fan, towards the place where it unites to the coracoïd. It grows narrow, and curves to re-ascend towards the back, and it has at its anterior edge a prominence to support the extremity of the clavicle.

In the foss, which the omoplate and coracoïd form by their union, is articulated a humerus, gross and short, swelled and rounded at its upper head, a little more slender in the middle, and, finally, flatted and dilated to support the bones of the fore-arm. These two bones are broad, flat, and united together, and with those which follow, so as to enter truly into the composition of the fin, or paddle. Thus, many anatomists have failed to recognize them, and have believed that the fore-arm was wanting in the ichthyosauri. This is not so, but it actually appears to form the first rank of the carpus.

The second rank, or the first of the true carpus, is formed of three bones, and is succeeded by two of four each, all flat, angular, and joined in a sort of pavement-like arrangement. Something like this is observable in the salamanders, and still more in the dolphins, but less complicated.

The rest of the paddle is formed by series of osselets, or little bones, which may be compared to the phalanges of the dolphin, but are still more numerous and crowded. Five or six of these series predominate for the entire length of the paddle, only becoming a little unequal towards the end to form the point; and a sixth or seventh of rounder and smaller osselets, prevails along a part of the anterior edge.

In the complete paddles twenty of these little bones may be distinctly reckoned in each series; and some still smaller and in disorder remain towards the extremity.

All these bones are flat, and their angles are adjusted quite like a pavement, so that they must have formed, as in the cetacea, a paddle whose parts had very little motion one over the other, and presented no external visible division.

Neither the researches of M. Cuvier nor those of his predecessors have been so fortunate respecting the pelvis and posterior extremity, as respecting those just described. It would appear, as we said before, that, in general, the hinder limb must have been feebler, and less strongly attached than the anterior, since it is so frequently wanting altogether, or sadly mutilated.

Two bones of the pelvis, but a little mutilated, were found in one of M. Cuvier's skeletons. One bone was slender, growing flat in front: the other more gross, triangular below, and a little more compressed at top. Articulated together by their two extremities, they intercept a hole of an elongated and elliptical form. M. Cuvier suspects then to be the pubis and ischion. Their posterior extremity is truncated and rough, and it concurred to the formation of the cotyloïd foss, probably with one of the bones of the ilia, which is lost, but the remains of which seem to have been found in another skeleton.

The femur is smaller and shorter than the humerus, but resembles it a little in form, being in the same manner triangular above and compressed below.

On its inferior edge it supports the two bones, the tibia and fibula, which, like those of the fore-arm, are flatted and almost confounded with the rest of the paddle.

After them comes a rank of three bones, then one of five, and five ranges of bones which grow more and more narrow in proportion as they approach the point of the paddle. The number of these little bones is not exactly determined, but it does not appear to be less than in the anterior paddle, and the arrangement is the same.

Thus we possess the skeleton of the ichthyosaurus in all its parts, and, if we except the form of the scales, and the shades of the colours, nothing is wanting to the complete representation of the animal.

It was a reptile with moderate tail, and a long pointed muzzle, armed with sharp teeth. Two eyes, of enormous bulk, must have given to its head an aspect altogether extraordinary, and facilitated its vision during the night, It had, probably, no external ear, and the skin passed over the tympanic bone, as in the cameleon, the salamander, or the pipa, without even becoming any thinner.

It respired the air directly, and not through the watery medium, like fishes, and was therefore often obliged to rise to the surface to inhale it. Still, its short, flat, and undivided limbs could permit it only to swim, and there is no appearance that it could have crawled on the shore, even as well as the seals. Had it the misfortune to be wrecked there, it must have remained immovable, like the whales and the dolphins. It existed in a sea where the mollusca, which have left ammonites behind them, inhabited, and which, according to all appearances, were species of sepia or pulps, which contained in their interior (like the *nautilus spirula* of the present day) those spiral and singularly chambered shells. Terebratulæ, and various species of oysters, also abounded in this sea, and many kinds of crocodiles frequented its shores, if even they did not inhabit it conjointly with the ichthyosauri.

We may assign with precision, at least in the species with slender muzzle (*I. tenuirostris*), the proportions of the parts. In a total length of three feet and a-half, the head and tail each take up a foot, and there remain a foot and a-half for the trunk, at the two extremities of which are the fins, for one can scarcely say that there has been a neck. The anterior fin, or paddle, was seven inches and a half long, with a breadth of nearly three inches. The hinder paddle was a little less both in breadth and length.

The head of *I. communis*, possessed by the Baron, must have been at least two feet and a-half in length: therefore, it announces an individual nine feet long, or thereabouts. A skeleton, discovered on the coast of Dorsetshire by Miss Mary Anning, has, however, been referred to this species, though only five feet long. In fact, the size may vary very considerably in

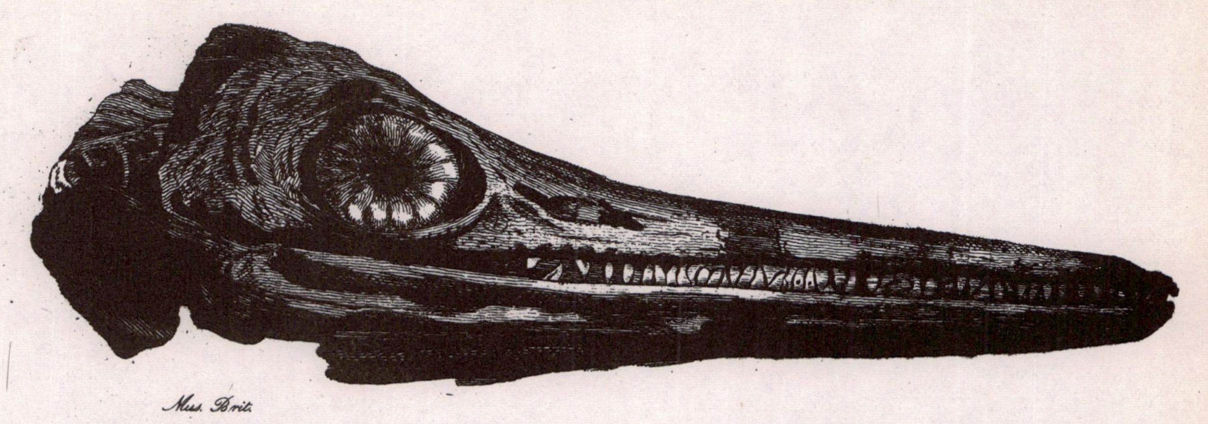

Ichthyosaurus tenuirostris.

reptiles, without the teeth affording us any indication of the age.

There are, however, ichthyosauri of much larger size, especially in the species *platyodon*, being twenty feet and upwards. A cranium in possession of Mr. Johnson, of Bristol, measured in breadth, behind, two feet six, and its longitudinal diameter was fourteen inches. M. Cuvier has vertebræ six inches in diameter, which he refers to individuals of at least one-and-twenty feet in length. One found near Bath, in the oolite, measured nearly seven inches, and many portions of fins from Newcastle announced individuals of very large size.

On the whole, the ichthyosaurus did not fall far short, in size, of the *masasaurus* of Maestricht, already described, whose length has been calculated at five-and-twenty feet.

M. Gotthelf de Fischer has described, under the name of tooth of ichthyosaurus, a conical tooth, found on the banks of the Occa, nineteen inches long, and seven broad at its base, hollowed with a conical cavity about seven inches in depth. This, indeed, would indicate a reptile of most enormous dimensions, but M. Cuvier thinks, with great probability, that it is only the tusk of an elephant.

The Plesiosaurus.

This genus is also entirely English, and altogether due to the sagacity of Mr. Conybeare. Some vertebræ mixed with those of the crocodile and the ichthyosaurus in the lias of the neighbourhood of Bristol, appeared to him to differ from those of both these genera. A considerable portion of a skeleton, in the collection of Colonel Birch, confirmed him in his notions concerning the species from which these debris proceeded. He added, to complete them, some bones of the extremities found with these vertebræ, and thus was enabled to publish, in 1821, the characters of the new animal, in a memoir, conjointly with

Mr. de Labêche, which was inserted in the fifth volume, first series, of the Geological Society.

Still the head was wanting; but having continued his researches with Mr. de Labêche, and profiting by the acquisitions of Colonel Birch, Mr. Conybeare in the following year was enabled to describe a tolerably entire head, though a little crushed, and a large under jaw, which he considered referable to this species. He also added many other bones. (See Geol. Trans. vol. i. second series.)

Lastly, in 1824, in the month of January, a skeleton, almost entire, was found, by the before-named Miss Anning, at Lyme Regis. This confirmed or rectified the conjectures which Mr. Conybeare had formed on the parts which he had examined. But he also learned from it a particularity altogether new, and of which he had not entertained the slightest suspicion. This was, that the neck of this animal had been of a most disproportionate length, and composed of many môre vertebræ than are seen even in the birds, which have the most, and particularly those in the swan, which, in this respect, surpasses every other animal.

This astonishing specimen was purchased by the Duke of Buckingham, and placed at the disposal of the Geological Society.

Of all the inhabitants of the ancient world, this animal is, perhaps, the most anomalous, and the most deserving the name of monster, if we could, indeed, dare to characterize any of the specific productions of Nature by such an appellation. The name *Plesiosaurus*, given by Mr. Conybeare, means akin or approximating to lizards, because he conceived that it more resembled this genus than the ichthyosaurus.

After these notices of Mr. Conybeare, the Baron examined anew many vertebræ, and some other bones from Honfleur, to which he had before turned his attention, and which he had proposed to describe as belonging to an unknown saurian reptile. These, he was now convinced, were debris of the plesiosaurus. Hence there is no doubt but that this animal existed also on the French side of the channel, and was accompanied, as in England, by the ichthyosaurus and crocodiles of various kinds. Even from the interior of France, from the neighbourhood of Auxonne, in the department of the Côte d'Or, M. Cuvier procured similar relics. There are some also found in the interior of this country, at a great distance from Lyme, for M. Brogniart obtained some fragments at Newcastle-upon-Tyne.

The magnificent specimen from Lyme is composed of many stones, which fit well to each other. The only doubt that can possibly be attached to them relates to the narrowest part near the base of the neck. But even if this neck did not belong to the same individual, it is not less extraordinary by its excessive elongation, and most assuredly belongs to the same species.

The animal lies on its belly, and its length in the state in which it is seen is nine feet six inches, from the end of the muzzle to the extremity of the tail.

The head is in advance a little before the rest, with six vertebræ, in a continued series. Then come four vertebræ, a little displaced; but the series is again renewed, and exhibits eighteen vertebræ in their natural order. There are seen twelve, more or less deranged, some of which may, perhaps, belong to the back. The following six are nearly in their places, and conceal the humero-sternal apparatus under them. Then come two crosswise, and then three considerably detached from their natural position. The rest of the vertebræ of the back, as far as the pelvis, eleven in number, is tolerably continuous, but altogether out of the direction of the spine, and thrown on the left side, which permits us to see the arrangement of the abdominal ribs. The pelvis is also, in a great measure, discovered.

Behind the pelvis, twenty-five vertebræ can be counted, forming the tail, pretty nearly in line, except the sixth and seventh, and still furnished, in a great measure, with their little chevron-formed bones.

The remains of the four limbs are tolerably entire. The anterior on the right side, and the posterior on the left, scarcely want anything to complete their description.

These vertebræ, by which the plesiosaurus was first distinguished, are easily recognized by two small oval fossets, which they all have at their lower face, and by the faces of their bodies, which are very little, if at all, concave, and the middle part of which is even a little convex.

In general, also, and only excepting a part of the cervical vertebræ, their transverse diameter is greater than their axis, though the difference is less than in the ichthyosaurus. Their annular part is articulated with their body by a suture, and is easily detached from it. It has, in almost all, a spinous apophysis rather elevated, and articular apophyses, of which the posterior are higher than the anterior, and rest their facets almost horizontally on the anterior ones of the succeeding vertebræ.

According to the first observation of Mr. Conybeare, it appeared to him that at least forty-six of these vertebræ constituted a part of the neck and back; but it was afterwards found that the number so doing was much greater.

The anterior vertebræ are a little longer than the others. The only lateral inequality which they show on each side, are two fossæ of no great depth, very near each other, placed very low, and which give insertion to the two tubercles of a small cervical rib.

Between these fossæ, and at the lower face, are two small fossets or dimples. These two small holes characterize all the vertebræ of the plesiosaurus, and the cervical as well as the others. In proportion as we proceed to the vertebræ farther back, these fossets are seen to approach to, and be confounded with each other. The portion of the vertebra where they are hollowed, becomes a little salient, assumes a figure vertically more oblong, and remounts by degrees, so as to belong, in part, to the annular portion of the vertebra, and not merely to the body.

The lateral prominence thus changes by little and little into a true transverse apophysis.

In the vertebræ which follow, this apophysis is tolerably large, obliquely directed towards the top, and belongs entirely to the annular part, so that when this part has fallen no trace of apophysis remains in the body of the vertebra.

The vertebræ of the tail are distinguished, as usual, by the small facets which they have underneath, for the chevron bones.

These bones in the plesiosaurus, as in the crocodile, are articulated under the juncture of the two vertebræ, so that there are two facets for each of their branches, and each vertebra has itself four facets, two at its anterior edge, and two at its posterior.

These caudal vertebræ have also two transverse apophyses, which, as in the young crocodiles, are attached by a suture; the impression of which remains visible on the body of the vertebræ, below the suture which unites the annular part to it.

The more we proceed in the examination of the tail, the more we find these apophyses diminish in length and thickness, and the marks left by their sutures diminish in proportion.

These forms of the vertebræ of the plesiosaurus, however peculiar, and notwithstanding the length of their axes, incontestably resemble those of crocodiles, and especially of certain fossil crocodiles, such as those of Caen, and the second found at Honfleur, much more than those of the ichthyosauri, or even of lizards. Mr. Conybeare was, therefore, perfectly justified in considering the plesiosaurus as approaching the crocodiles in many points, at the same time that, in its lineaments, it shows a relation to the ichthyosaurus.

Without the surprising discovery of the skeleton which we have mentioned, it would have been impossible to determine the number of vertebræ which this animal possessed in each of the portions of its spine.

Mr. Conybeare, after his first researches, had calculated that there might have been, in the neck and back, a total of forty-six, and even this much surpasses the number in all known saurians, and even in the ichthyosaurus.

The skeleton of Lyme exhibits in their places, thirty-five evidently cervical, and supporting only small ribs articulated by two tubercles, and terminating in a hatchet-form, like those of the crocodile, in the same part. Then come six, whose little ribs are elongated, and assume by degrees the form of the dorsal ribs. The dorsal and lumbar vertebræ are a little in disorder, so that it is impossible to say if their number be complete or not. One-and-twenty have been counted.

Then come twenty-three caudal vertebræ, and three appear wanting towards the end, which would make them twenty-six. This makes eighty-eight vertebræ in all, and Mr. Conybeare adding two sacral vertebræ, makes ninety.

In front of this series of vertebræ, is, in this skeleton, a head, so small, that taking it as unity, the neck is five times its length, the trunk four times, and the tail three times. Thus the head does not make a thirteenth of the whole length. On examining, too, the state of this trunk, and the length which the vertebræ belonging to it ought to occupy, if they were in line, there is reason to believe that the shoulder and pelvis have been more approximated than in nature, and the ribs a little mixed up, so that the trunk must have been rather longer than it appears.

It is, however, quite certain that, in the living state, the plesiosaurus must have exhibited the true neck of a serpent attached to a trunk whose proportions differed little from those of a common quadruped. The tail from its shortness has little analogy with that of reptiles, and the form of this animal must have been the more extraordinary, inasmuch as its extremities, like those of the ichthyosaurus, were true fins or paddles similar to those of the cetacea.

In the back, or in most part of it, the ribs have but one

head, or at least the number of those which have any besides a tubercle, must have been very small. This head of the rib articulates with the extremity of the transverse apophysis, which is sometimes concave, sometimes convex, though it is not possible to assign the place of the vertebræ which have these separate conformations. In the groups examined by the Baron, near the vertebræ in which the end of the transverse apophysis is convex, ribs were found whose heads were concave, and *vice versâ*.

These ribs, in the greatest part of the back, appear to have been composed, each of two parts, a vertebral and a ventral, and it is judged, from the skeleton of Lyme, that the ventral part of one rib was united to that of the opposite rib, by an intermediate cross-piece. So that each pair of ribs (the sternal, if any excepted) surrounded the abdomen by a complete cincture, and this cincture was composed of five pieces. The cameleons, the marbrès, and the anolis, have also the belly surrounded by complete circles, which would lead us to conjecture that the lungs of the plesiosaurus, like those of these three subgenera, were very much extended, and, perhaps, like them, unless the scales were very thick, it changed the colour of its skin according to the greater or less force of its inspiration.

Mr. Conybeare, in his restored figure of the plesiosaurus, makes the simple ribs, not terminated in the hatchet-form, to commence at the thirty-seventh vertebra. He marks seven on each side which go on increasing in size, but have no ventral part. Then he gives fourteen with this ventral part—then three which want it—and further back he places four lumbar vertebræ without ribs.

The humero-sternal apparatus was, in a great measure, reestablished by Mr. Conybeare, at the time he wrote his first notice of this animal.

What is most remarkable here, is the coracoïd bone, which is more dilated into a fan-like form than in any other saurian, so that its dimension, from front to rear, is nearly triple the transverse measurement. Its anterior edge does not appear to have had the emarginations, which are remarked in most of the saurians, and it also wants the hole which is usually seen in the disk.

The omoplate, in the designs of Mr. Conybeare, is long, narrow, elevated by a not very salient crest, and divided transversely into two parts. In front, from one omoplate to another, is a transverse crest, in the form of a crescent, whose convexity, directed hindwards, would unite to the anterior extremities of the two coracoïd bones. This Mr. Conybeare calls the sternum, and gives it no longitudinal apophysis, so that the two coracoïds would unite through almost the totality of their internal edge. In the skeleton of Lyme these parts are concealed, by vertebræ and portions of the ribs, and probably their osteology is not yet completely made out.

This skeleton reveals the pelvis much better. It appears that its ventral part, composed of the pubes and ischia, somewhat resembled that of the land-tortoises, that is to say, the ossa pubis joined each other, and the ossa ischia joined each other by a symphysis, and the posterior extremity of the first joined the anterior extremity of the second, so as to make, on the total, a suture in the form of a cross, and to leave on each side a round hole analogous to the oval foramen in man, and the majority of the mammifera.

In the greater number of reptiles this union of one pair of bones with another does not take place, and the two ovalary foramina unite in the skeleton in one large common aperture.

The pubis appears to have been larger, and especially more broad towards the cotyloïd cavity, than the ischion. This last is wide and fan-shaped.

The ilia, of which only one remains, and is displaced, were narrow, and not voluminous.

The extremities of the plesiosaurus are more elongated than those of the ichthyosaurus, and the hands and feet constitute more pointed fins. The humerus and femur are at first cylindrical, terminated above by a convex head, without neck or tuberosities, flatted and widened below. Nevertheless, the humerus is distinguished from the femur, because it is more flatted towards the bottom, and its external edge forms a more concave curve.

The bones of the fore-arm and those of the leg are short and broad, and almost alike in both limbs. One of the two thickest of these bones is narrowed in the middle; the other is flatted, and its external edge presents the arch of a circle. In the leg this flat bone represents the fibula, and is there a little emarginated at this external edge. In the fore-arm it represents the radius.

We then find some flat and round bones, which represent the carpus and tarsus.

To the carpus there are but four bones in the first rank, one of which, a little outwards, is the *os pisciforme*. There are three in the second. To the tarsus, it appears that there were in all but six, the two largest of which, probably, represent the astragalus, and the calcaneum of the lizards.

All the rest of the fin or paddle is formed by the metatarsians and the phalanges, very obviously disposed in five longitudinal series which represent the five toes; but the phalanges, as in the fin of whales, are in much greater number than usual.

There are, at least, seven in the second and third of the fore-toes, which are the longest; and, at least, ten upon the third toe of the hinder extremity. The absolute numbers of all are, however, difficult to state, because some small phalanx may be lost, especially on the lateral toes. The shortest appears to have been the thumb, which does not seem to have had more than four or five articulations, comprising the metacarpian or metatarsian.

All these little bones are united by synchondrosis, as in the cetacea, rather than by articulations, admitting freedom of motion. They are all a little flatted, truncated, and dilated at the ends, and narrowed in the middle. The last terminate in an obtuse point.

In this individual, which furnished the skeleton of Lyme, the anterior limb, taken from the head of the humerus to the end of the longest toe, was nearly twenty-three inches long. The hinder limb was two feet. Thus they exceeded, a little, one-sixth of the total length.

These forms and proportions were confirmed by a less complete skeleton, discovered at Lyme, by Captain Waring.

The head is that part of the plesiosaurus which is least known.

The muzzle of moderate length, the form of the parietal, the disposition of the bones which surround the orbit, and the temporal foss, exhibit analogies with the iguana. But the teeth adhere in distinct alveoli, as in the crocodile, and Mr. Conybeare thinks that the nostril is near the anterior edge of the orbit, as in the ichthyosaurus.

In a fragment of muzzle presented to M. Cuvier, by Dr. Buckland, there is no trace of nasal aperture. It would appear, therefore, that in these two genera of reptiles, the ichthyosaurus and plesiosaurus, as in the cetacea, to which they approximate in so many other respects, the nostrils were situated towards the summit of the head.

The teeth of the plesiosaurus are slender, pointed, arched a little, and longitudinally channelled. They are unequal. The anterior below, and the posterior above, are longer and thicker than the others. It is not easy to determine the number of the upper teeth, in consequence of the state of the materials, but in entire dentary bones of the lower jaw, on each side, twenty-seven alveoli were distinctly visible. The first six on each side are the largest, and in this part, which makes nearly the third of the length of the bone, the jaw is somewhat swelled out. These dentary bones were far larger than those of the Lyme skeleton, and are referred, by M. Cuvier, to an animal of, at least, thirty feet in length. Many other bones have been found announcing a gigantic size in plesiosaurus, but M. Cuvier thinks it probable that they belonged to a different species from the Lyme skeleton.

In fact, there are several species of plesiosaurus as well as of ichythosaurus. Mr. Conybeare has characterized one by the vertebræ found in the argilla of Kimmeridge. They are much shorter from front to back, than those of the common plesiosaurus, and as flat as draught-pieces, or as the vertebræ of the ichthyosaurus, although their faces are not so concave. They are recognized by their sutures, their facets, and, particularly, by the two small holes of their lower face.

Mr. Conybeare has named the species which furnished the skeleton of Lyme, *plesiosaurus dolichodeirus*, or long-necked plesiosaurus, and that whose vertebræ came from Kimmeridge, *plesiosaurus recentior*.

But other species appear to have existed. M. Cuvier received a cervical vertebra which was found at Boulogne, and apparently in the oolite. It is distinguished by a blunt longitudinal crest in its lower face between the two small holes, and certainly must have come from a different species from the two former. It is provisionally named by M. Cuvier, *plesiosaurus carinatus*.

Some other cervical vertebræ from Honfleur were in the Baron's possession, longer in proportion, and flatter underneath, than their correspondents of Lyme. But the difference is not much to be rested on.

But the Baron considers certain vertebræ of the tail, which he received from Auxois, as belonging to a distinct species. Their body is not cylindrical, but *pentagonal*.

He comes to a similar conclusion relative to a vertebra from the coast of Calvados. It is triangular, like some of those of the animal of Maestricht, that is, flat and broad below, growing more slender towards the top, and presenting on the sides of its lower face, its transverse apophysis.

The species to which these two kinds of vertebræ may be referred, the Baron names *plesiosaurus pentagonus*, and *trigonus*. These appellations, however, he has left open to alteration.

We have seen that the head of the plesiosaurus was remarkably small, being less than a thirteenth of the entire body. In the ichthyosaurus the head is one-fourth. Supposing these animals, therefore, to have come in contact, a thing by no means improbable, as they inhabited the same waters, and from their conformation and analogies were evidently fierce and rapacious reptiles, the ichthyosaurus must have been an overmatch for its antagonist, unless the long and flexible neck of the latter gave it some advantages on the score of activity. The plesiosaurus in its movements, and even, in some degree, in its figure, must have resembled the chelonian reptiles, or sea-turtles. Supposing the turtle to be stripped of its shelly armour, the resemblance would be tolerably exact. There can be no controversy respecting the plesiosaurus having been an aquatic animal, from the nature of its paddles, and that it was marine is equally to be concluded in consequence of the debris by which its remains are invariably accompanied. It is probable that, like the turtle, to whose extremities there is a strong analogy in the plesiosaurus, it may have occasionally visited the coast. Still its mode of loco-motion on *terrâ firmâ* must have been exceedingly awkward. Neither was it by any means so well fitted for swimming as the ichthyosaurus. Its long neck must have presented a considerable impediment to its progress through the watery element. It is the conjecture of Mr. Conybeare, that as it breathed the elastic air, and had frequent need of respiration, it generally swam upon or near the surface of the water, arching back its long neck like the swan, and plunging it downwards at the fishes that passed within its reach. He also thinks that it may have lurked in shallow water near the coast, concealing itself among the weeds. Thus raising its nostrils to the surface, like the cayman, it might have found a

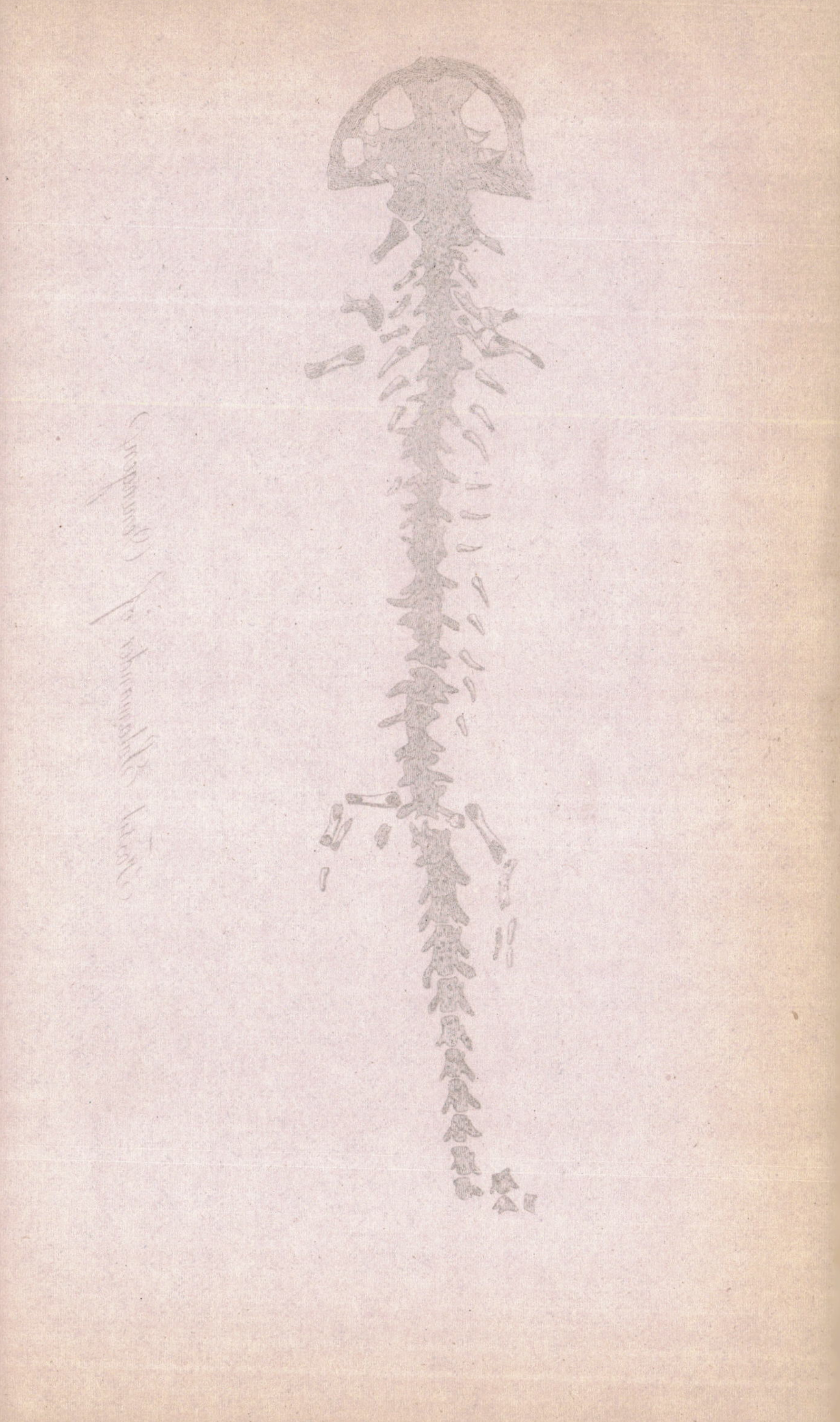

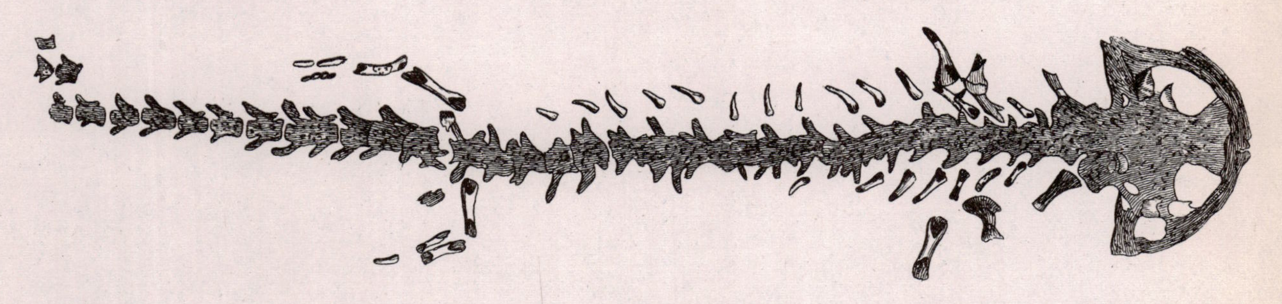

Fossil Salamander of Oeningen.

secure shelter from its enemies, and a place of ambush from which to dart upon its prey. By the suddenness and quickness of its attack, it must have proved a formidable foe to all less powerful animals, and more especially to those of the finny tribe.

Since the above historical details of plesiosauri successively found at Lyme Regis was written, a still finer specimen of this fossil has been found there, which is now (and very properly so) in the British Museum. A description and figure of it is, we understand, likely to be presented to the Geological Society by Mr. Conybeare or Dr. Buckland. We shall not, therefore, attempt to anticipate the observations of these gentlemen by any minute or lengthened account. We may, however, be permitted to state that this plesiosaurus is eleven feet long,—the series of the vertebræ is complete, except ten or twelve at the lower part of its enormously elongated neck,—the lower jaw appears to have slipped from its proper position, thereby, however, exposing to view the interior of the mouth; the sternum, bones of the pelvis, and ribs are in good order; the head is extremely small, not exceeding perhaps three inches in width, but the neck is as long as the body and tail together.

This very perfect and highly interesting specimen of the remains of an ichthyosaurus was discovered in February, 1829, by Miss Anning; nor can we suffer the present opportunity to pass by without bearing testimony to the arduous and zealous exertions of this female fossilist in her laborious and sometimes dangerous pursuit. It is to her almost exclusively that our scientific countrymen, whose names have been already mentioned, owe the materials on which their labours and their fame are grounded, nor, we are persuaded, will they be unwilling to admit that they are indebted for some portion of their merited reputation to the labours of Mary Anning.

We shall terminate this long account of fossil reptiles with the notice of one discovered by Gideon Mantell, Esq. in the sandstone of Tilgate. That gentleman has named this reptile, which seems to have been herbivorous, the *iguanodon*, and has given, in the Philosophical Transactions for 1825, a most interesting article on its teeth and bones. The sandstone of Tilgate is a part of the iron-sand formation, which forms a chain of hills stretching through Sussex in a west-north-west direction from Hastings to Horsham. In various parts of its course, but especially round Tilgate, it contains a quantity of organic remains of various kinds. Mr. Mantell thus enumerates those which he considers to be characteristic of Tilgate Forest. (See Geol. Trans. 1826, vol. ii. part 1, second series, p. 134.)

Stems of vegetables allied to the genus *Cycas*, and, perhaps, Euphorbium.

Leaves of a species of fern.

Plates and bones of turtles.

Teeth and bones of crocodiles, and other saurian animals, of an enormous magnitude.

Bones of birds.

Teeth and scales of fishes.

Teeth of an unknown herbivorous reptile (the *iguanodon*), differing from any hitherto discovered, either in a recent or fossil state. Teeth of an animal of the lacertian tribe, resembling those found at Stonesfield, near Oxford, and figured by Lhwyd.

So great is the difference between the teeth of the crocodile, the megalosaurus, and plesiosaurus, and so much do they differ from those of the other lizard tribes, that it is scarcely possible to commit an error in their identification. But some other teeth were discovered in the summer of 1822, in the sandstone of Tilgate, which, with an obvious indication of herbivorous characters, exhibited other peculiarities of so remarkable a kind, as to arrest the attention of the most superficial observer, and announce something of a very novel and interesting description.

Mr. Mantell made a comparison of these teeth with those

of existing lizards in the Museum of the Royal College of Surgeons. The result of this comparison proved most satisfactory. He found in the iguana teeth decidedly analogous to the fossil, in conformation and structure. He has figured one of these teeth of iguanodon, the largest and most perfect specimen which he could find. The surface of the tooth is worn down obliquely by mastication. The edges are serrated. The fang is broken, and the hollow filled with sandstone. There is a cavity or depression in the base of the fang, occasioned by the absorption produced by the pressure of a secondary tooth.

Like the teeth of the existing iguanas, the crown of the fossil tooth is acuminated. The edges are strongly dentated. The outer surface presents ridges, while the inner one is smooth and convex. These teeth, like most others, appear to have been hollow in the young animal, and to have assumed solidity with advancing age.

From the character of the fossil remains, which more immediately surrounded those relics of the iguanodon, it is concluded that if this animal was amphibious, it was a native of the fresh water, and not of the ocean. Calculating on the proportions of the living animal, and supposing the same relative dimensions in the fossil, as to the teeth, the individual which possessed the tooth we have been describing must have been upwards of sixty feet in length. A similar deduction has been made by Dr. Buckland respecting the size of the iguanodon, from a femur and other bones in the possession of Mr. Mantell.

It would appear, from the researches of Mr. Mantell, that the iguanodon bore on its head a remarkable horny appendage, as large, and similarly formed, as the smaller horn of the rhinoceros. What he discovered of this is, externally, dark brown. Some parts of the surface are smooth, and others furrowed, as if for the passage of vessels. Its structure is osseous, and there is no internal cavity. But it does not appear to have been joined to the skull by a bony process, like some horns of mammiferous animals. The horned Species are by far the most abundant among the existing iguanas. The *cornuta* of St. Domingo is like the common species in magnitude, colours, and general forms; but upon the front of the head, between the eyes and nostrils, are found four large and scaly tubercles. Behind them rises an osseous and conical horn, which is enveloped by a single scale. The fossil horn of which we have been speaking was, beyond all question, a dependency of this description. There were even found upon its surface impressions of the tegument by which, in all probability, it was connected with the cranium.

We shall now close this description of the reptile inhabitants of the ancient world, by a quotation from a book to which the author of this imperfect sketch of fossil remains has been most deeply indebted, and to which he will always be both ready and proud to acknowledge the extent of his obligations. The Baron Cuvier thus expresses himself in the conclusion of his immortal work on the "Ossemens Fossiles:"—

"It will be impossible in future not to recognize as an established truth, the multitude, the magnitude, and the surprising variety of the reptiles which inhabited the seas, or which covered the surface of the globe, at that ancient epocha in which the strata were deposited, commonly designated by the too restricted appellation of the formation of Jura. Also, that they inhabited immense tracts of territory, where not only man had no existence, but where, if there were any of the mammiferous tribes, they were so very rare, that not above a fragment or two can be cited as authentic.

"This variety, this magnitude, and this number, are still further announced, independently of the undetermined pieces of which I have spoken in the article on the megalosaurus, by many of those collected by Mr. Conybeare, and which, at first, he imagined to belong to the plesiosaurus, but which do not find their representations in the skeleton of Lyme.

"In his second memoir, for example, pl. XXI., is seen a portion of a lower jaw, and a bone which appears to me to be one of the bones of the ilia. Also, two other bones, which I imagine to have come from different pelves.

"Time will, in all probability, lead to the complete restoration of these beings, whose existence is already conjectured from the remains in question; and judging from the zeal and ardour with which such researches are now prosecuted in every quarter, we may conclude that the era of their resuscitation is at no very remote distance.

"I have no doubt, but that in proportion as the discoveries already commenced shall be completed, new discoveries will be multiplied, and that, perhaps, in a few years, I shall be obliged to confess, that the work which I this day terminate, and to which I have dedicated so much labour, will appear but a superficial view, a first and hasty glance cast over the immense creations of ancient ages."

From this last conclusion of our illustrious author, given as it is with all the modesty of true greatness, we must be permitted utterly to dissent. At all events, whatever opinion he may himself entertain upon the subject—that of the scientific world must ever remain unaltered. As long as a profound acquaintance with all the relations of organized beings—an unrivalled acuteness of discrimination and comparison—a sober soundness of deduction, united with an expansive and philosophical genius—a union as rare as it is admirable;—as long, in fine, as the most patient and laborious research — the greatest candour, and the most luminous eloquence shall be entitled to the reverential consideration of mankind—so long will the "*Ossemens Fossiles*" remain an imperishable monument to the memory of Cuvier.

THE remains of Fossil Fish are found in the strata anterior to the chalk, in that substance, and in strata which are more recent. These remains consist of bones, of spines, and scales. Sometimes they are converted into a substance either calcareous, siliceous, or pyritous; but for the most part they have not changed their nature.

These fossil remains have belonged to distinct genera, some of whichare new. Our limits will not permit us to enter very largely into their enumeration.

It is not necessary, nor perhaps altogether possible, to explain how these different species of fossils have been formed, in localities of different degrees of antiquity, from the oldest *zootic* strata, to those which are being formed daily under our actual inspection; but it is not difficult to conceive why the fish have left a greater quantity of those buried remains than any other class of vertebrated animals. It is sufficient for this purpose, to recollect that the fish, constantly living in the water, and often in the muddy bottom, can, when they die, be deposited without coming in contact with the air, and, in consequence of their form, which is most frequently extremely flatted, in such a position as is very favourable to their conservation. Their carcasses, carried along by the currents, are deposited in some still water, where the liquid element easily abandons the calcareous molecules which it held in suspension, and which then envelope either the skeleton or the entire fish. Accordingly it is well known that, on the coasts of Iceland, genuine ichthyolites are formed every day in a sort of bluish mud, which hardens by exposure to the air. After this, it is not astonishing to find fossil fish in every species of strata, at whatever depth, or at whatever height, compact or loose, freshwater or marine; but it is more especially in the schistose and calcareous fossil depositions, that they are observed in the greatest abundance, disposed, as it were, like plants in an herbal, and accumulated in the greatest quantity; as we find them, for example, in the strata of Monte Bolca or Vestena Nuova.

The traces which the fishes have left of their existence in the bosom of the earth are very different in their nature. Rarely enough do these consist of actual pieces of their skeleton; that is to say, the true fossils of this class are rare in nature. For the most part they are nothing but impressions: the bones, after having existed in the midst of the substance which has enveloped them, becoming by degrees, in the long end, decomposed, have concluded by disappearing more or less completely. At other times, their impressions may have been filled, so to speak, too late, and the fish appears in relief, or very much compressed; but nothing is seen but its external form, or that of its scales,—nothing of the skeleton, properly so called.

It is rare to find fish in an isolated state, and especially so in the coarse limestone; where, nevertheless, the frequent presence of the calcareous osselets of the ear, proves that some existed there when the waters of the sea washed the strata where such remains are found. We may also conclude, that as in the strata where there is no crystallization or petrifaction, as at Grignon, no skeletons of fish are to be found, and yet calcareous osselets are found, proving the existence of these animals there formerly. Petrifaction has been necessary for the preservation of the skeletons which the petrified strata contain.

Such fish as die naturally must become the food of other fish, or of crustacea; so that we should not be surprised at not finding them very often in the fossil state, in places where we are assured that they must have existed in abundance. They are more usually to be found, as we have seen, in great numbers, in one and the same locality, where they were evidently destroyed by a volcanic eruption, or some other sudden and violent catastrophe. At Monte Bolca, for instance, there can be no sort of doubt that the revolution was sudden, and that the fish must have been covered some instants after their death, by the deposition in which they are found. One of those fossil fish, to be seen in the galleries of the Paris Museum, and which is supposed to be a blochius, had not time before its death to let go another fish, which it was in the act of swallowing.

In some climates, when a fish, and particularly one furnished with an air-bladder, dies in summer, it remains at the bottom of the water for three or four days; then rises to the surface, even before it begins to be offensive, and does not sink to rise no more, but when the parts which constituted it are disunited by putrefaction. Most assuredly if some days had passed between the death of the blochius just mentioned, and its involvement in the crystallization where it was found, it would have mounted to the surface of the water, and would have been separated from the fish which it swallowed, when it was surprised by the catastrophe which destroyed it.

If we had not this example evidently proving the rapidity of this catastrophe, we might mention other fishes found in the same place, in the bodies of which were seen the skeletons of those which they had swallowed. This would prove that they had died suddenly, after having satisfied their appetite.

It is not, therefore, astonishing to find so few fossil fish in the shelly strata which have been formed in the bottom of the sea and without catastrophe; and those which are found there must have been covered, shortly after their death, by a stratum of sand, which concealed them, and hindered them from rising to the surface.

The remains of fish differ in so remarkable a manner, according to their localities, that a person well exercised in this sort of investigation can determine, from the nature of the trace, the place from which an ichthyolite has come. The same observation nearly holds good respecting the substance into which they may have been converted. Thus the bones of fishes are calcareous, siliceous, or pyritous, invariably according to each kind of locality. But, as we said before, their remains, however altered as to form, or, to speak more correctly, however removed from the consistence of bones properly so called, have suffered little change in the nature of their composition.

In a zoological point of view, we may say, that a certain number of these fossil remains have belonged to distinct and new genera; but the majority evidently appertain to all the divisions of the existing ichthyological series. It would appear, however, that the remains of abdominal fishes are generally more abundant than the rest.

Before we speak of the genera and species to be met with in the fossil state, we shall notice the principal localities which have presented them.

The first which we shall remark are the ichthyolites of Glaris. The only place in which they are found is about five or six miles to the south-east of Glaris, in the bottom of a small valley, called the Sernft, nearly three-quarters of a mile above the village of Lengi, in a part of the mountains which border this valley, and which has received the name of Plattenberg. The substance which contains them is a black or blackish schistose fissile rock, containing mica in distinct spangles, and limestone, which presents itself in small beds, parallel to the stratification.

These traces of fish are only parts, more or less complete, of the skeleton. These schists, however, sometimes present the impression of fishes, in scales, fins, and other external forms. Such remains are rare, and are never found, as it would appear, accompanied by shells.

The rock containing the fossil fish forms one or many banks in a nodulose steatite, in which the valley of Sernft is hollowed. This rock is thought to appertain to the transition strata of German geologists.

Mount Pilate is situated in the canton of Lucerne, a little to the centre of Switzerland: it commences to the west of the lake of Lucerne, and extends from north to south almost into the canton of Berne. On the most elevated point of the mountain, and somewhat below it, a considerable quantity of ichthyolites are found in the slate rocks, which detach easily by foliation. In almost all of them a fish may be found. The projecting part is reduced to dust, but the impression is left. A great quantity of teeth are also found here. It appears that Mount Pilate is not at all of the same nature as the strata of Glaris, as to geological structure.

It is especially in the county of Mansfeld, in Thuringia, Voigtland, and the Palatinate, that the most remarkable depôts of a species of ichthyolite are found inclosed in the metalliferous slate. The places where they are more particularly found are: — In the territory of Hesse, Riegelsdorf, Thaliter, &c.; in the Mansfeld county, Rothembourg on the Saale; in Thuringia, Eisleben, Sondershausen, Sangershausen, Kamsdorf, Bottendorf, Saalfeld, Ilmenau, &c.; near Magdebourg, Alvensleben; in the Palatinate, Munster-Appel, and the environs of Kreuznach. They are also found in France, near Autun, in the department of the Haute-Saone, three leagues from that city, in a mountain called *La Muse.*

It appears that the substance of their flesh has penetrated the stone, which replaces it, and has modified the latter.

In some cases, the impression of the fish scarcely occupies any thickness. It is represented by scales, fins, and the head, all flatted. The stones, containing these fish, are divided into two parts, so that this image is found on each of the two pieces.

These fishes, which may have been about three feet in length, are for the most part on the back, or in violent and bent positions, and the head is usually disfigured. The substance in which they are found, according to the agreement of all mineralogists, is a coppery, marly, bituminous schist, sprinkled with argentiferous pyrites, and sometimes with mercury in the state of cinnabar. These schists must be very ancient, since above them are found strata containing ammonites, belemnites, and entrochites.

In the secondary strata, ichthyolites are found. They exist at Grammont, four leagues from Beaune in France, in a hard, grey calcareous stone, which appears to form a part of the ancient limestone, containing gryphites and belemnites.

Fortis has found fossil fish in a fissile calcareous stone, partly forming the high mountain of Pietra Roya, a portion of Mount Matès, in Italy. They are couched flatly, in relief, and their ridges are converted into silex. On cutting the stone, the fish, instead of being divided more or less equally between the two parts, remains attached altogether to one of them. Unluckily, Fortis has not described these fishes.

At Stabia, in Italy, on the borders of the sea, in a place called the *Tower of Orlando*, to the west of Castellamare, are found fossil fish, in a limestone coarsely fissile, fetid, and of a grey, bordering on bluish, which has very great relations with that of the Apennines.

In the chalk in the neighbourhood of Paris, of Beauvais, of Mount St. Pierre, of Maestricht, of Perigeux, and of Gravesend in this country, the remains of fossil fish are found; and it is probable that they exist in many other places where the same strata prevail; but these remains are so badly preserved, that it is difficult to recognize to what genera they may have belonged.

In the coquillaceous coarse limestone, or, as we call it, crag-limestone, remains of fossil fish have been found in the quarries of Nanterre, in those of St. Denys near Paris, and in almost all the strata of the craig-limestone. But these remains are in the same predicament as the last-mentioned.

In Pappenheim, Solenhoffen, Aichsted, Ruppin, and even Anspach, ichthyolites are found. The most remarkable of the quarries of this locality is one which we have mentioned before, situated between Aichsted and Solenhoffen. The remains are usually impressions, or reliefs, of the skeleton, sometimes with a small part of the scaly portion, or of the trunk. The stone in which they are imbedded is calcareous, tolerably hard, of a whitish yellow, and evidently fissile. Authors are not agreed on the degree of its antiquity. These fishes are found with crustacea; and among others, with a species of limula and with asteriæ.

The most celebrated of the localities of ichthyolites is that of Monte Bolca, or Vestena Nuova, on the confines of the Veronese and Vicentine territories. The mountain of Vestena Nuova is volcanic, and raised more than a thousand feet above the limestone quarry. The part containing the fish is rather hard, and backed against high calcareous mountains, with coquillaceous strata; but these are evidently more ancient.

The mountain is composed of two kinds of stone. One is but a very hardened marl, forming thick strata, and not containing, as would appear, any organic remains. The second is a fissile fetid marl, which divides in foliations. The fossils do not appear to be found except in a stratum of two feet in depth; and M. de Blainville is of opinion, that they all appertain to species at present existing in the Mediterranean.

These ichthyolites consist of skeletons, sometimes in a perfect state of preservation, placed, without having undergone any violent disturbance, generally on the side. Bones themselves are to be found there, a little friable. Sometimes there is only a hollowed impression. Scales are rarely seen, but a coloured trace is often observable, indicating the form of the fish. Most of them appear to have been caught, or rather deposited in a perfect state of integrity in the stone, while it was forming; but some are found which appear to have been more or less destroyed previously to their engagement. The geological affinities of this rock do not appear to be very clearly known, and in all probability it owes its existence to the volcano in its neighbourhood.

Other organized bodies are very rarely found with these fish. Shells are extremely scarce, and but one species of large lobster has been found there, and a crab related to *C. Mænas*, Linn., of which M. Desmarest has spoken in his work on fossil crustacea.

Ichthyolites are found in the Vicentine territory, about a hundred paces from the town of Schio, in gross spherical nodules, a little compressed, contained in the great calcareous strata, composed of a greyish stone, mixed with argilla, and quartzose sand. At Monteviale, a league and a half distant from the road to Vicenza, some are also found in a brown, bituminous, argillo-calcareous schist, attaching to a coal-mine, which is worked between heaps of madrepores above and below. Similar fossil fish are also found at Salzeo, twenty miles to the north of Vicenza, at the foot of that part of the Alps which unites itself to the Tyrol, in a fissile, black, pyritous, and fragile schist, eight feet in thickness, under a bluish, foliated schist, hard, or slaty, at the summit of a volcanic mountain.

At Tolmezzo, a small borough of Frioul, are some very small species of fossil fish, in a fissile stone, like that of Vestena Nuova. M. Faujas, who mentions these, has favoured us with no details which could lead to an appreciation of the genera to which they belong.

A stone, containing a fossil fish, was taken from a quarry which is on the declivity of a mountain, six hundred feet above the level of the sea, and at the distance of two miles from it, at Antibes.

In Dalmatia, fossil fish have been found, with marine plants, corallines, and muscles, in a whitish fissile marble, which the inhabitants use for covering their houses, in the gulf of Jukowa, island of Lesina, near a small hamlet called Verbager. They have also been found in the island of Cerigo, in a stone analogous to that of Vestena Nuova.

In Asia, ichthyolites have been found in Mount Libanus, near Gibel, in a calcareous stone, somewhat argillaceous, usually white, but occasionally brown. These ichthyolites are tolerably numerous, and have many relations with those of

Pappenheim. They are, like those, skeletons almost complete, couched flatly, and very seldom with scales.

Le Brun, in his Voyages, tells us, that petrified fishes are found in a mountain of Syria, at some leagues from Tripoli. They are very probably the same as those of Mount Libanus.

The islands of Malta and Sicily contain a considerable quantity of the fossil teeth of fishes, and especially those of squali. Impressions of fish are found at Melliti, near Syracuse, in the Val di Noto.

Shaw, in his Voyage to Barbary, tells us that, in different parts of the coast there, fossil fish are found, but he gives us no details concerning their locality. Barrere, also, in his "Observations on the Origin and Formation of Figured Stones," tells us that, on the coast of Oran, certain limestones are to be found, exactly expressing the figures of fishes.

The ichthyolites of Iceland are in the midst of a sort of marl, or hardened mud, of a bluish colour, usually forming narrow and elongated masses. They are found in the bay of Patriksfiord, where, it is said, this sort of fossils is every day being produced. The entire skeleton, and often the greatest part of the scales, are found in the centre of a kind of nodule, which remains floating, or, at least, not adherent to the main body of the muddy substratum.

The lias at Lyme contains various remains of fish, and especially of what appears to have been defensive radii, as in the balistes and palates.

We insert figures of two of these fish from the "Geological Transactions." The one has been named by Dr. Leach, *Depedium politum*, from having rectangular scales, with a projection on one side, which fits into a groove in the opposite scales. The other fish has not been named.

We have now to speak of ichthyolites of fresh-water formation. At Aix, in Provence, three-quarters of a league from the town, is a plaster quarry thus composed :—1st. a stratum, many feet in thickness, of a schistose, or argillaceous marl,

Dapædium politum. from Lyme.

Geol. Tr.

Fossil Fish from the Lias at Lyme.

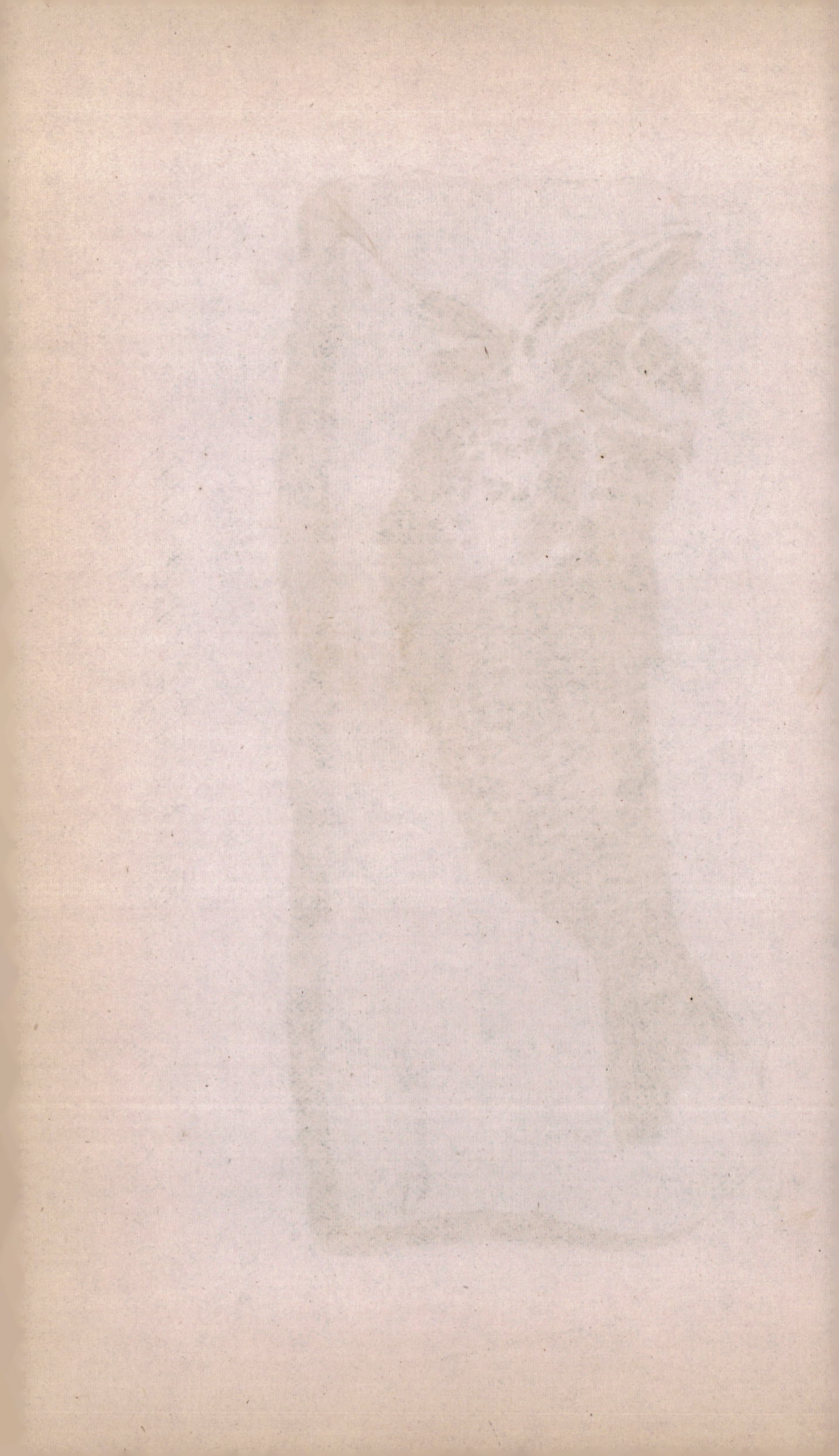

in thin leaves; 2ndly, a white calcareous stone, tolerably compact, and containing argilla; 3rdly, another calcareous stone, pretty hard, also containing argilla; 4thly, a schistose marl, very thick, and highly coloured, containing crystals of gypsum; 5thly, at fifty-six feet in depth, a fissile stone, mixed with limestone, argilla, and a little bitumen, which contains fossil fish, and below this is the gypseous stone.

The remains of fishes belonging to this locality most usually consist in portions more or less considerable of skeletons, regularly disposed in a flat position, pretty nearly like those of Vestena Nuova.

Fossil fish are also found in the plaster quarries in the neighbourhood of Paris, at Montmartre, Mount Valerien, &c., in the fissile marl interposed between the beds of stone and plaster, or in the latter. They are seldom well preserved, and exhibit decided traces of alteration previous to their deposition. They are usually portions of the skeleton, like those of Aix.

In Italy, at Scapezzano, three miles from Sinigaglia, in the marches of Ancona and of Monte Alto, very small fossil fish are found, in an argilla, a little calcareous, bluish, of no great hardness, of an uniform grain, and somewhat fissile. These fishes lie in skeletons, often exhibit scales, and even evident traces of soft parts.

At the promontory of Forcara are found, without order, and, as it were, moulded in an hardened argilla, mixed here and there with true porous lava, fossil fish of different dimensions. Passeri tells us that, about two miles from Pezzaro, in an argilla similar to the preceding, they are also to be found. M. Brocchi also informs us that some of them exist in the Monte Volterrano, which M. Leman believes to be ichthyolites, of fresh-water formation. Faujas also mentions fossil fish found at Alessano, at the very extremity of Italy, opposite Corfu, and which are small, and moulded, as it were, in a very white calcareous sort of clay or mud.

In a multitude of other places ichthyolites have been found.

In the Vivarrais, at a league from Privas, a department of the Ardeche, on one side of a mountain, are found impressions of skeletons, and skeletons themselves, of fossil fish of a single species, in a marly, fissile, greyish earth, so light that it will float in water, and situated below more than two hundred feet of different kinds of lavas, which are surmounted by vast basaltic causeways. Fossil fish are known which come from the environs of Cadiz, and which are contained in a fissile marl, extremely light, a little partaking of steatite. They have even been found in China, according to the testimony of Lebrun; in Bohemia; in Nottingham, and other parts of England, in certain quarries. They have also been found at Saarbruck, in the upper part of the formation of pit-coal, and also in the coal-mines of this country.

At Elve, near Villefranche, a department of the Aveyron, ichthyolites have been found in a limestone containing many fossil shells. The geological affinities of this rock are not known. It is a hard, marly, bituminous limestone, by no means schistose, but very fetid. The place where the fish is found is bluish, and forms a sort of globule surrounded with white.

M. C. Prevost found the head of a fish, in a fossil state, near Villers-sur-Mer, in Normandy, in a bluish marly limestone, which, in its geological position, corresponds to the upper lias of our geologists, and to the middle layers of the limestone of Jura of the French. It is this formation, as we have before had much occasion to notice, which yields such immense quantities of remains of ichthyosaurus. Near Caen, also, in a stratum of the same character, M. de Magneville found an ichthyolite apparently of the same species as one found in our lias by M. de la Bêche, and called by him *Dapedium*.

We are very far indeed from having given a complete enumeration of the different localities in which ichthyolites have been found. To do so would be impossible within the limits to which we are of necessity confined. Imperfect, however, as our sketch has been, it is sufficient to convey to the reader an idea of the immensity and the extension of the remains of this class of vertebrated animals. We must now give a notice of the principal genera and species to which these remains have been supposed to belong, craving indulgence also for the imperfection of our attempts in this way, which must be attributed as well to the actual state of the subject as to our own limited opportunities of observation. The great work on fish, on which Baron Cuvier has been so long employed, will, doubtless, when given to the world, shed much additional light on the fossil as well as the living genera and species of this most important division of the animal kingdom.

In the locality of Glaris the remains found are referred by M. de Blainville to the following genera :—

1. Anenchelum. Species, *A. Glarisianum*. De Blainville. This was, for a long time, referred to the eel kind; but M. de Blainville having examined the remains, which consisted of a part of the head and a large portion of the posterior extremity, did not hesitate to pronounce that this fish had a caudal fin, quite distinct from the anal and dorsal; and, moreover, that the latter, much longer than the two former, had much more expanded radii, and, consequently, much less numerous than in the eel. The vertebræ, also, are much narrower, much longer, and consequently less numerous in a given space. M. de Blainville makes, therefore, a distinct genus of this fish, under the name above cited.

2. Palæorhyncum, another genus provisionally formed by M. de Blainville. The species is *P. Glarisianum*. This was taken to be the *Thornback*, but, as it would seem, erroneously. The remains represent the muzzle very much elongated and pointed.

3. Herring. *Clupœa*. The impression of this was figured by Scheuchzer, in his *Querelæ Piscium*. It indicates a fish whose body was narrow and elongated. The caudal fin is bifurcated, and the dorsal is situated between the pelvian and

anal fins, the latter of which is pretty long, and has from ten to twelve radii. M. de Blainville has named this species *C. Scheuchzeri.* At first he took it for a carp, but finally referred it to Clupæa.

Another species attached by the same author to this genus is that named by him *C. elongata.* This was a fine ichthyolite from the same locality, and figured by Knorr. The impression offers nothing but mere traces of the pectoral fins. The pelvian seem thrown very far back under the abdomen, which is very long. The dorsal fin, on the contrary, appears to have been considerably in front of the pelvian, and therefore not at all opposite to the anal. The tail is terminated by a fin deeply bifurcated, and each bifurcation is rather pointed. The head, though disfigured, appears to have been elongated, and the body very evidently so. The number of vertebræ is forty, and the ribs are slender and very numerous. This ichthyolite has much of the pike form, though M. de Blainville refers it to the herring.

A third species is *C. megaptera,* De Blainville. The body is much shorter in proportion than the preceding. The vertebræ are very numerous, and diminish gradually in length from head to tail. The abdominal cavity is pretty large. The head is much disfigured, and unsatisfactory in its indications. The pectoral fins, very long, and probably narrow, are attached very low. The pelvian are rather small, and situated a little behind the moiety of the trunk. The dorsal has eight or nine radii; the first of which, very short, and the second longer, are simple, while the following five have, at their extremity, a sort of divided ray. The anal fin is very far back, rather low, and formed of eight or ten gradually diminishing radii. The tail is deeply notched, and terminated by a fin of eighteen radii.

4. ZEUS. To this genus are referred *Zeus Regleysianus,* De Blain. An impression evidently formed by the skeleton of a spinous and probably a thoracic fish. It is with doubt, however, that M. de Blainville speaks of the genus of this

ichthyolite, as the head is entirely wanting. There remain of the spinal column twenty crowded vertebræ, with strong spinous apophyses. The tail has about fifteen: the abdominal cavity is small: the pelvian fins have seven or eight radii. The upper fin is single, occupies almost the entire of the back, and is composed of twenty spinous rays. The anal fin occupies nearly the entire space between the anus and the caudal fin. It has three large spinous rays, and about twelve more smaller. The caudal fin is incomplete.

Z. platessa, De Blain. Another ichthyolite, in the collection of M. Brogniart, and probably appertaining to this genus. It is a skeleton indicating an oval, elongated fish, somewhat like *pleuronectes platessa.* The head would appear, though very mutilated, to have been pointed. The vertebral column is composed of twenty-two articulations. The terminal fin is deeply bifurcated: the dorsal has but one spinous ray remaining: there is no trace of the anal. Though referred to Glaris, the locality from which this fossil came is unknown.

Zeus spinosus, De Blain. Another ichthyolite, in possession of M. Brogniart, and apparently of a different species. The head is almost entirely effaced. The vertebræ seem to have been about twenty. The tail is deeply bifurcated, and has about eight radii in each lobe. There are no traces of pectoral fins. A long spinous ray, in nearly the middle of the belly, may have belonged to a pelvian. The dorsal fin seems to have been divided into two portions: the anal fin is extended to within a small distance of the tail. The determination of this species is liable to doubt; and the locality of the fossil, though arranged with those of Glaris, is not known. It is, however, certain, with respect to the eight species now enumerated, that the formation in which all have been found is marine.

Nothing of any certainty is determined respecting the ichthyolites of Mount Pilate; still M. de Blainville does not think that they have any relation with those of Glaris, because the disposition of the fossil remains seems very different; because a great quantity of teeth have been found separate, which is not the case at Glaris; and because Mount Pilate differs entirely from the other locality in geological structure.

The species of fishes in the metalliferous schists of Mansfeld, Thuringia, &c., are not numerous, though the ichthyolites are so abundant. They are referred to the following genera:—

1. PALÆONISCUM. Species, *P. Freieslebense.* De Blain. M. de Blainville thinks that this impression, from Eisleben and Mansfeld, should form a distinct genus, probably approximating to the sturgeons. This fish has the form of a small dolphin in the anterior part of the body and head. It is evidently abdominal. The pelvian fins are very near the anus. There is but a single dorsal fin, very large, intermediate to the central and anal, and raised on a sort of pedicle, like that of the sturgeon's. The tail is short, but very broad, and terminated by a very large fin, the upper lobe of which, longer than the lower, appears to have been furnished with scales. The radii of all these fins are very numerous, fine, and not divided, as is the case with the sturgeons.

2. PALÆOTHRISSUM. Under this generic name, M. de Blainville includes several species of fish found in great abundance in these schists, and hitherto found nowhere else. The essential characters are—abdominal, malacopterygian, or soft finned; a single upper fin situated before the anal, and between it and the pelvian fins; bifurcated tail, the upper lobe of which is usually much longer than the lower, and covered with scales in all its upper half.

P. macrocephalum, De Blain. This species is evidently abdominal; has but a single dorsal fin, intermediate to the ventral and anal. Tail bifurcated, and the two lobes nearly equal. The head is very thick.

P. magnum, De Blain. This ichthyolite is eighteen inches long; in other respects it differs little from the preceding.

P. inæquilobum, De Blain. A small species; the remains of which, more or less complete, indicate a fish about six inches long. The lobe of the tail is very sharp. A fossil of the black schists of Autun.

P. parvum, De Blain. Many relations with *P. magnum*, but the head is much less thick.

3. HERRING (*Clupæa*). *C. Lametherii*, De Blain. The deep bifurcation of the tail of this ichthyolite determined M. de Blainville to refer it to clupæa. The pectoral members are terminated by a fin of eighteen rays. The pelvian are extremely small, and have but five; the dorsal has about fifteen, and the anal ten or twelve.

4. PIKE (*Esox*). *E. Eislebensis.* Another species, also abdominal. The lower fins are disposed as in the preceding, but it is much larger, being sometimes twenty-seven, and never less than eighteen inches long. The dorsal fin, moreover, is exactly opposed to the anal, which constitutes it a pike.

5. STROMATEUS. *S. major*, De Blain. In Hesse, a fossil fish has been found, exhibiting much analogy with *Zeus* or *Chætodon*. It is referred by M. de Blainville to this genus.

S gibbosus is the *rhombus minor* of Scheuchzer, and distinguished by M. de Blainville from the preceding. The form is greatly elevated and compressed ; there are traces of a very large head, an operculum very marked and open, but no traces of pectoral fins. The caudal fin is deeply bifurcated.

S. hexagonus. A fine impression, figured by Knorr. The dorsal line is convex as well as the ventral. The head is moderate.

Many other remains are found in the same localities as the preceding, but nothing determinate has been concluded concerning them.

Among the ichthyolites of the compact limestone come the following species :—

In France, *Elops macropterus*, De Blainville. At Grammont, four leagues from Beaune in Burgundy, was found an ichthyolite, thus named by M. de Blainville. It projects from the surface of the stone. It was a squamous, and probably an abdominal fish. The body is fusiform and thick: the head moderate and well proportioned; the jaws are armed with very fine teeth; the eyes are large and placed high; the pectoral members are remarkable for their length and scythe-like form. This fossil has great affinities with some species of salmon, especially with those called *Elops* by Bloch.

Esox incognitus, De Blain. A specimen in the French Museum, from the same limestone, but the locality is unknown; clearly a fish, abdominal, and pretty long. From the disposition of the dorsal and anal fins, this ichthyolite may be regarded as belonging to the pike genus; still M. de Blainville puts this in the form of a query.

In the chalk formation, near Brussels, among the remains of fossil fish, there is but one species about which any approaches to certainty have been made. M. de Blainville thinks it may belong to *pleuronectes*, and that it approximates to *pleuronectes maximus* (the turbot). But even this seems questionable.

In the crag-limestone, below the gypsum, in the quarries of Nanterre, has been found

LABRUS *Julis?* An impression or relief, with some remains of skeleton. It indicates a fish evidently normal, fusiform, exhibiting but a single dorsal fin, which commences immediately from the nape, and extends to the posterior third of the body. It appears to have thirty and odd radii. The number of vertebræ is from twenty-four to twenty-six. The head is pretty large; the orbits moderate; the opercle large, and greatly cleft. There are traces of but one member, which M. de Blainville thinks the pelvian. As its position is pretty forward, the fish may be sub-thoracic.

The catalogue of Davila speaks of fossil fish found in the quarries of St. Denis. But the author gives neither description nor figure on which to found any specific characters.

Ichthyolites, from Pappenheim and Solenhoffen, are referred by M. de Blainville to the following genera:—

Herring (*Clupæa*). Sp. *C. sprattiformis*, De Blain. This is the most common species, and evidently belonging to this genus. The impression indicates a small abdominal fish, from four to five inches long. The head is moderate, the eyes large, the opercle very much cleft; the radii of the pectoral fins, ten in number, are very fine. The pelvin fins are attached about the middle of the length of the body, have fourteen or fifteen radii, and at their base a sort of long scale, as in certain herrings. The dorsal fin is single, and opposed to the pelvian; it is small, and has ten or a dozen radii. The anal fin, equally small, and having nearly ten radii, is in the middle of the space which separates the ventrals from the caudal. The tail is terminated by a fin with very fine radii, and deeply bifurcated.

C. dubia, De Blain. Pretty evidently belonging to the same genus, but a little larger than the preceding.

C. Knorrii, De Blain. A wonderfully perfect impression. The principal difference between it and the foregoing is in the more backward position of the dorsal fin; which, instead of corresponding with the pelvian fins, occupies the space between them and the anal.

C. elongata, De Blain. Distinguished from the preceding species by a more elongated form. The dorsal fin appears to be exactly opposed to the anal.

C. Davilei, De Blain. The remains of this fish indicate a form rather short and thick. The head, moderate, has no scales. The pieces of the opercle are very large, and so is its aperture. The branchiostegous rays are rather slender, and from seven to nine in number. The vertebræ are very short, and consequently numerous. M. de Blainville counted sixty-four. The abdominal cavity is long, and formed by a great number of tolerably fine ribs. The pectoral fins are pretty large, and placed very low. The pelvian or ventral are in the middle of the abdomen, and have nine rays. The dorsal fin is large, arched towards the back, commencing much before the ventral, and extending nearly to the anal. It has twenty-five rays, all soft. The anal fin is also very much raised, and has but seven rays. The pedicle of the tail, which seems thick and short, is terminated by a very ample and deeply bifurcated fin. All the anterior part of the body is covered with scales, which are large and rounded.

PIKE (*Esox*). Sp. *E. acutirostris*, De Blain. A fine impression, regarded as belonging to the same locality, and figured by Knorr: general form that of sturgeon, but it is more probably a species of pike. The body is fusiform; the head, moderate, is prolonged in front into a pointed muzzle, which has some resemblance to that of the sea or gar, pike (*Esox Bellone*), though much less slender. The opercular cleft appears to have been moderate. The pectoral fins are rather small, as well as the ventral, which are triangular, and situated towards the middle of the total length. The anal fin, at the middle of the space which separates the ventral fins from the caudal, is exactly opposite to the dorsal, which is single, and of the same form. Finally, the tail, tolerably long, is terminated by a moderate fin, rather deeply bifurcated, the lobes of which are equal.

In the little which is seen of the skeleton, the vertebræ are rather numerous, and the apophyses short and weak. The skin was also covered with scales, pretty large, and rounded; and, according to the remarkable observation of Knorr, a substance is visible in some places, which, he thinks, proceeds from the flesh; and which, like dried isinglass, detaches, and falls of itself.

Another impression, represented by Knorr, exhibits, still better, the form of the body of a pike, inasmuch as it appears considerably more elongated and cylindrical; unfortunately, it is not complete.

Knorr himself regards these two relics of fish as having belonged to *Esox Bellone*.

STROMATÆUS. Must we regard, says, M. de Blainville, as

belonging to this locality, a fine impression of a fish of the genus Stromatæus, which is figured by Knorr, but without any designation of country? The nature of the stone, which appears of a yellowish grey, and especially the black dendrites, by which the impression is bordered, and the ferrugineous colour of the impression itself, lead this gentleman to the conclusion, that the specimen in question belongs to the locality of Pappenheim. But the resemblance of this fish with that of the metallifcrous schists already mentioned, casts some doubt on this conjecture. However, be all this as it may, it is quite certain that the fish of which we write belongs to the genus Stromatæus. This is proved by the position and form of the dorsal and anal fins, which appear altogether similar, and exactly opposed to each other; and by the character of the caudal fin, which is very large, and deeply bifurcated. The general form of the body is almost hexagonal. This fish has some relations with *Chætodon rhomboides*; and it is difficult to conceive how Knorr could have unhesitatingly decided that it was a plaice. M. de Blainville refers it to the *S. Hexagonus*, which we have already mentioned in our account of the ichthyolites of the metalliferous schists.

Pœcilia. Sp. *P. dubia*. Under this name, M. de Blainville designates an impression which he observed in the cabinet of M. Drée. It is on a hard, calcareous, fine-grained stone, without any other organic remain. The colour is reddish, and it is said to come from Anspach. This impression is merely external; that is to say, there is no trace of the skeleton, but merely marks of a lozenge form, which indicate the insertion of the scales, and all the fins are visible. This impression indicates a fish short, and tolerably thick. The head is defaced. The mouth is without teeth, and very much cleft, as is also in all probability the opercle or gill-covers; the pectoral members are moderate, and have about ten rays; the ventral, about ten or eleven; the dorsal, rather incomplete, exhibits only six rays, and is nearly opposed to the ventral fins; the anal is hardly visible; the caudal is large, probably bifurcated, but incomplete.

The shortness of the body has caused M. de Blainville to approximate this fossil to the genus Pœcilia of Bloch. The position of the dorsal fin is not the same, and approaches more to that of the same member in the carps. M. de Blainville queries, after all, whether it may not be of the fresh water?

We shall now speak of the ichthyolites of the basin of the Mediterranean, beginning with those of Monte Bolca, or Vestena Nuova, in the Veronese.

We may premise, that the principal work which has ever appeared on the subject of these ichthyolites was published by a society of Savans of Verona, of which Count Gazzola was one of the most distinguished members. Unhappily, the scientific execution of it was confined to Dr. Seraphin Volta, brother of the celebrated natural philosopher, who, besides having no collection at his disposal, seems to have been very imperfectly acquainted with this class of animals. In this work are to be found the description of the locality, and of the nature of the stone, and a determination, but a very faulty one for the most part, of the species and their analogues. The figures, however, are extremely good.

The species of fish hitherto recognised in this locality are extremely numerous, though doubtless all that exist there are very far from being yet enumerated, and the number will greatly augment in proportion as they are better studied. We shall consider them in their genera and species, as we have done the others.

Qualus. Sp. *S. innominatus*, De Blain.; *S. carcharias*, Lacépède and Volta. This is rather an incomplete impression, in which no trace of teeth is visible, nor of dorsal fin, nor even the termination of the tail, so that it is not easy to pronounce a judgment upon it. Still the form of the head, the muzzle not much prolonged, the form of the pectoral fins, which are very pointed and arched, that of the fins which appear to be ventral, and perhaps the entire assemblage of the parts, all led M. de Blainville to the belief that this fossil belonged to Squalus, and to that species which he calls *innominatus*, which has been confounded with *carcharias*, and which is found in all the seas of warm climates, as well as in the Mediterranean.

Sq. glaucus. This is after a much better preserved impression than the last, and in all probability the species is fossil at Vestena Nuova. The teeth are visible, triangular, pointed, scarcely denticulated, but notched or curved on the posterior edge; the form of the muzzle, pectoral fins, and tail, approximate this fossil to the Squalus glaucus, or some neighbouring species, both inhabitants of the Mediterranean.

Sq. catulus? Volta. This impression is complete enough to decide that it represents a species of the genus Squalus.

Raia. Sp. *Trygonobatus vulgaris?* The body is evidently round or oval; the pectoral fins are united anteriorly, and there is no prolongation of the muzzle; the pelvian or ventral fins are entire; the tail is very long and very slender, and furnished with a denticulated point, or sting; the teeth are very small and graniform;—so that it is not possible to doubt that it is a sort of *pastinaca*, or *trygonobatus*. It is a mistake, however, to regard it as analogous to the *Sephen ray* of Forskaël, which is totally different. The impression is very perfect, and exists in the French Museum.

Trygonobatus crassicaudatus, De Blain. The general form of the body, the disposition of the pectoral fins in front, the absence of all cephalic prolongation, the integrity of the ventral fins, all denote a fish of this subgenus. The tail, which appears truncated, is remarkable for its great appearance of strength, and its breadth at the base: towards its posterior extremity there is an indication of a sting. This fish was most likely a native of the Mediterranean.

Narkobatus giganteus, De Blain. This impression, though

considerably defaced, yet very evidently represents a torpedo This is easy to perceive from the general form of the body, and especially from that of the tail, the fins of which exactly correspond with those of this subgenus; but it is remarkable for its very large size, infinitely superior to that of all known existing species, and perhaps for its oval and elongated form, if that appearance be not the result of its peculiar position. Volta also regards it as the analogue of the torpedo of the Mediterranean.

BALISTES. *Balistes dubius*, De Blain. The general sublozenge form of this impression, the forward mouth, two or three very strong stings at the upper angle for the first dorsal fin, and one equally sharp projection at the lower, and the position of the second dorsal fin very far back and opposed to the anal, are characters which have determined M. de Blainville to refer this species to Balistes, and not to *Ostracion;* still less to the *Ostracion turritus* of Bloch, which is a native of the Indian Ocean and the Red Sea.

An impression which M. Volta regards as analogous to *Cyclopterus lumpus* evidently appertains to the same species; but this species, as the epithet *dubius* indicates, has not been determined by M. de Blainville, though he decidedly assigns it to the genus Balistes, and to the habitat of the Mediterranean.

TETRAODON. *Tetraodon Honckenii.* This is an impression of a bulky fish, swelled out vertically; the vertebral column exists in its entire length, but there are scarcely any traces of fins. It is about two inches and a half long, to one broad; the body is oval, the muzzle tolerably elongated, the lower jaw is the longest, without any trace of teeth, and the body is covered with small prickles. Many individuals have been found; one of which, by far the largest, was in the brown or ash-coloured stone of Vestena Nuova. M. Volta at first regarded it as analogous to the *Tetraodon ocillatus* of the Mediterranean; but, afterwards, without giving his reasons, he thought proper to refer it to the above named species, which exists in the seas of Japan.

Tetraodon hispidus. An impression of the same nature as the preceding, but still smaller, more orbicular, with the head less pointed, and the forehead more plane. The figure and position of the pectoral fins, and the number of rays of the tail (ten), induced M. Volta to regard this ichthyolite as the *Tetraodon hispidus* of the Mediterranean; but it is a query if it differ much from the preceding.

DIODON. *Diodon reticulatus?* Volta, in the work which we have mentioned, gives the description and characters of a true *Diodon reticulatus*, under this head; but it is very doubtful, to say no more, that the fossil which he refers to it, and of which but a single specimen has been found, appertains in reality to this species.

PALÆOBALISTUM. It appears indubitable, according to M. de Blainville, that an impression figured in the same work, very imperfectly terminated, by no means represents the *diodon orbiculatus*, as M. Volta will have it; nor, in fact, a diodon at all. The form of the tail, and especially some thick teeth which remain near the mouth, though a little in disorder, clearly prove the contrary. It is much more like a species of Balistes, as was the opinion of M. Faujas. M. de Blainville thinks it should form a small distinct genus, in consequence of the form of the teeth; he names it, provisionally, *Palæobalistum*, and the species *P. orbiculatum.*

CENTRISCUS. *C. longirostris*, De Blain. This is certainly a species of this genus, very much approximating to the *velitaris* of Pallas, but differing in the greater proportional length of the muzzle, which equals one half of the body, and is also much more slender. The first sting of the first dorsal fin is also considerably longer.

C. aculeatus, De Blain. M. Volta, not having found in authors, and especially in Bloch, who was evidently his guide, any figure which resembled this ichthyolite, has thought proper, after long reasoning, to make it a lost species of the genus URANOSCOPUS. This allocation, however, is totally wrong; it is only sufficient to compare the figure which he has given with that of the *Centriscus scolopax*, to see the close affinity between the two species; it should, however, perhaps, be separated from the latter in consequence of the form of the head, supposing that it was entire, and especially because the large sting of the first dorsal is much longer and more forward, without a trace of any other, and the ventral fins are also nearer to the head.

SYGNATHUS. *S. typhle*. Respecting this impression, about a foot in length, in which no traces of fins (except the caudal, which has all the characters of the genus) are to be found, no doubt can possibly exist; it is evidently a species of Sygnathus, and most probably the one above-named, which is found in very great abundance in the Mediterranean. The anterior part of the head, terminated by a long point, formed by the vomer, proves that this fish must have undergone a considerable alteration before it was engaged in the stone.

S. breviculus. Evidently a species of Sygnathus. The muzzle is greatly elongated, the mouth very small, the lower jaw longer than the upper, and the body polygonous and cataphracted; no trace of fins is to be seen except that of the tail, and an indication of very small pectoral fins. This fossil has been considered identical with the *Pegasus volans;* but the body of the latter is much more elongated, and the form of the muzzle totally different.

M. Volta has referred to Pegasus volans another fossil impression, which, as far as it can be made out, bears no sort of resemblance to that fish. So little is this the case, that it must be concluded that this author never saw the fish to which he has compared the fossil, nor even a figure of it. The specimen, however, is far too much mutilated to permit any determination of its genus or species. His *Pegasus lesiniformis* is not exactly in the same predicament, because the impression is more perfect. It is still, however, more than probable that it is no Pegasus; nor is there any genus in which it can be placed without difficulty.

LOPHIUS. *Loph. piscatorius*, var. There can be no doubt of the identity of this ichthyolite with the frog-fish; but as it is extremely small, it may perhaps be the variety called *ganelli* by M. Risso, which is always smaller than the common *Loph. piscatorius*, and which is very common in the Mediterranean.

FISTULARIA. *F. Bolcensis.* This impression is very common at Vestena Nuova, may be about half a foot in length, and seems really to belong to this genus. A very small vertical mouth is visible, at the end of a very long and very broad muzzle; the pectoral fins are scarcely to be distinguished, and the ventral not at all; but at the extremity of the body, above, there exists a single dorsal fin, very low and very short, opposed to an anal, of the same form. M. Volta supposes that this fossil is analogous to the *Fistularia chinensis;* but it is evident, even in the state in which it is seen, that there are numerous differences, in the proportion of the muzzle, in the total absence of stings in front of the dorsal fin, in the separation of the dorsal and anal fins, and in the general form.

F. dubia, De Blain. A species very probably of the same genus, but far too incomplete to give any assurance of its identity. It is represented in the Veronese work, under the name of *Fistularia petunha;* the head, however, proportionally larger than in the preceding species, seems to forbid this approximation.

ESOX. *E. longirostris*, De Blain. An incomplete fossil, referred by Volta to *E. Bellone*, but its identity with which is more than doubtful. Its muzzle is proportionally longer, and seems widened towards the extremity, which is certainly not the case with Bellone.

E. syphræna. A very fine impression, about a foot in length, and to all appearance perfectly analogous with the Sphyræna of the Mediterranean. The form of the head, of the muzzle, and of the lower jaw, the size of the eyes, and the place of the fins which exist, are all exactly similar in both.

The impression, which M. Volta refers to *Esox vulpes*, has assuredly no relation to that fish. It appears to approximate more to certain species of *clupœa*, or rather, perhaps, to *salmo murœna.*

E. falcatus. A very incomplete impression, extremely confused, and exhibiting scarcely any portion of the skeleton. It indicates, however, a cylindrical and very elongated fish. The head is very short, and the jaws more particularly so. The latter are equal, and armed, according to M. Volta, with robust and granulous teeth. The branchiostegous rays seem about six in number, and are very robust; the pectoral fins are very small. Beyond the middle of the body, a ventral fin of eight rays is visible; and two or three inches beyond that an anal of seventeen, but not very distinct, to which a single dorsal is opposed, consisting of nineteen rays. Finally, after a caudal pedicle, which is pretty long, comes a terminal fin, very large, and deeply bifurcated. In the most ancient, as well as in the most recent authors, nothing is described or figured resembling this fossil.

The impression given by M. Volta, as an analogue of the *Esox lucius*, or common pike, is so singularly defaced, that it is astonishing how any person could venture to decide in a case of so much difficulty, and on such scanty materials. Nothing remains, in the least degree, to be recognized, but a portion of the lower jaw, which has some resemblance to that of a pike, and has been erroneously represented as the upper jaw. It is impossible to tell to what genus of fishes this impression most approximates.

Esox macropterus, De Blain. M. de Blainville regards, as belonging to this genus, a very fine ichthyolite, preserved in the French Museum, and which may be about twenty inches long, and three or four broad, which indicates a very elongated fish. The body is attenuated at both extremities; the head, which is small, is terminated by a mouth, not much cleft, with thick labial bones, but no trace of teeth; the tail is rather pointed; the pedicle is long and narrow, and supports a bifurcated fin, remarkable for its smallness, and having thirty-six rays. The number of vertebræ is extremely considerable, being upwards of seventy-five, thirty-four of which are caudal; they are, consequently, very short. The abdominal cavity is very large, and the ribs are very fine; the branchiostegous rays, of which there are ten at least, are very fine; the pectoral fins are remarkable for their length and narrowness, and especially for the thickness of their first ray, which borders them through their entire length: the others, on the contrary, seem very slender; the ventral, situated a little beyond one-half of the body, are very small, but five or six rays have been counted in them. There is but a single dorsal fin, placed very low, a little more raised in front, with about twenty rays, and which exactly corresponds to the anal, the form of which is absolutely the same.

LORICARIA. M. de Blainville denies the existence of a fish of this genus among the fossils of Monte Bolca, though M. Volta is desirous of referring to it a tolerably well-preserved and well-figured impression. It appears evident to M. de Blainville, that it rather comes from a fish with gross head and depressed body, or from some jugular fish, than from loricaria. It may be the same as *gobius Smyrnensis*.

The same is pretty nearly the case with the genus SILURUS, though the author of the Veronese Ichthyolithology admits four species. His *Silurus Bagre* cannot belong to this genus, because the impression referred to it, besides being totally different in its general aspect, presents no indication of the first dorsal fin, which is so long, nor of the barble. His *Silurus cataphractus* has no character of this species, and it may even be doubted if it be a silurus at all, though something like barbles are to be distinguished on each side of the mouth.

The impression under which the name of *Silurus cattus* is put is so incomplete, that it is impossible to determine anything concerning it, except that it is a fish, and may perhaps be *exocetus exiliens.*

As for the *Silurus ascita* in the Ichthyolithology, the impression is sufficiently complete to prove that it was an abdominal fish ; but the general form, or *facies*, has evidently but little relation with this species of Silurus.

HERRING *(Clupæa). C. murænoides*, De Blain. This impression is entirely in profile. The muzzle, which is truncated, belonged to a regular fish, a little compressed in its form. It has been referred by Volta to *Salmo muræna*, but there is nothing whatsoever about it that can characterise a salmon; but, on the contrary, the size of the opercle, the great bifurcation of the tail, and the shortness of the body, appear to indicate a species of herring.

C. cyprinoides, De Blain. A fine impression of a fish, with the head rather small. The aperture of the opercle, or gill-cover, seems to have been very large ; the pectoral fins are placed immediately after; the ventrals are exactly opposed to the dorsal, which is single, narrow, rather high, arched like a scythe; but its first ray is by no means very long, which is a very distinct character of *Salmo cyprinoides*, to which this fossil has erroneously been referred. The caudal fin is remarkable for its size, its deep bifurcation, and the sharp form of its lobes.

C. evolans, De Blain. Though, on the first view, this impression bears some resemblance to an *exocœtus*, or flying fish, in consequence of the extent of its pectoral fins, this notion will not bear the test of close examination. The pectoral fins are far from being so large as those of exocœtus; the same is true of the ventral. The head also appears to have been too thick ;

and the caudal fin especially does not present the very singular character of that genus, in having the lower side longer than the upper.

EXOCŒTUS. *Exocœtus exilens.* This impression may, perhaps, belong to a species of the flying fish; nevertheless, the pectoral fins are very small, and the head very thick. Probably, after all, it does not differ much from the last.

MUGIL. *M. Brevis*, De Blain.; *Polynemus quinquinarius*, Ichth. Ver. This impression, about six inches long, which was unique at the time in which Volta wrote, has evidently belonged to a spinous fish, with two separated dorsal fins. It very probably approximated to *mugil*, and consequently to *Polynemus;* but nothing is less certain than that it is a species of this last genus. The portions of what M. Volta considers as analogous to the fine kinds of separated rays of the dorsal fin, would seem to be placed under a vertical line through the eyes, whereas, in the *quinquinarius*, they are much further back. The head, moreover, of the fossil is very stout.

TRIGLA. *Trigla Cyra.* An impression is referred by M. Volta to this fish, which is very abundant in the Mediterranean, but the impression does not appear sufficiently complete to establish this identity; indeed the smallness of the pectoral fins, and the total absence of any indication of decomposed rays, would lead us to the belief that it was not a trigla, but rather, perhaps, a true abdominal fish.

SCOMBER. *Scomber pelamis*, Volta. An impression, ten inches long, complete enough to identify it with the genus in question, but not sufficiently so to determine the species.

S. altalunga, Volta. The great length of the pectoral fins seems to justify this analogy.

S. thynnus. A fine impression, twenty-eight inches long, pretty well preserved as to the general form, though the head has no character that can be recognised. There is, however, much of the tunny in the general appearance. This fossil is very common at Vestena Nuova; as we find also that the tunny is very common in the Mediterranean.

S. cordyla, Icht. Ver. An impression sufficiently well preserved to exhibit many relations with *Scomber cordyla*, from which, however, it differs in having the first dorsal fin and the body more elongated.

S. trachurus, Icht. Ver. An incomplete impression, which M. de Blainville is more inclined to refer to the *Scomber pneumatophorus* of De Laroche.

S. Klenii, Icht. Ver. This impression is sufficiently well preserved to warrant the conclusion that it belonged to a species of the subgenus trachurus, but that it is the scomber of Klein is somewhat doubtful.

S. ignobilis, Icht. Ver. In the general form of this ichthyolite, which is tolerably entire, nothing very analogous with the species of this genus is to be found. In fact, the belly is very convex; the head very small; the ventral fins extremely large, and also the first dorsal, which is evidently much more raised than the other. M. Volta was induced to regard it as the analogue of the species above named, only by the existence of a little point between the two dorsal fins.

S. speciosus, Icht. Ver. It is very probable that this fine impression belonged to a species of this genus, but not quite so certain that it belonged to that just named. Forskaël tells us expressly, that Scomber speciosus has no teeth, while there are small ones to be seen in the fossil. But the number of rays in the two dorsal fins appear to be alike; and the scales, moreover, seem to have been pretty large in this fossil. The impression referred by M. Volta to *S. glaucus* is more likely to have belonged to this species.

S. pelagicus, Icht. Ver. Though this fossil has evidently belonged to a species of the genus of Scomber, with an elongated body, and which, probably, had but a single dorsal fin, we cannot by any means be certain that it belonged to *Scomber pelagicus*.

S. chloris. This, again, is a species of Scomber, which may, perhaps, have some relations with *chloris*, as the body is short, and very much raised; as there are three spinous rays in front of the anal fin, and as the dorsal fins seem united; but the lower jaw is indubitably shorter than the upper, exactly the reverse of what takes place in the true *chloris*, which, moreover, belongs to the African seas.

Scomber orcynus. The general form and assemblage of the parts indicate a Scomber, and very probably the *Scomber orcynus* of Rondelet, inasmuch as there is a single dorsal fin rather short, but elevated; an anal almost correspondent to it, and eight spinous rays both above and below. This species is found in the Mediterranean, and Linnæus has confounded it with the tunny.

We may observe, in general, that the number of species of this genus, Scomber, is much more considerable in the Mediterranean than was believed at the time when the Veronese Ichthyolithology was published. This we find from the works of MM. Risso and Rafinesque, who have discovered species there very much approaching to many hitherto supposed to exist only in the Red Sea or the Indian Ocean. So much is this the case, that there is scarcely any reason to doubt that all the fossil species of Monte Bolca, which are very numerous, are to be found in the Mediterranean.

Perca (Perch). *P. formosa*, Linn. *P. Americana*, Icht. Ver. This fossil skeleton, in which no traces of ventral fins are visible, has certainly some relation with the general forms of the perch, in the size of the head and mouth. There is but a single dorsal fin, situated in the middle of the back, as in *Perca formosa.* But the anal fin is much longer and nearer to the tail; and this, with other differences, takes away all certainty respecting the species, and even respecting the genus.

Sciæna. Of this genus the impression referred to *Sciæna jaculatrix*, by the author of the Veronese Ichthyolithology, has no relation whatever to that species, and is, moreover, extremely imperfect. His *S. Plumieri*, which is well preserved, has more analogy with the genus Sciæna. The head is very small; the teeth appear strong; the gape of the mouth is not large; the jaws are equal; the body is moderately elongated; and the vertebræ are from twenty-four to twenty-five in number, twelve only of which are thoracic. The back, which is a little convex, has two dorsal fins nearly equal, the first of which has seven stings or spines, and the last nearly as many ramified rays. The anal fin, which is smaller, has a single spinous ray. The pectorals appear large, and the caudal is scarcely semilunar. M. de Blainville, however, does not regard the specific identity as proved.

LUTJANUS. *L. Lutjan*, Icht. Ver. M. de Blainville is inclined to regard this as the same as *Scomber cordyla*, above cited, for the following reasons:—The impression represents a form attenuated at both extremities. The vertebral column is composed of twenty-four or twenty-five large vertebræ, and terminated by a large and deeply bifurcated caudal fin. There is no sting to the fins. The fish, however, appears to have been covered with pretty large scales, and to have had the teeth tolerably strong.

L. ephippium, Icht. Ver. This ichthyolite is yet better preserved than the preceding. It represents a fish much shorter, more raised, almost convex, and whose mouth is armed with very strong teeth, the anterior two of which are the longest. The head is very long; the vertebral column consists of twenty-four or twenty-five vertebræ, six of which, at most, are abdominal. The dorsal fin, very long, is scarcely excavated at the place which separates the spinous part from that which is not so. The anal is large, with two or three spinous rays; the caudal is large and rounded. The analogy between this fossil and the species to which it is above referred, seems pretty probable.

HOLOCENTRUS. *H. calcarifer*, Icht. Ver. This is a species which has great affinity with the preceding, and in which all the fins are disposed absolutely in the same manner. The head, however, if it has been well represented, is much smaller, and the jaws appear to be without teeth. The analogy with *Holocentrus calcarifer* is more than doubtful. As to the *Holocentrus lanceolatus* of Volta, the impression is too incomplete to afford any ground for decision; it may, however, belong to this genus. It is also more than probable that his *Holocentrus maculatus* belongs to it also; but, again, this specimen is too imperfect to allow of a positive determination.

H. macrocephalus, De Blain. There are certainly some relations between this fine and well-preserved impression and the *Holocentrus sogho*, which exists in the rivers of North America; but it is equally evident that it should form a distinct species, in consequence of the thickness of its head, the convexity of the forehead, and even from the nearly bifurcated form of the tail.

SCORPÆNA. The impression referred to *Scorpæna scrofu*, Icht. Ver., is altogether too much defaced for any cautious naturalist to pronounce a positive judgment on its identity.

SPARSU. *S. vulgaris*, De Blain.; *S. dentex*, Icht. Ver.; and *S. sargus*, ibid.; and *S. macrophthalmus*, ibid. Three impressions equally well preserved, which M. de Blainville unhesitatingly refers to one and the same species. The head is large, partly covered with scales; the opercle equally large; the mouth is furnished with hooked teeth, pretty sharp. The body is very much raised, almost in the proportion of one to two. The vertebral column is nearly straight, and composed of twenty-four vertebræ, with strong spinous apophyses, nine of which are abdominal. It is terminated by a very large and somewhat semilunar caudal fin. The pectoral fins are small, and attached pretty high; the ventral are nearly sub-abdominal; the dorsal is single, commences immediately behind the nape, and is formed of seventeen rays, nine of which are spinous: the anal has nine rays, the first two or three of which are spinous. Though the disposition of the teeth is pretty nearly the same as in *Sparus dentex*, the proportions of the parts will not permit us to regard this fossil as the analogue of that species.

There are other fossil species attributed to Sparus in the Veronese work, but they are either identical with some of the foregoing, or established on materials too insufficient to afford a safe foundation.

LABRUS. An impression referred, in the above-mentioned work, to *L. merula*, Linn., is so completely defaced, that though it does somewhat indicate the form of Labrus, it is impossible to give any assurance concerning it; it may, perhaps, be *Labrus ciliaris*.

L. turdus, Icht. Ver. A very fine impression, indicating a fish of this family, of a tolerably elongated form, a character sufficiently applicable to this species. The head is remarkable for its length, and especially for the protraction of the muzzle, which presents no trace of teeth. The scales were very large, and there were some on the gill-covers: the caudal fin is very thick. Notwithstanding these differences, which, perhaps, are attributable to the greater or less degree of preservation, there does not appear to be much doubt concerning the identity of this species.

L. punctatus. Very probably another species of this genus, remarkable for the bulk and shortness of the body. The head is very gross, but entirely decomposed; the dorsal fin is single, and commences at the nape; it is composed of seven spinous rays, and fourteen or fifteen others not spinous; the middle ones of which, being very long, cause the fin to attain to nearly one-half of the caudal. The anal is pretty nearly of the same form as the last part of the dorsal, and appears to have but a single spinous ray; the caudal fin is large and entire; the number of vertebræ is from twenty-three to twenty-four.

Though this fossil is most likely to belong to the genus, yet it presents no trace of the ray of the ventral fins extended into a long filament, which is a character of *L. punctatus*

L. rectifrons, De Blain.; *L. ciliaris*, Icht. Ver.; and *Sparus Bolcanus*, ibid. These two ichthyolites must be approximated to the foregoing species, for it is probable that they belong to one and the same, though Volta has placed them in two separate genera. The form of the body, and even the proportions of the parts, are almost alike in both to the last; but it would seem that the dorsal fin commences still nearer to the nape in these specimens, and that the number of spinous rays is more considerable, since there are at least ten to be reckoned. It also appears that there are two in front of the anal; the head, in particular, has a different form, for its front, almost straight and declined, is terminated by a very small mouth. This species is as yet unknown in the living state. *L. bifasciatus* and *L. malapterus*, of the Veronese Ichthyolithology, cannot, from the very incomplete state of the specimens, be possibly determined.

CHÆTODON. It is of this genus, beyond all doubt, that the greatest number of species, remarkable for their fine preservation, are to be found in the locality on which we are now writing.

Ch. pinnatiformis, De Blain.; *Ch. pinnatus*, Volta.—This ichthyolite, which is frequently found at Monte Bolca in the best possible state of preservation, has certainly many relations with the *C. pinnatus* of Linnæus; but still it appears to differ from it very sensibly in the following particulars:—the point of the greatest breadth or elevation is considerably in front of one half the body, at the articulation of the ventral fins, while in *pinnatus* it is in the middle, and towards the anus; the upper jaw is shorter than the lower in the one, and the reverse in the other; the form of the dorsal and anal fins is different; in the fossil they are directed much more vertically, and, moreover, the size of their rays decreases much quicker, so that the latter half forms a very narrow fin.

C. subvespertilio, De Blain.; *C. vespertilio*, Volta.—Though this fossil skeleton is in a still finer state of preservation than the last, and especially more resembles the living species to which it has been compared, some differences are, nevertheless, observable; thus the body is less elevated, proportionably; the dorsal and anal fins are larger, and instead of being concave at their posterior edge, they are very convex; the ventral fins too, admitting them to be entire, which is very probable, are much shorter, since they do not reach to the anus; whereas, in *C. Vespertilio,* they proceed considerably beyond it. The figure is excellent.

C. substriatus, De Blain.; *C. striatus,* Icht. Ver.; and *C. asper,* ibid. M. de Blainville regards these two incomplete fossils, as belonging to one and the same species, and as presenting characters distinguishable from *C. striatus.* Without speaking of the proportional height which is visibly less, the form of the head, and especially that of the muzzle, are totally different.

C. subarcuatus, De Blain.; *C. arcuatus,* Volta. A skeleton, as complete as if it proceeded from the hands of a good preparer, coming from a species evidently approximating to *C. arcuatus,* but distinguishable from it by a very sufficient number of characters: thus, the front-head is formed by a curve line, but perfectly even; whereas, in *arcuatus,* there is a very sensible boss over the eyes. Although the dorsal and anal fins are very much alike, yet they both present the very singular character of being, as it were, divided into two, pretty nearly towards the middle: still, the last ray of each part is not much more prolonged than the others, though it is so in *arcuatus;* and the longest ray of the anal is the last of its first part, while in *arcuatus* it is the first. Finally, the pectoral and ventral fins are much longer in the *arcuatus* than in the fossil.

C. Argus, Icht. Ver. The identity of this species with the fine ichthyolite, figured by Volta, is much more difficult to deny than the preceding, if we except the little difference of the body being at least as much elevated in front of the pedicle of the tail, as behind the gill-covers, while the contrary evidently takes place in the fossil. The upper jaw, too, is much shorter in the latter than in the former, which may be accidental. There is also a slight difference in the form of the first dorsal fin; but, as to the rest, the resemblance is perfect.

C. rhombus, De Blain.; *C. mesoleucus?* Volta. This fossil skeleton is equally well preserved, to enable us to decide that the species from which it proceeded was not the analogue of *mesoleucus*. In fact, it has not the prolonged muzzle; and the gill-covers, which are perfectly preserved, present no traces of spinous radii.

C. nigricans, Volta. The case is the same with the species; though there may be some approximation, the form of the dorsal fin is entirely different.

C. canescens, Volta. Nor can this approximation of Volta's be admitted. The first two rays which follow the spines are infinitely more elongated in the living than in the fossil species, and the form of the anal fin presents very notable differences.

C. saxatilis, Volta. Here there is more analogy than in the majority of the foregoing. Perhaps, indeed, the identity is perfect: the living species is found in Egypt.

C. chirurgus, Icht. Ver. There is, perhaps, still more probability that this is the analogue of the living species. The fossil, however, is more orbicular, that is, proportionally more raised. Its dorsal fin is not so long, nor placed so low; it has fewer spinous rays; and the anal fin, also a little different in form, appears evidently situated further back. There is no trace of the curved sting or spine in the fossil.

C. ignotus, De Blain.; *C. macrolepidotus*, Icht. Ver. There is the utmost evidence that this approximation by Volta is totally wrong. There is scarcely anything in common between the fossil and the living animal. In fact, the latter has ten dorsal spines, the fourth of which is longer than the body; while, in the fossil, the rays of the dorsal fin, as well as those of the anal, proceed, gradually diminishing, from the first to the last.

C. lineatus, Icht. Ver. The general form of this fossil fish indicates a species very much approximating to *lineatus;* but, as the head is entirely wanting, it would be too much to pronounce decisively on its identity.

C. canus, Icht. Ver. A small species, pretty like the preceding, that is, considerably elongated; but it differs a little in the form of the fins, and especially in that of the anal.

C. triostegus, Volta. There is really some analogy between this fossil and the living species, in the general form of the body, and especially in the declivity of the front-head. But the anal fin is strikingly longer in the living animal; and probably there is some difference in the proportion of the spinous rays of the dorsal: the body is also more elevated.

C. rostratus, Icht. Ver. A specimen far too much deteriorated to attempt any approximation between it and living species.

C. orbis, Icht. Ver. Very imperfect; but there is quite enough to prove the non-affinity to *orbis*.

C. subaureus, De Blain.; *C. aureus*, Volta. There is, without doubt, some approximation in this fossil skeleton, which is well preserved, and the figure of which is very good, to the *C. aureus*; but the differences are still more apparent, and prove that the fossil belonged to a distinct species. The head of the latter is much larger, and has a totally different form; the dorsal fin commences much less in front, and the first radii are the longest, as is also the case with the anal; neither do they reach to the tail, whereas they pass it considerably in *aureus*. The form, too, of these fins is not the same. The *Zeus gallus* of Volta indubitably belongs to the same species as this fossil.

C. papilio, Volta. This species, which Volta has not found to approximate to any living species, is effectively distinct from all hitherto known. The body is very much raised, and nearly of a lozenge form; the anterior half of the dorsal fin, which is almost square, is remarkable for its great height; the dorsal fin is preceded by four small spines, and terminated in contact with the caudal fin; the anal, though equally prolonged, is much less raised, though a little more behind than in front. Four or five vertical bands are to be seen on this ichthyolite, of a deeper colour than the rest.

C. velifer, De Blain. *Kurtus velifer*, Volta. This is one of the most common and best preserved ichthyolites of Vestena Nuova; all the characters of Chætodon are easily recognized in it. But this species differs from all that are known, in the magnitude of the first half of the dorsal fin, which is triangular, and which commences immediately above the nape, and especially in the magnitude of the ventral fins; the anal fin, though tolerably long, is very low, and yet proceeds decreasingly from the first ray to the last: the pedicle of the tail is very narrow.

Zeus. The *Zeus triurus*, and the *Z. vomer* of Volta are indubitably one and the same species, and M. de Blainville thinks that they should form a distinct species of *Chætodon* rather than of *Zeus*. The most remarkable characters consist in the convexity of the ventral line, much more considerable than that of the back; in the dorsal fin, which has but a small part more elevated in front; and finally, in the anal, which is very long and very low through its entire extent.

Zeus platessa, De Blain. In this genus must be placed the rather incomplete fossil figured by M. Volta under the name of *Coryphæna apoda*, to which it is quite evident that it cannot belong. Its general contour is that of a chætodon, or of a Zeus: the dorsal fin commences a little behind the nape; the first part is more elevated, and has six or seven tolerably long and simple rays, and the fin is afterwards continued very low, as far as the pedicle of the tail, which is very narrow; the anal fin has nearly the same form as the second part of the dorsal, and there are neither pectoral nor ventral fins; this, however, must be attributed to the changes which the ichthyolite has undergone; and we may remark here, that in ichthyolites generally, the traces of the ventral fins are rarely to be found.

Z. rhombeus, De Blain. This is the *Scomber rhombeus* of Volta, which in reality belongs to this group. It is a fish remarkable for its breadth, which is almost equal to its length, owing to the great projection of the belly, which considerably exceeds that of the back; the dorsal fin appears to have been divided into two parts, the anterior of which is somewhat high, and formed of a great number of simple rays; and the posterior, much lower, is equally composed of very small rays, producing a fan-like form; the anal fin, much longer than this second part, in consequence of the great curve of the belly, has in other respects the same form and composition; the pectorals are very short; the ventral fins are composed each of a single ray, almost cylindrical, and much longer than the body.

Monopterus. *Monopterus gigas*, Volta. This genus is established on rather a defaced ichthyolite, about a foot long and one half broad: the head is high, rather short, and has a truncated appearance; the back is extremely gibbous, and covered with some traces of small scales; there is no indication of eyes, of gill-covers, or even of pectoral or ventral fins, but solely a dorsal fin, pretty far back, rounded, of moderate size, and apparently formed of rays entirely soft; the anal is formed in the same way, but differs in being notched a little behind, the first rays being much longer than the others; the caudal, at the end of a pretty long and conical pedicle, is remarkable for its length, the narrowness of its lobes, and the depth of its emargination.

Pleuronectes. *Pleuronectes platessa*, Volta. *Pleuronectes quadratulus*, Ib. Both belong to one species: the fossil skeleton in the first figure is far from complete, especially in the anterior part, where the summit of the head is wanting. The second is much better, and to a certain extent the *quadratulus* of Belon may be recognized in it.

Cottus. Under the name of *Cottus bicornis*, Volta has established a new species in this genus, on an ichthyolite in such a state, that it would have been much better to have said nothing about it.

Gobius. *G. Smyrnensis* of Volta, is exactly in the same predicament as the last.

G. barbatus and *G. Veronensis*, Volta, are two ichthyolites in a skeleton state, nearly perfect, indicating very probably a species of this genus, and very probably the same; the general form of the body, head, and caudal fin of Gobius are to be recognized, but it is very difficult to pronounce upon the species.

Blochius. *B. longirostris*, Volta. One of the most singular fishes of this locality, and found, as it would appear, in great abundance. It is remarkable for its great length, sometimes attaining to nearly three feet; for its slenderness, and its general form, which approximates to *ammodytes*, with this essential difference, that it appears to have small ventral fins under the throat; its head is terminated by a very long and pointed muzzle, and there is no trace of teeth. Volta makes a genus of it.

Callionymus. *Call. Vestenæ*, Volta. This fossil appears evidently to belong to a different species from all the rest in this locality. The body is narrow and elongated, and terminated by a caudal fin, with a pretty deep bifurcation: this character is never found in the genus to which it has been referred. There is but one dorsal fin, pretty long, and opposed to the anal, which is of the same form and parallel with it; there are scarcely any traces of pectoral or ventral fins in the figure, but on the ichthyolite they may be observed, but very small. It is difficult to determine the genus of this fossil, but assuredly it does not belong to Callionymus.

Gadus. *Gadus merluccius*, Volta. This may be a Gadus, but it is difficult, if not impossible, to determine the species.

Blennius. *B. cuneiformis*, De Blain. *B. ocellaris*, Volta.

This ichthyolite, which is not perfect, represents an extremely singular fish, having somewhat of the form of B. ocellaris, as to the thickness of the head, but in other respects there is hardly any approximation: in fact, the body is much shorter, and altogether cuneiform; the head is much more thick; the first dorsal fin, too, is infinitely higher, and commences sooner. There appears but little doubt that this is a new species, whose analogue is yet unknown.

Ammodytes. *A. tobianus*, Icht. Ver. This fossil, very much defaced, indicates an elongated fish, but one infinitely less so than the *Ammodytes tobianus*.

Ophidium. *Ophidium barbatum*, Volta. An anguilliform fish, with the head very small; the back is furnished, in its entire extent from the nape, with a fin with very numerous rays, augmenting from front to rear; the anal is of the same form, commencing towards one half of the animal, and uniting at the extremity of the body to the dorsal, and there assuming a rounded form, which is quite different from what takes place in *O. barbatum*. The whole body is spotted with brown.

Muræna. *Muræna conger*, Volt. This fossil very probably represents a species of this genus, but it is utterly impossible to determine what that species may be.

Murænophis and Synbranchus. Three ichthyolites represented under the names *Murænophis, Muræna cæca,* and *Synbranchus immaculatus,* appear to belong to one and the same species of anguilliform fish, very much elongated, and without any trace of fin; but this is nearly all that can be said about it.

It is easy to see, after this brief view of the ichthyolites of Vestena Nuova, which we have given from M. de Blainville, that the majority of the analogues admitted by Volta are more than doubtful, and that the number of species is much less considerable than what he has calculated them to be. He has reckoned a hundred and five, whereas, at most, there are not above ninety-four: seven he attributes to the fresh-water, while, in fact, there does not appear to be a single one which belonged to that element. His allotment of the species to the different quarters of the globe seems to be equally destitute of proof, for, in the great majority of his analogical references, he is decidedly erroneous.

We must now dismiss with great rapidity the ichthyolites of the remaining localities, respecting the great majority of which nothing certain has been determined. Nothing positive, for instance, as to species or genera, is known respecting those of the Vicentine, or of Friuli. At Murazzo Struziano, however, a very complete impression was found of a fish evidently abdominal, which M. de Blainville refers to Clupæa, and has named *Clupæa dentese;* it exhibited thirty-six or thirty-eight vertebræ; the head is rather small, and the mouth is armed with very strong teeth in both jaws.

Two species of Clupæa, from Mount Libanus, have been named and described by M. de Blainville. The first is *C. brevissimus;* rather short in proportion to its breadth; the vertebral column, concave towards the back, is composed of thirty-one or thirty-two vertebræ, nearly equal, rather small, and with weak spinous apophyses; the mouth is large, and very much cleft; the lower jaw a little the longest, and no trace of teeth; the gill-covers and the eyes are large; there are at least eight branchiostegous rays, pretty broad; the pectoral fins have twelve or fifteen, and the ventral, somewhat in front of the middle of the body, are small, and composed of six or seven rays at most; the dorsal fin, pretty nearly approached to the head, is low, composed of sixteen rays, gradually diminishing, and sustained by weak apophyses; the abdominal cavity is moderate, and the anal fin, which commences a little before the conclusion of the dorsal, is weak, very low, and also long; the caudal is moderate in size and furcated.

The other species is *C. Bernardi.* It is more elongated than the preceding; the vertebræ are thirty-six; the ribs are from twenty-two to twenty-four, and very slender; the tail is longer than the abdomen; there is no trace of pectoral fins, and the ventral are very small; the dorsal fin is very low, and the anal commences considerably beyond the end of the dorsal, leaving a large abdominal cavity; the caudal is like that of the preceding species.

From Mellili, near Syracuse, M. de Blainville has seen the impression of a fish, which he is inclined to refer to the genus *Cyprinus*. To the same he is also inclined to attribute a species discovered at Scapezzano, a locality which we have already pointed out. Near Aix, in Provence, the following species appear to have been discovered: *Mugil cephalus, Perca minuta,* De Blain.; *Cyprinus squamosseus,* De Blain. These are all ascertained by M. de Blainville, and are fresh-water fish; but Darlac has cited others, such as *Mullus barbatus, Coryphæna hippurus, Anarrhycas lupus,* many species of Gadus, and *Trigla cataphracta,* which M. de Blainville supposes to be the same with his *Cyprinus squamosseus.*

In the celebrated plaster quarries of Montmartre, &c., are enumerated the following;—Sparus? Perca? (it is not decided to which of these the impression, which is very imperfect, is referable;) *Amia ignota,* De Blain.; *Mugil,* Lacepède; *Pæcilia Lancetherii,* De Blain.; *Anormurus macrolepidotus,* De Blain.; *Salmo,* Cuvier; *Perca,* De Blain.; *Cyprinodon,* Cuv.; *Cyprinus minutus,* De Blain.; and *Cyprinus squamosseus?* De Blain.

Among the ichthyolites of Œningen, the following species are reckoned;—*Esox lucius,* Knorr; *Cyprinus jeses,* Scheuz.; *Capito?* Scheuz.; *Cyprinus bipunctatus,* and many others which are not clearly ascertained.

We shall terminate this account of fossil fish with a brief notice of such remains of this class as are to be found scattered in almost all countries, in strata of various kinds, and which consist, for the most part, of vertebræ and teeth.

In almost all the ancient works on oryctology, the name of *ichthyospondyles* is given to the vertebræ of different species of fish which are found in the bosom of the earth. All the books which treat of petrifactions contain a greater or less number of these, figured more frequently than described: the best works on this subject are those of Scilla, Knorr, and Walch.

These fossils have been met with in all zootic strata, as it would seem, from the most ancient to the most modern; they are found in the schists, the compact limestone, the chalk, the shell-limestone, the gypsum, the diluvial, and the alluvial formations: some are found which have even evidently been rolled.

The vertebræ thus found have been preserved as far as regards the form; the substance, probably, has been pretty frequently converted into that of the stone which contains them. No author, we apprehend, has as yet attempted to refer these vertebræ to known or unknown species: the task, though, perhaps, not impossible, especially if we had a great number of specimens at our disposal, would yet be no very easy enterprise; for these vertebræ are never accompanied by their apophyses, they are for the most part insulated specimens, and there is so strong an inter-resemblance in the vertebræ of a great number of living fish, that it is with some difficulty that they can be distinguished. All that can be said, at present, respecting the vertebræ in question, is that they seem to have belonged to large species, and that sometimes under this name have been confounded the coccygian vertebræ of cetacea, which are, however, easily enough to be recognized by the absence of those deep and regularly disposed holes which are remarkable at the surface of the true vertebræ of fish.

The fossil teeth of fish are the parts most frequently to be found in the bosom of the earth, because they are much less subject to decay and alteration; accordingly, we find them in great abundance in all collections.

In old works, very frequently, we find these fossil teeth spoken of under the generic name of *ichthyodontes*, and they are justly divided into two groups,—the *glossopetræ* or *lamiodontes*, that is, teeth more or less flatted, which have belonged to the family of Squali, &c.; and *bufonites*, *batrachites*, &c., meaning teeth more or less rounded, and which are still very generally regarded as having proceeded from some species of Sparus, or Anarrychas. The denomination of *glossopetræ*, or petrified tongues, seems to have been derived from an erroneous notion, formerly entertained concerning the form of the tongue of the serpent; and especially from another ludicrous prejudice, that the apostle Paul, in passing by Malta, had destroyed all the serpents of that island, and that the fossil teeth of Squali, found there in such abundance, were the petrified tongues of those reptiles.

These fossils often preserve their native composition in the bosom of the earth, and very often pretty nearly their natural colour; some are, however, changed in this last particular, being coloured of a blackish blue, or of an ochreous red, according to the character of the strata in which they have been buried: the colour thus proves an indication of the locality. Many of them have been altered in their chemical composition; some are mentioned, from the locality of Sienna and Placenza, which are converted into turquoises. The places in which they are most frequently found, are the islands of Malta and Sicily, which appear to be almost sown with them in certain parts. They are equally common in Calabria, in Tuscany, in the territories of Sienna and Placenza, and very probably in all the sub-apennine hills; also, in the neighbourhood of Brussels, in the mountain of St. Pierre, near Maestricht, in the environs of Montpellier, those of Paris, of London, in the Isle of Wight, &c., they are in great abundance.

The nature of the strata in which the numerous tribe of Squali, which are in various collections, have been found, does not appear to have been sufficiently studied; there are none which seem very certainly to have proceeded from the schistose or transition strata, nor even from the compact limestone;

it is in the chalk, and the shell or crag limestone, (*calcaire grossier*, of the French,) that some begin to appear; and, as we proceed upwards, they increase in greater and greater proportions.

Though a very great number of these fossil teeth are found and of very different forms, yet we must not suppose that they proceed from an equally great number of species; in fact, a very moderate study of the teeth of Squali is sufficient to prove that, in the same species, there are sometimes not two on the same side exactly similar, and yet their differences are so marked, and so constant in their recurrence, that the Squali are easily characterized by the consideration of this part of their organization alone.

The fossil teeth seen by M. de Blainville, either in their natural state or figured, are by him referred to the following species:—

1. *Squalus cornubicus*. To this animal are referred a great number of teeth from Sicily, Brussels and its environs, and the neighbourhood of Montpellier, &c. All the fossil teeth named by oryctographers *subulati, cuspidati subulati, ophioglossæ, glottidæ, ophiodontæ*, &c., and by Lluid, *ornithoglossæ recurvirostres*, must certainly belong to this species. They are, in general, slender, narrow, elongated, and pointed, with entire edges, somewhat trenchant, flat within, and a little convex without. This kind of fossil teeth appears to be by far the most common, since they are found in all parts of the earth, and at all depths; accordingly we find this species of shark very common in all our European seas.

2. *Squalus ferox*. Teeth, more or less broad, sometimes rather elongated, altogether straight, or but little curved backwards, and which, without denticulations on their edges, are accompanied at their base by a very evident point on each side, belong to this large species of shark, which exists in the Mediterranean. They are found at Boutonnet, near Montpellier, and in different parts of England.

3. *Squalus tricuspidens.* We may very probably distinguish from the preceding, and regard as belonging to a species of Squalus not yet known, certain teeth with three points, but which are straight and very high, or very slender. They come from the neighbourhood of Brussels, and are figured by Bartin.

4. *Squalus vacca, columbinus,* or *grineus.* A kind of teeth are found in Sicily, with a very broad basis, almost straight, and the trenchant edge of which presents a point, not much raised, not denticulated, compressed, a little curved behind, and accompanied with five or six very strong points decreasing backwards, and three or four much smaller ones in front. They belong, indubitably, to the species called by the Sicilians *Squalus vacca.*

The other species of Squali, to which the fossil teeth are referred, are *Sq. pristodontus, Sq. lamia, Sq. auriculatus,* De Blain., and *pristobatys* dubius?

There are also found pretty frequently, in the bosom of the earth, in localities, and, as it would seem, in strata of different kinds, the teeth of *aetobates,* or *raia aquila,* separated or united, in greater or less numbers.

These teeth are sorts of parallelopipeds, of forms a little different—sometimes entirely straight, sometimes curved in a chevron form, or thus ᴧᴦᴦ; one of the faces of which is smooth, and more or less hard and varnished, and the other, which was adherent to the skin of the mouth in the living animal, is traversed by lines, parallel and perpendicular to the length of the dentary bone.

It is also to these sorts of teeth, to which the name *mylio-dontes* may be given, that certain little lozenges or cubes belong, whose structure is the same, but which are much smaller.

All these pieces, united in greater or less quantity, form a broad plate, usually elongated from front to back, and which is adherent to the skin of the palate and of the place of the tongue. Sir Hans Sloane, in a paper in the Philosophical Transactions, was the first who compared these fossils with the palatine teeth of *Raia aquila*.

The chevron-formed teeth are found, in a fossil state, in the marly strata of Placenza, and sometimes in calcareous rocks, as in the mountain of Antelaus in the department of the Piave, where there is a great quantity of them incorporated in the stone, either isolated, or sometimes united three or four together. These fossil teeth are considered, by M. de Blainville, as probably having belonged to a ray, approximating to the *Narinari* of Marcgrave.

The spines, which arm the tails of some of this genus, have also been found in different localities, and in sufficient abundance, in this country.

Under the name of *bufonites*, or *batrachites*, we find figured, in the works of some oryctologists, a great number of fossil bodies, more or less rounded and shining, which are evidently portions of the teeth, or dentary palates of fishes. The above names were given, because it was imagined, for some reason not easily conceived, that they were engendered in the heads of toads or frogs. The substance, which most usually contains them, seems to be the compact limestone; which would lead us to conclude that they belong to tolerably ancient formations. The species from which they come are very far indeed from being determined.

Many other parts of fishes have been found in different places and at different depths, but the examination of them is not likely ever to lead to any important results, either geological or zoological.

The remains of invertebrated animals are so excessively numerous in species—we should say, perhaps, innumerable—that it cannot be expected that we should enter into minute details concerning them, or pretend to present our readers with any thing like a complete enumeration of them. We must content ourselves, therefore, with some general observations on the most interesting points of the subject, and a notice of the localities, &c. of the most remarkable genera, and such as are of most frequent occurrence.

We cannot commence this department of our labours more appropriately, than by some remarks on the nature, circumstances, and varieties of petrifactions in general.

It has been asserted that the petrifaction of organized bodies was a mechanical operation, in which the stony matter replaced, molecule by molecule, the substance of those bodies. But this supposition is far from being clearly demonstrated. Some bodies, in passing to this state, have preserved both their external and internal forms: others have preserved the first only; and there are more, such as certain of the fungiform polyparia, in which the part approaching to the edge has preserved its original contexture; while that which is found towards the middle is only a confused petrifaction, or crystallization, in which no organization can be discovered.

Some ancient authors, who have written on oryctography, would consent to class among petrifactions those only whose analogues were already known: the remainder of such productions was considered as nothing but figured stones—the work of simple chance. At the present day, however, there are none who devote their attention to this very interesting department of natural history who entertain the slightest doubt that the bodies, which are found in the strata of the globe, have belonged to beings once endued with life, and whose remains have been buried, some, after their natural death, by slow depositions—others, by sudden revolutions. But there is one point which, assuredly, we shall never be able to ascertain, and that is, the period of time which was requisite for the deposition and petrifaction of each stratum. Certain substances can be preserved in the earth, and pass into a petrified state; but all are not alike capable of a similar conservation. Flesh, the bills of birds, claws, horns, soft fruits, or other soft substances, do not appear to have been ever found in the fossil state. Teeth and horns are sometimes petrified; but, for the most part, they are found preserved—a portion of the gelatinous substance even remaining in the latter, which are also not unfrequently penetrated by mineral productions.

Specimens, which have been found in the coarse limestone, as well as in the superior marine sandstone, and which have been considered as ribs of *manati*, are changed into a very hard and sonorous calcareous stone, although the strata in which they existed are not petrified.

It is remarkable enough, that, in many localities, such as Nice, Gibraltar, Cette, Aix, and Corsica, the fossil bones are imbedded in a stony stratum, the colour of which, in all those places, is a brownish red.

Amber, and the different organic bodies which it contains, are preserved, but never in a state of petrifaction. Wood and ligneous cones are very often changed into silex, and have sometimes altogether disappeared in the petrified strata, leaving nothing but their external mould or impression.

In general, that body which is found in the completest state of preservation, is the calcareous covering, or shell, of the mollusca. It is often found to be penetrated by different mineral substances; and our knowledge of certain shells is owing to their having been penetrated by a chalcedonious matter, which has kept them in a state of preservation.

The study of fossil organized bodies has taught us, that, after the crystallizations took place, which are observed in granite, porphyry, and other primitive substances, which contain no vestiges of bodies that once were animated, the waters covered these crystallizations; if, indeed, they had not already been formed in their bosom, as every thing leads us to believe; for we pass from these last, without any sensible interval, to strata which contain organized bodies, which most assuredly have existed in the waters.

We are informed, indeed, by some scientific men, that above certain strata, containing the remains of organized bodies, crystallizations have been found similar to those of the granite; but circumstances of this kind, which may be referable to volcanic action, are so rare, and the places where they have been remarked, of such confined extent, in comparison to all the surface of the globe whence nothing of the sort has been found, that no satisfactory inference can be drawn for the establishment of any general principle respecting such phenomena.

Our observations can never enable us to ascertain whether or not the primitive substances which we behold have been preceded by one, or many other worlds of greater antiquity, which may be covered by them; but, admitting that they have been preceded only by other substances similar to themselves, we see that life originally commenced with aquatic animals, of species and genera very different, for the most part, from those which exist at the present day.

In the most ancient strata are found trilobites, orthoceratites, ammonites, belemnites, encrinites, terebratulites, and many other genera, the greater portion of which no longer exist in the living state. Of those, too, which still exist, some, such as the encrini, which at present, in our actual seas, are of the rarest occurrence, were once so common, that their debris, connected by a calcareous cement, constitute, of themselves alone, strata of very considerable extent.

If any doubts can be raised relative to the crystallization of primitive substances in the waters, scarcely any can exist respecting that in which organic remains are found, and which evidently appears to have been operated in that liquid medium. On this hypothesis, if it do not deserve a better name, it is probable that the waters, which contain the elements of those crystallizations, contain little or nothing of them any longer at the present day, since we do not find that genuine petrifactions are formed now, as formerly. Nevertheless it appears, as we shall show hereafter, that certain crystallizations, which have taken place after a preceding one had engaged the bodies which we find in a fossil state, must have been operated subsequently to the retreat of the waters.

We may believe that certain strata, such as those of the phyllades and of the chalk, have been deposited in fluids which had the property of destroying or dissolving certain calcareous substances which existed in these strata, but where no further traces of them are to be found at the present day.

If we have nothing but analogy to lead us to such a conclusion respecting the phyllades, it is not so with the chalk, which exhibits phenomena, conferring a character of certitude upon our inferences to this effect.

In the strata of phyllades, we find, in general, only trilobites, and other turbinated bodies, such as ammonites, and the shell of which exists no longer; but these strata must have contained a much greater number of marine bodies which have been destroyed. What leads us to this belief is, that, in the time when those trilobites existed, there also existed a very great number of other marine animals. The proof of this is to be found in many localities; and, among others, at Dudley in this country, and Chimay in the Netherlands.

Since, then, it is clear that, at the period when the trilobites existed at Dudley and Chimay, there also existed, in the same places, a great quantity of marine animals, why should we not extend the same conclusion to the formations of phyllades and transition limestone, and believe that, with the trilobites of these latter strata, there existed an equal quantity which have disappeared? This conjecture is corroborated by the almost positive certainty that a great number have been dissolved in the superior chalk without leaving any traces behind. Moreover, the trilobites, and other turbinated bodies, which are found in the aforesaid depositions, lived on animals, and probably on testaceous animals—the traces of which would be found, had they not been dissolved or destroyed in the course of the time in which the schistose phyllades were being formed.

It is probable that the absence or presence of organic bodies in the strata of the phyllades, has occasioned certain of these depositions to be ranked among primitive substances, and others to be considered as intermediate; for the superposition of the primitive rocks can no longer be a guide in this case, since the example of the granite of Christiana, which rests on a stratum with orthoceratites; but the organic remains being already very rare in certain strata of phyllades, may it not be possible that they are still more rare, or have disappeared altogether, in those which are ranged with primitive substances?

Certain families of mollusca, such as oysters and gryphites, in passing to the fossil state, have preserved their shelly covering in all localities and all strata. Others, such as those of the volutæ, cyprææ, crassatellæ, and others, have disappeared in almost all the places where crystallization or petrifaction has taken place. The terebratulites are preserved almost every where; still, in certain ancient strata, such as those of Vologne, Coblentz, Tenior, in the Alleghany Mountains, and in Virginia, they have disappeared, leaving only their internal and external moulds.

The polyparies, the serpulæ, and, in general, all such shells as adhere to certain bodies, are better preserved than the others.

The solid parts of the stellerideæ, echinodæ, and encrini, in passing to the fossil state, are changed into calcareous spath, which breaks into rhomboïdal laminæ; and it is always easy to ascertain if these bodies are fossil, by the spathic state in which they are found. The testa of these animals is very often preserved in the chalk, where so many other bodies have disappeared; but in some places, as in the Alleghany Mountains and some parts of this country, the stems of encrini have disappeared, and have left nothing but their impression.

It must be admitted that the testa of some shell-animals may, in certain strata, be changed into an irregular crystallization. Were it not so, we might believe that the bodies which have the most exact form of shells, both univalvus and bivalvus, which are found in the environs of Caen and Bayeux, in an oolitic bed, lower than the chalk, and which are often disengaged from their crust, were not genuine shells. It appears that the testa of those which these moulds represent, after having disappeared, must have been replaced by a crystallization which assumed precisely all its forms. It is quite certain that, on breaking them, instead of any fibrous testaceous substance, we find that those bodies are composed of nothing but crystals. The different species of pleurotomaria, the ammonites, the modiolar cypricardiæ, the testa of which is very thick, and some other shells of the same strata above mentioned, are in this predicament.

The belemnites do not seem ever to disappear in this way, and they are met with even in the chalk, and in localities where all soluble shells have disappeared. On breaking them, they are always found composed of a sort of crystallization, consisting of needles, or sharp lines, radiating from the centre to the circumference; but, as they have never been found except in the fossil state, we cannot be assured that they were not organized in this manner, previously to having passed into that state; and we cannot form the same supposition for them as for the radiated echinodoemes and encrini. What, however, appears very certain is, that, before they passed into the fossil state, they were of a solid and calcareous matter; since some are found which were pierced and inhabited by pholades, and to others serpulæ are found adherent.

In disappearing from other strata, as well as the chalk, the testaceous covering of the mollusca has left the mould of its forms, both internal and external. This mould is so exact, that it represents, in all its parts, the lines, the striæ, or the slightest asperities, which belonged to the original production. Such of these moulds, as derive their origin from the animal kingdom, have been called helmintholites, entomolites, ichthyolites, amphibiolites, ornitholites, zoolites ; such as are derived from the vegetable kingdom are named phytolites.

The external moulds being entire, and often without the slightest fracture, the testæ on which they have been found cannot possibly have left them in any other way than by solution, after the soft matter, in which they were imbedded, had undergone a crystallization or petrifaction, which thus preserved all the communicated forms.

In certain strata, such as that of the Mount St. Pierre of Maestricht, for example, we find that some external moulds of univalve shells are filled, only through one-half of their length, with a substance similar to that of the stratum, as if this substance had not been in sufficient quantity to fill the entire mould. We cannot, however, be certified that this was the case, since we see that, in these external moulds, there remain very well-formed portions of the internal mould. It seems more reasonable to attribute the cause of this singular fact to a partial solution of the latter, after the formation of these impressions.

Although we are unacquainted, at the present day, with any agent capable of producing a similar dissolution, without also attacking the calcareous mould which surrounds these bodies, still we cannot attribute their disappearance to any other cause than the action of the waters, and other fluids, which are continually traversing the earth, from its surface to its profoundest abysses.

If the waters have the power of dissolving the calcareous

substance which is no longer found in the mould of the testæ of mollusca, they must have carried it into places at a greater depth, or, perhaps, they may have formed new crystallizations*.

We are informed that, in the environs of Amberg, a very considerable quantity of the alveoli of belemnites was found; while the exterior envelope of those fossils was extremely rare, and scarcely ever was found in an entire state.

M. de Rance, from whom the substance of our observations on this subject is borrowed, not having seen those alveoli, does not venture to pronounce on their nature and origin. He thinks, however, that the alveoli have been preserved, only because they were seized by a petrifaction, which filled the cavity of the belemnites at the same time that it enveloped their bodies, and formed, in all probability, the stratum in which they are found. If they alone exist at present, it must be because the shells, which contained these alveoli, have been dissolved subsequently to the petrifaction. Still, even supposing this to be the case, we ought to find the mould of their external forms.

In certain localities, as at Montmartre, are found, in the marly strata, *models* or moulds, in marble, of marine shells and of crustacea, without any appearance of an external mould having existed of a different nature from the model. In breaking the marl, these models become detached from the rest of the mass, and represent exactly the external forms of the shells and crustacea. They are covered with a yellowish sort of

* It may be as well to explain here the sense in which the word *mould* is used by naturalists. The *external mould* is the vacancy left by the testaceous covering of the fossil body which has disappeared in the localities where petrifaction has taken place. The *internal mould* is the name given to the paste or substance of the stratum which is moulded and petrified in univalve or bivalve shells. The word *model* might, perhaps, be used with more propriety to designate whatever has filled these shells before their dissolution; though, strictly speaking, it is a model not of the animal, but of the place which it occupied.

stucco, or plastering, which appears to be the cause of their thus detaching themselves from the mass.

If we do not admit that the shells and other bodies become changed into marl, it is very difficult to explain the formation of these models—both the internal and external mould being the same substance. Had there been a disappearance of the testa of the shells, as in the other localities, it would be necessary to suppose that a petrifaction had seized the bodies; that afterwards the marl was moulded; and, finally, the mould itself was changed into marl. But it must be confessed that these transmutations are not very easy to be comprehended; and, perhaps, without straining at any better explanation, we may be contented to conjecture that all the calcareous substances contained in the stratum have been converted into marl.

When the hipponyces are found in a stratum, where there has been a disappearance such as we have been speaking of, they present a very singular phenomenon. Their upper shell, which is composed of a substance analogous to that of the cyprææ volutæ, and other soluble shells, has disappeared, leaving only its mould; while the under, which is of a foliated texture, like that of oysters, remains untouched, with the exception of that part where the adductor muscle is found. This organ, which is displaced, or at all events extended, in proportion to the growth of the animal, has furnished, on the side of the under shell or support, the same soluble matter which it furnished to the extremity by which it was attached to the upper, or shell proper; so that, when this last, and its thick support, were in circumstances proper for their dissolution, the shell, and the place of the support where the muscle was attached, have alone disappeared, and the rest of the support has remained untouched.

The lirostrites (Lamarck), as well as the spherulites, exhibit, in like manner, very singular facts in their petrifaction. Their testa, or at least that of the lower valve of the first, the contexture of which is analogous to that of the oyster, is preserved. An internal mould, petrified and free, is found in this valve, but does not fill it altogether. A pretty large empty space is found on one side; and this space must necessarily have been occupied by a body, or by a soluble portion of the shell, which has disappeared after the petrifaction of the mould.

The internal moulds of the spherulites, or of shells analogous to them, are still more singular, in that, independently of two considerable depressions which come forward in this mould, there are two large holes which traverse it from one part to the other. Some of these moulds are, as it were, foliated. It would seem that the interval between each foliation must have been filled by solid and soluble bodies, which have disappeared since the petrifaction of the mould. Nothing that is known in the living state can assist us in conceiving what may have been the organization of animals which have left similar moulds behind them.

Neither do we know if the petrifaction which has seized these bodies has been rapid.—We might suppose it to have been so, on considering the moulds just mentioned—which might lead us to believe that certain soft parts of the animals may have been destroyed by it, or before it took place; and that others, such as more solid muscles, capable of resistance, have disappeared since: but it is very difficult to form satisfactory conjectures on this point. It appears certain, however, that in some cases, relative to these moulds, the soft matter was insinuated into, and petrified in, very narrow vacancies; and that, what surrounded them having disappeared, there remained very slender laminæ.

The baculites have not been observed hitherto, except in strata analogous to the chalk, or near that substance, and where the testa has disappeared. Frequently the interior moulds of their numerous partitions do not adhere one to the other; so that some portions of this singular shell, composed sometimes of more than thirty of these moulds, which hold together by their parts dove-tailed, seem to be articulated. They are never invested with crystals, like the ammonites, of strata more ancient than the chalk. In the paste, or substance of the stratum which fills the last of their partitions, there are found a prodigious quantity of little shells, or debris of polyparies and other marine bodies. The same is the case with the other partitions when the mould is not perfect; which would lead us to believe that, in this instance, the testa has been destroyed on one of its sides. But, with respect to the moulds which are perfect in their circumference, and which may be supposed to have been formed in complete shells, that of each partition is composed of very fine paste, without any mixture of organic remains—the marginal siphon having been too narrow to admit them to pass.

These remarks are equally applicable to the ammonites, which are often found with their testa, though still more frequently without it. In the first instance, it often happens that the last compartment is filled with the paste formed by the stratum in which they have been deposited; and that the other compartments are filled with a fine paste, or merely invested with crystals. In this case, we see that the fluid in which this stratum has been found contained two distinct substances—namely, the opaque matter of the stratum, and that which, being filtrated through the testa, or through the siphon, has formed the crystals, and furnished the crystallization which has hardened the stratum. It may be supposed that the animals which inhabited these shells were able to live in the waters which held in solution the substance of the crystals; for, when they were abandoned, this did not fall, or remain at the bottom of the sea, but after having been filled with the water which surrounded them; and it is difficult to believe that this water was expelled by any other which could have deposited the crystals in question.

Certain ammonites having been filled by quartzose sand, their internal mould is formed of sandstone, and what remained of the testa has been changed into silex. That of certain shells, found in the green sand of Blackdown, are in like manner changed into this substance. Shells are often observed which have been seized by the silex, or internal moulds, wherever formed of it; but it may be remarked, that the testa is but rarely changed into this substance.

We may believe that the matter which forms the siphon of the ammonites is not exactly the same as that of the rest of the shell; for it has sometimes resisted, while the other parts have been dissolved.

In certain localities, both in France and England, in the strata of the lower chalk, ammonites are found, the testa of which, in the last turbination, after having been filled with the matter which composes the stratum, seems to have disappeared—while that of the partitions, the siphon, and all that was internal, have been preserved; so that, in those particular parts, these shells are seen with their testa, such as they were when abandoned by the mollusca which formed them. There is reason to believe that the waters in which these shells, as well as the baculites, existed, and with which they were filled, did not contain substances proper for the formation of crystals—as in the more ancient depositions of Nevers, Caen, and others. Doubtless, the absence of these substances is the cause why the chalk strata are not found petrified, like the last mentioned.

Respecting the ammonites whose testa has disappeared, there remains only the internal mould; and the mould of the exterior, in concave, and very generally all the turbinations, have been soldered together after the disappearance of the testa. It was also after this disappearance that the vermicularia, which were attached to the testa, and which did not disappear with it, became, as they are at the present day, adherent on the internal mould. The proof that they did not become attached on the mould already formed, is found in many specimens of the internal moulds of ammonites; in which the folded edges of the last partition are soldered, without any intermediate substance, and form a whole, with the mould of the whorl, or turbination, which serves them as a support; although, before the petrifaction, there existed in this place a double thickness of the testa —namely, that of the interior of the last whorl, and that of the exterior of the preceding one.

The great and small oysters, which constitute the stratum by which the environs of Paris are covered, have been preserved untouched, with the balanæ, serpulæ, and flustræ, with which they are often charged, while shells of another genera, found along with them, have only left their mould; as may be remarked at Montmartre, Pontenai-aux-Roses, and other localities.

We have seen that the belemnites never disappear: it is not so, however, with their partitions, which appear to be of a substance different from that of the shell. They are preserved in some strata more ancient than the chalk; but as yet, we believe, no example has been found of their preservation in this last. When they are preserved, they are found either totally filled with crystallizations, or with a paste which has a tendency to separate between each partition; or, in fine, some partitions alone are crystallized, and others filled with a fine petrified paste. In no case, however, has the substance with which they are filled any relation with the contexture of the singular shell on which they depend; and what fills the alveolus does not entirely resemble the substance of the stratum, except when it has been filled after the partitions had been destroyed, either in consequence of their fragility or solubility.

It is very remarkable that, in the chalky mountain of St. Pierre, near Maestricht, the claws of crustacea are commonly found not referable to any other genus than that of pagurus, and yet the shells, in which those crustaceous animals must have lodged, are not to be found; while, on the other hand, in the strata of the Placenza, where there has been no petrifaction, and nothing has disappeared, univalve shells are often observed, covered with a polyparium; the presence of which, as well as the form of their aperture, proves evidently that they have been inhabited by paguri—and yet these last are not found. But, perhaps, it may be that, if any are found, their fragility does not permit them to be gathered, for they are never seen in collections.

It may be asked, if the oolites, which are met with in strata presenting ammonites, were already formed when the shells existed, or if they were formed at the same time in which the petrifaction of the stratum took place. The state in which these shells are found, as well as the belemnites, may assist us in the resolution of this question.

Ammonites are found, whose partitions, especially the more recent, are filled with oolites; but whether the testa of these shells were in a perfectly entire state, does not appear to be clearly ascertained. Considering its fragility, we may suspect that this was not the case; since that of some of them, which is filled, and which is six inches in diameter, is not much thicker than a sheet of paper. But in those species, whose testa is thicker, and which are well preserved, oolites are seen only in the last compartment, which is always open, and the others are filled with crystals. From all that has been observed, we may conclude that the oolites formed a part of the deposition in which they are seen, before the shells were filled; and, if it can be imagined that they were formed contemporaneously with the petrifaction, we may believe that the fluids, which deposited the crystals, did not contain the elements of the oolites, or that it lost them by filtration through the testa, or in passing through the siphon. It is the same with the alveolus of the belemnites, which is filled with oolites when the partitions have been destroyed, but in which they are never found when the latter are entire.

There are oolites which differ very considerably one from another. In some localities, as in the environs of Caen and Bayeaux, they are round or ovoid; their surface is shining; their colour ferruginous, and their strata concentric. A little point, of a clearer colour, appears to serve them as a centre, and in some of them two such points are distinguishable. In some localities of the same countries, they are smaller than the last, flatted, and some, which are of a larger size, exhibit varieties of flatted forms. These oolites, in general of regular forms, are met with in strata which, from the preservation of the fossils which they contain, appear to have been tranquil, and seem to differ essentially from those of other places,—for example, of Nevers and Auxerres. These last are found in white beds, of which they constitute the major part. They are accompanied by debris of shells, of polyparia, and other marine bodies. It appears that these depositions have been exposed to considerable convulsions, for there remain only of certain univalve shells, extremely thick, some very short and mutilated portions. These we may recognize by their spathic and shining brilliancy, and by their form, the remains of the stems of Encrinites, some flatted fragments, and some which appear to be debris of bivalve shells, though they have not the contexture of the latter. Others, which are rounded, are filled with crystals in their centre. The surplus of the mass is composed of oolites of different sizes, from the bulk of a poppy seed to that of a small pea. Some of the larger ones seem to be formed by an agglomeration of the smaller. The whole is cemented by a white and transparent crystallization.

These oolites are white, and appear to have been formed by the bruised matter of shells, and other marine bodies, mutilated debris of which are found along with them. Considering the state of disorder in which they are found, we may be justified in attributing to them a different origin from that of the others which we first mentioned.

What we remark in certain marbles which contain marine bodies, would lead us to the conclusion that they had undergone petrifaction at several successive times. The first petrifaction which probably took place in the waters would have formed the ordinary coloured stratum, which surrounds them in all their parts. From some cause, with which we are not acquainted, longitudinal fissures have taken place in this stratum in all directions, breaking the shells and other marine bodies thus fixed, and leaving a certain interval between the broken parts. A second petrifaction or spathic infiltration of a white colour, supervened to fill exactly not only all the fissures, but also the concave mould of the shells which had disappeared, as may be remarked in certain black marbles.

A third sort of petrifaction seems to have taken place in tho *breccie.*—For, in the debris of which they are composed are found fragments which appear to have been divided and then rejoined by a spathic crystallization which has no analogy with that which unites together all these fragments. Certain marbles seem to have broken twice in the same places, since the same clift is found filled with two parallel infiltrations, one of which is white and the other yellow.

M. de France mentions an orthoceratite in his possession, found in the stratum of brown marble of Valognes. This fossil is traversed, in different directions, by sinuous veins of calcareous spath from half a line to two lines in breadth ; and, what is singularly remarkable is, that one of these veins traverses in their diameter some partitions, whose separated parts do not correspond one before the other as before the separation. This fact seems to prove that the marine body, filled with paste, must have been cleft after its petrifaction, and that the calcareous spath then succeeded, and was crystallized in the cleft. Still one can scarcely conceive, according to what is seen to take place in our own days, how two clefts could have taken place within half a line distant from each other. There is nothing analogous to a fact of this kind, except the clefts produced by humidity in a chalky stone, or dried potters' earth. How, again, is it possible to explain certain spathic veins nearly parallel, and sometimes very closely approximated to each, which are found to traverse some of those shelly fragments, and which, without destroying them, cut most exactly all the shells and other marine bodies of which these marbles are composed? Simple desiccation could never have produced an effect like this, or divided into such small parts shells or polyparia, as we witness in the specimens alluded to. These shells themselves are even sometimes cleft, and filled with spath, while the paste or stratum surrounding them is not so.—Facts like these have perhaps not been sufficiently studied, and assuredly they merit every degree of attention.

It appears that it is more rare to find in the strata anterior to the chalk localities in which marine bodies, which have disappeared, have left their place empty, than in the strata posterior to that substance.

M. de France, without disputing the reasons which have determined geologists to give to the strata posterior to the primitive rocks, the names of intermediate or transition, secondary and tertiary, thinks that he is justified in making three different divisions of those in which organic fossil bodies are found;—namely, the strata anterior to the chalk, those of the chalk itself, and those posterior to the formation of that substance.

According to this arrangement of M. de France, we find that the strata anterior to the chalk contain forty-seven genera of polyparia, seven of echinodæ, five genera of crustacea, one genus of annelides, three of scapulæ, one of cephalopodes monothalami, one of cirrhipedes, forty-four genera of bivalve shells, one of phyllidiæ, fifteen of univalve shells, ten genera of partitioned shells, three genera of marine bodies little known, three genera of reptiles, eleven genera of fish, and twelve of vegetables.

In the strata anterior to the chalk, univalve and bivalve shells are found in a proportion, the difference of which is not very remarkable. In the lower strata of the chalk, univalve shells are still to be found; but this is not the case in the upper chalk. There we scarcely ever find any univalve unilocular shells, such as cerites, volutæ, and other soluble shells; and the marine bodies met with there belong to families which resist dissolution in the localities where the others have disappeared.

It is extremely probable that, in the upper chalk, univalve shells did exist, as well as in the preceding strata; and that they have disappeared, leaving no trace behind, because this substance has not assumed a consistence, or crystallization, capable of preserving the forms of the shells, or other marine bodies which it contained, and which have been dissolved. This we cannot avoid believing, when we find, as before instanced, the supports or bases of the hipponyces in the chalk, without uniting the shells which they sustained; and when we see that the oysters, the lower valves of the craniæ, those of dianchora, the spirorbes, and other adherent shells, which are found in the chalk remote from all other bodies, bear the traces of polyparia, and other testaceous marine bodies on which they have adhered; and yet we do not find any of those said bodies in the same formation.

In one specimen of the chalky substance of Mount St. Pierre, in the possession of M. de France, and which was sufficiently solid to preserve the external and internal mould of a species of large cerithium, on which some oysters had adhered, the testa of the univalve shell has disappeared, while that of the oysters has remained untouched.

Similar examples are found in the green sand in this country, which is under the chalk. The oysters are perfectly well-preserved, and the stratum is of such a consistence, that the form of the univalve shells, on which the oysters adhered, is preserved also, but the testa has disappeared.

In some localities the chalk acquired so much consistence, previously to the dissolution of the shells and other marine bodies which it contained, that their forms are still to be found, and may be seen with the *belemnites macronatus*, and other bodies which essentially characterize this substance, a prodigious quantity of pectunculi, of baculites, of gervilliæ, of ammonites, and other shells which are never seen in the upper chalk of the environs of Paris. But it may be remarked that the bivalve shells are there in a much greater proportion than the others.

The silex, which is found in the chalk, has seized upon shells and other marine productions, and the echinodæ, in particular, are often filled with it; but it is very remarkable that it has only seized on bodies belonging to the class of those which are usually preserved.

Some authors tell us that the silex found in shells has been formed by the animal matter which they contained; but it is easy to be convinced that it could not have had such an origin in the echinodæ which are filled with it, and the exterior of which is covered with oysters, with craniæ, and with other bodies, which could not have lived there but after the death of the animal on whose testa they are found. If these examples are not deemed sufficient, many others may be adduced, which prove that the siliceous moulds have been formed in the ananchites after the interior of the testa had been invested by crystals of calcareous spath.

The siliceous moulds of galerites, and other echinodæ, which are found at the surface of the earth without testa, have very probably been covered by silex while they remained in the chalk strata from which they have been derived. The testa was probably destroyed when the strata were washed and carried off by the rains, and the shells thus exposed to the injuries of the weather and accidental shocks; for, in the chalk strata, no siliceous moulds are found without being accompanied by the testa in which they were formed.

By rubbing with a brush a piece of the chalk found at Meudon, in France, after having moistened it, it appears sonorous when it is dried; and it is observed to be traversed in all directions by veins, like the marbles of which we have already spoken; but no shells in it appear to have been traversed by those veins, as we have seen to be the case in the marbles.

The state in which they are found would prove, on the contrary, that those veins have not broken or traversed them.

The chalk contains nineteen genera of polyparia, two of stellerides, or rather debris, which may belong to four genera established in this family, eight genera of echinodæ, two of crustacea, one of annelides, three of serpulæ, twenty-five of bivalve shells; the genus *planospirites*, little known; twelve genera of partitioned, or comparted shells, two of fish, two of reptiles, one of vegetables, and (which is remarkable for the smallness of the number) four genera of univalve shells.

It would seem that the chalk strata have not been placed in circumstances proper for the production of marbles, for it does not appear that any have been recognized in them.

Silex is found abundantly in the chalk, and in the more recent strata, but it is more rare in the ancient strata. The wood which is found in the latter is not so generally siliceous as that which is found in more recent depositions. It is very rarely met with in a calcareous state.

Some varieties of that species of silex called *petrosilex molaris*, (vulgo, millstones,) contain shells, while others do not; but there is reason to believe that all those, which are in circumstances analogous to those of this silex which contains the shells, did formerly contain some, which have disappeared.

Some of the shelly strata, posterior to the chalk, as that of Grignon, in France, are not petrified, and their compactness is comparatively trifling. In others, as at Dané, a department of the Marne and Loire, and at Saillencourt, a department of the Seine and Oise, the shells, polyparia, and the debris of other marine bodies, are deposited lightly one over the other, and connected by their point of contact with a crystallization almost imperceptible, so that the mass is porous and incapable of retaining the waters. At Sainteny, a department of La Manche, a similar stratum exhibits all the marine fossil bodies, and their debris, invested with a slight brownish-coloured crust.

In the strata of coarse, or, as we term it, crag-limestone, in the environs of Paris, as at Grignon and Chaumont, department of the Oise, the univalve shells are filled to the top with shelly sand. This sand is frequently found loose, without any adherence to the shell; but in certain shells it is found petrified in those turbinations of the spire which are most remote from the aperture, although it is not so in this last part.

In the Bahama Islands, in our own days, a stratum similar to that of Dané has been found. The debris of bivalve shells belonging to that part of the world have been found in specimens of this stratum, with colours. There is an aggregation of the bodies which compose it, but no petrifaction.

Certain fragments, which probably come from the Mediterranean, are composed of a very hard and cavernose limestone, which impastes the valves of the little moulds or modioles of the vermicularia, or serpulæ, the interior of which has remained empty, and invests in like manner the debris of the same testacea. On the exterior of these fragments are polyparia of different genera, serpulæ, with some portions of coral, and the lower valves of craniæ. Other pieces, which are found in the sea, on the coasts of the department of Calvados, are composed of grains of quartzose sand, of shells, of moulds, with their red colour above and mother-of-pearl inside, of debris of balanæ, of asteriæ, and other polyparia. As the debris of marine bodies which are contained in these pieces, and especially in the first, do not appear to differ in any thing from those which are not fossil, we may believe that these petrifactions are not so ancient as those which are found in the strata of the earth. With respect to the polyparia found in the second, they are strangers to our climate, and have been detached from the cliffs on the sea-coast, and depend on the stratum with polyparia in the neighbourhood of Caen, so that even if it were well proved that those petrifactions are recent, it is yet clear that they are composed of the debris of beings which have existed at very remote periods. It is remarkable that the first fragments, which we have just mentioned, are not pierced by shells or perforating animals like the ancient limestones found in the Mediterranean.

In some localities there are beds of the coarse limestone which are composed only of miliolites, and other very small marine bodies, either entire or in debris, without any other mixture, and always without any adherence. A stratum of this kind is found at Beyne, near Grignon. The strata anterior to the chalk exhibit nothing analogous, either in the smallness of the marine bodies or in the want of cohesion between them. Where these last were found, the species were not equally numerous, nor do they appear in general to have been so small.

It may be remarked, in general, that in the strata anterior to the chalk, bivalve shells are to be found with their two valves very often united, or the internal mould of these two valves, which proves that they were so at the moment of petrifaction. This is not the case in the other strata, and especially in that of the coarse limestone, where it is very rare to find bivalve shells entire, and there is, perhaps, no exception to this but in the stratum of the Placentine territory.

It has been remarked that organic bodies, found in the fossil state, differed from those now existing in proportion to the antiquity of the strata in which they were discovered. This remark is fully confirmed by a summary view of the living and fossil genera, and the peculiar circumstances in which the latter are placed. We find, that out of five hundred and two genera of polyparia, stellerides, echinodæ, annelides, serpulæ, cirrhipedes, and shells, eighty-nine genera are not found in the fossil state; two hundred present themselves both in the living and fossil states, and one hundred and fifteen in the fossil only. On uniting these two last numbers, and observing in what particular strata they are met, we find one hundred and thirty-four genera in the most ancient strata, seventy-five in the chalk, and two hundred and three in the strata posterior to that substance. If we examine in what strata are to be met the genera which are found in the living state, and also in the fossil, we shall ascertain that the most ancient contain sixty-five, those of the chalk forty-two, and the most recent, one hundred and seventy-two. A different proportion obtains respecting the bodies which are found in the fossil state alone; the most ancient strata contain sixty-three genera, the chalk contains one-and-thirty, and the most recent strata thirty only.

It has been said, that ammonites were found in the London clay, but this report does not seem to rest on very sufficient authority.

M. de Humboldt, in his work on the Independence of Formations, has stated, that among the fossil shells the univalves predominate, as they do at this day, in the living state, under the tropics. The following is the result presented by a tabular view of them:—The number of the genera of univalves exceeds that of bivalves by eleven for those of the living state only; twenty-four for those found equally in the living and the fossil state, and five for those in the fossil state exclusively. It is less by sixteen for those found in the strata anterior to the chalk, and by nine in twenty-five for those found in that substance: but it becomes greater again in those found in the most recent strata, for there the genera of bivalves rise only to fifty, while the univalves amount to eighty-nine.

With respect to species, the number of univalves in the living state exceeds that of bivalves by eight hundred and fifty-eight; and in the fossil state, the number of univalves exceeds that of bivalves by four hundred and thirty-three.

It was supposed that there had been remarked, at Orglandes and Hauteville, in the department of the Manche, a bed of coarse or crag-limestone, analogous to that of Grignon, and situated under a stratum of chalk foundation. But this could not have taken place without some catastrophe, which displaced the strata, or, perhaps, the beds of chalk formation having left empty spaces between each other, into which the composition of the lower stratum might have introduced itself.

Were it otherwise, we should be obliged to withdraw our assent from the fine observation of the Baron, on the ever-increasing analogy which subsists between the beings which exist at the present day, and those which are to be met with in the strata, in proportion as they are more recent.

The *falunières* * of Orglandes and Hauteville contain a great number of genera, which are found at present in the living state, and which have no analogy with the fossils of the chalk, nor with those which are more ancient. If the last, which contain ammonites and belemnites, was more recent than the falunières, how could it happen that the shells of these genera are never to be met with in the falunières, since it is only in the chalk that they have disappeared? There appears, on the contrary, every reason to conclude, that the debris of beings contained in the falunières are of an epoch more recent than the chalk.

We know not what would happen if some genera of animals existing at the present day became extinct—but we may well imagine that that would be a very remarkable era for fish and insects, in which they ceased to be devoured by sharks and

* In the province of Touraine, in France, they give the name of *falun* to a loose sandy stratum, composed principally of the debris of shells, which, in consequence of its nature and easy disaggregation, is employed as marle or manure. It is considered by the French geologists to belong to the formation of the lower layers of the coarse limestone, or that with cerithia, of the neighbourhood of Paris. *Falunières*, accordingly, is the name given to strata composed of shells and other marine bodies, broken in part, and in which there is little cohesion, such as those of Touraine, which are of very great extent; those of Hauteville, above-mentioned, of Grignon, in the department of the Seine and Oise, of Courtagnon, department of the Marne, and some other localities. These *falunières* depend, as we said before, on the coarse marine limestone formation, and not on more ancient strata. Some of them, such as those of Touraine, which are composed only of debris, the angles of which are worn down, appear to have been exposed on some shore to the action of the waves. But, in all the others, the most fragile things are often preserved entire, and the angles of the broken bodies are very sharp. This seems to prove that the last-mentioned strata were placed in different circumstances from those of Touraine.

swallows. Now, as we are certain that a very great number of genera found in the fossil state exist no longer, we may conjecture that, after their disappearance, some changes must have taken place in the situation of the beings then existing—the nature of such changes we cannot appreciate, but we observe that the number of genera has augmented in the coarse limestone.

It is stated that in this stratum, at Grignon, more genera and species are found than could be found on any of the French coasts. This M. de France believes to be the case, in consequence of the temperate character of the actual climate of France. But he does not hesitate to believe, that between the tropics, where the seas contain a much greater quantity of mollusca, the coasts or bottom of those seas abound as much in debris of testaceous marine bodies as the stratum of Grignon, and that it cannot be doubted that this stratum was formed under a climate analogous to that of the tropics. The nautili, and many other fossil genera of this locality, not found living except in the hot climate, establish the certainty of this point.

There does not appear to be any remarkable difference between the fossils of Europe and those of America. At the mouth of the Alleghany river, and on the banks of the Mohauk river, near Utica, in the state of New York, are found trilobites, encrinites, terebratulites, and other shells, which must have come from very ancient strata, and in which the testa has disappeared. A specimen of sandstone from the summit of the Alleghany mountains was found filled with internal moulds of debris of the stems of encrinites. Beyond the rivers of Genesée, in proceeding to the falls of Niagara, the internal siliceous moulds of shells are found, both univalve and bivalve, which may be suspected to belong to the coarse or crag-limestone. Specimens have been seen from Virginia, which appeared evidently to have been derived from a bed of this last-mentioned formation. They contained petunculi, arcæ, mactræ, selenes, bound together with a coarse and quartzose sand. These shells can scarcely be said to differ in any respect from those of the same species which came from the same formation in our own parts of the world.

In North Carolina are found naticæ, large pernæ (*perna maxillata*), venericardiæ, pectines, and other shells, which have many relations with similar species which have been met with in the Placentine. These shells are free, filled with a yellow quartzose sand, and seem to depend on the coarse limestone, or on other formations less ancient than the chalk. But M. de France states, that he has never seen any fossil from America which exhibits any relations to the strata of the last-mentioned formation.

In the coarse limestone, which is certainly a marine deposition, shells are to be found, whose genera are no longer in existence, except in the fresh water:—such are the ampullariæ, the melaniæ, and the cyclostomata. These genera, too, with the exception of the last, are found, at the present day, only in climates warmer than our own. Indeed, respecting the subject of climate, in relation to fossils, every thing is calculated to excite astonishment, and nothing admits of explanation. The case is very nearly the same, in relation to the genera which formerly inhabited the sea, and which are no longer found except in fresh waters, unless we admit a difference in the degree of saltness in the former. This, indeed, must be more considerable at present, than it was before the number of ages, since the sea occupied our continents, had elapsed;—during which, the rivers and streams have been, and still are, incessantly supplying it with saline particles, which it receives to return no more. But, if we admit that the sea, having been less saline, might have allowed of the existence of certain genera in its waters, in former days, but which can exist there no longer, we must also admit that such genera as live there in the present day, and which lived there at the era in which the coarse limestone was formed, have been enabled to support a higher degree of saltness. Now, the fact is, that one supposition is no more easy to be conceived than the other.

On the whole, however, it must be observed, that it is not the presence of very precise characters which enables us to distinguish marine and fresh water shells. What usually fixes the judgment respecting the last, is the recognized identity of certain genera, or species, which have never been met with in the living state but in the fresh water, and which have never been found as fossils in marine formations.

As to fossil remains in general, they have a greater or a less analogy with what is now existing; and this analogy is more or less easy of verification.

In plants, for instance, we easily distinguish the fossil wood, the family of monocotyledon trees from that of dicotyledons; but it is not so with the genera. This difficulty, however, may proceed from the contexture of woods, in the living state, not having been sufficiently studied.

The study of fossil stems, leaves, and fruits, has led to the recognition of many genera, which have been distinguished by M. Adolphe Brogniart. But there are many vegetable debris which, as yet, no botanist has been able to refer to any thing analogous in the existing state of the vegetable kingdom.

We have already amply seen that the long labours of the most illustrious comparative anatomist of the day have made known a wonderful number of species of cetacea, of reptiles, of birds, and of mammalia, many genera of which have disappeared from the surface of the globe, and, perhaps, have never been known in the living state to man; but of *his* remains, as we formerly remarked, nothing has been found, nor yet of those of the quadruma; and, hitherto, it has only been in strata more recent than those of the coarse limestone that any fossil remains of the mammifera have been discovered.

We have also seen, with respect to fishes, that, being in a great measure composed of soft organs, which have been destroyed before petrifaction could have seized them, they have left most frequently only their skeleton, scales, or impression. Their remains may sometimes conduct to the knowledge of the genus, but rarely, if ever, to that of the species.

The case is different with the testa of aquatic or terrestrial animals, which has often been preserved untouched in the sands or in the rocks, or which has not disappeared from the last without leaving the trace of its internal and external forms.

This preservation permits us to recognize the genera and species, and to appreciate the degree of analogy which they may possess with what is found at present in an existing state.

But, in general, it is difficult to pass decided judgments in this respect: to do so, it would be necessary to determine what it is which constitutes the species, and to fix the line of demarcation between that and the variety, if any do exist.

Observation has demonstrated, in reference to the testa of animals which are provided with one, that very sensible differences exist—1st, between individuals of the same species, taken in the same locality; and, 2dly, between the same species, taken in different localities, either in the living or in the fossil state.

These differences consist in the size, in the absence, presence, or number of ribs, of tubercles, or of striæ; or rather, in some localities, these characters are hardly visible, while, in others, they are sometimes very strongly marked.

There are sensible individual differences between the *cardia rustica*, taken on different coasts, such as those of Rochelle, of Cherbourg, of Normandy, and of Dunkirk. It is the same with the fossils; and a species in the possession of M. de France (*pleurotoma dentata*), taken in ten different localities, varies in its forms according to all the localities. Specimens from the Placentine, for instance, are in general much longer and less bulky than those of the environs of Paris.

Among the shells, in the living state, many species being distinguished only by their colours, and this character being wanting in the fossils, these last should exhibit, and do in fact exhibit, much fewer species in certain genera.

We can clearly recognize the identity of some fossil species with the living: in other cases, we only find an analogy; and there are some in which the relations are much more remote. These three different states, or circumstances, are well expressed by M. de France, in the terms *identical, analogous, subanalogous.*

With the exception of a trochus, and two or three species of terebratulites, which are derived from strata anterior to the chalk, and which have some analogy with species now existing,—and, again, of one species of the last-mentioned genus, found in the chalk, and which appears to be identical with *terebratula vitrea,* it is only in the strata more recent than the chalk that identity or analogy can be observed.

It is very remarkable that the greatest number of identical or analogous species are found in the strata of the Placentine, and of Italy generally—for, to two hundred and forty there are one hundred and sixty in this predicament; and of these, one hundred and thirty-nine have been marked as identical by M. Brocchi. It is possible that this distinguished naturalist may have a little overrated the number, or not placed the strictest bounds between identity and analogy.

The upper marine sandstone formation, of the neighbourhood of Paris, appears to contain a less number of marine fossil bodies than that of the coarse limestone; although some species are found in the first which are not to be met with in the other. Some, from both formations, appear to be identical. But the most part, very nearly the whole, present nothing more than analogy; for out of fifty species from the upper sandstone, M. de France found but three which perfectly resemble those of the coarse limestone. In the latter, some species are larger and others smaller than in the former; which is the case with other different species in the sandstone. In fine, some species, which are very common in the coarse limestone, are very rarely found in the upper marine sandstone.

The bulbiform fusus, which is found at Grignon, appears to

be also found in the stratum of upper sandstone of the environs of La Chapelle and Louvres, in the department of the Seine and Oise; but it is so much modified in the latter, that it has been classed in the genus pyrula, before it was discovered that this modification arose from the influence of a stratum, not the same with that of Grignon. This pyrula was not found in the last place, no more than the fusus in the other. This error had crept into the system of Invertebrated Animals, by M. Lamarck; but it was before the strata in question had been distinguished by the researches of MM. Cuvier and Brogniart. This species is also found in England, in Hampshire, but with additional variations of form.

Taking the environs of Paris, as it were, for a centre, we find, according to the remark of M. de France, that the genera and species of the coarse limestone have a tendency to analogy with those of Italy, as we proceed through Anjou, Touraine, and the environs of Bourdeaux; and in the same manner we find a growing analogy to certain English fossils in those of the department of the Oise.

A multitude of genera, found in the fossil state in Europe, are not only no longer existing, in the living state, in our seas and waters in general, but, for the most part, are found to inhabit at present only the equatorial regions.

The number of genera found in the fossil state, is superior to that of the living genera, in the polyparia, the echinodæ, the annelides, the tubicolæ, the bivalve, and comparted shells. It is inferior in the serpulæ, the cirrhipedes, the pteropodes, the phyllidi, the univalve shells, and the heteropodes.

The number of the genera of fossil crustacea, being scarcely more than a third of those now existing, we may believe that the latter have increased since the revolutions which have buried the remains of those now found in the fossil state.

Notwithstanding the greater number of some families of marine organic bodies found in the fossil state, we may believe that that of these bodies, now existing in the living state, is greater than it has been at any other era, if we consider that our era, geologically speaking, is one single period, while the fossil remains are the product of many successive ages.

The ammonites being met with in Europe, in America, in India, and, according to M. Lamarck, in all countries, exhibit a genus which has been able to exist under all climates, if climates were distributed over the earth in the same manner formerly as they are at present; or, on the contrary supposition, the universality of this genus would lead us to believe that, throughout the whole earth, the temperature was the same, or that it has undergone a series of successive alterations.

The nautili and spirulæ, being, among existing genera, those which have most affinity with the ammonites, and living only in climates the temperature of which is very elevated, we may believe that in which the ammonites lived was similar. If the presence of the ammonites, in the polar regions, could make us believe that in them the temperature was elevated to the degree in which it is at present in the equatorial regions—as we see no reason why the temperature of the latter might not have been augmented, as it is now, in a relative proportion to the entire quantity of heat over the globe—it will follow, by a necessary consequence, that the inteıtropical parts of the world must have been uninhabitable.

If such was the case in the first age of the world, and that subsequently the globe became cooler, life must then have commenced in the polar regions; and if the globe became more and more refrigerated with the progress of time, those regions must have been the first to become deserted. With this theory many facts accord; and it has the additional advantage of coinciding with the doctrine of those learned theologians who have fixed the site of the garden of Eden exactly at the north pole.

Certain it is, that the presence, in the northern countries, of the remains of animals and vegetables which, in consequence of the cold, could not exist there now, lends some foundation to this hypothesis. Still, however, a conjecture of this kind requires the support of a much greater number of facts before it can receive the full assent of any cautious philosopher.

There are certain genera, such as oysters, mussels, &c., which are discovered in a fossil state in all countries, like the ammonites. But we cannot pretend to deduce from this fact the same consequences as from the universality of the ammonites, inasmuch as oysters and mussels are found, in the living state, in all the climates of the earth.

Identical species, in localities different or much remote from each other, are rare among the fossil shells—especially if, by identical, we understand exactly similar. M. de France has met with but one genuine example of such identity, which is presented by the *bulinius terebellatus*, which is found at Grignon, altogether similar to that which has been collected from the strata of the Placentine. It is with other species as with the *auricula ringens*, which is found in the strata posterior to the chalk in this country, and in the environs of Paris, in Touraine, in the neighbourhood of Bourdeaux, and in Italy; but in each of its localities, this species is rather analogous than identical. However, if we only take into consideration the general modification of all the species taken in different localities, we may be justified in regarding them as identical; and none of the species, found in the strata posterior to the chalk, are to be found below the coarse limestone.

Above all the strata, whether marine or of fresh-water formation, which appear to have been deposited in waters more or less tranquil, another is found which presents itself in the neighbourhood of Paris, in the basins of the Seine, the Marne, the Oise, and the Loire, and doubtless of many other rivers; and in which are found the debris of all the other strata, mixed up with the bones of terrestrial mammiferous animals and cetacea.

This stratum, which shows itself immediately under the ordinary soil, and even sometimes at its very surface, is not petrified. It exhibits, however, in the plain of Grenelle, some siliceous agglomerations, which are found at a tolerable depth. Its thickness varies, and probably in proportion to the more or less elevated situation of the stratum on which it reposes. It commences on the road to Orleans, near the Grand Montrouge, by some inches of thickness, and proceeds, increasing, to above eighteen feet of depth, in the plain of Grenelle, near Vaugirard. Afterwards, it extends, ascending towards the north of the other side of the basin, as far as the forest of St. Germain.

All that is observed from this forest, as far as Montrouge, proves that the waters have filled this space; and they could not fill it without doing the same to very extensive distances, to east and west in the basin, the lowest parts of which are occupied by the Seine. Neither can it be doubted that the same is true of the basins of the Marne and Oise, since they are caused by a similar stratum.

It is extremely probable that these basins were formed when they were filled with water by the event, whatever it was, which deposited the stratum there, as it is equally probable that they were more deep and extensive, since there is a stratum deposited, the thickness of which, in some places, is more than eighteen feet; unless we may suppose that, in the commencement of the irruption, the waters had carried off some portions of the strata over which they had flowed, and that those were replaced by the bodies, carried on by the torrent, when it diminished in intensity. This supposition assumes some degree of probability, when we find that all these bodies are foreign to the place in which they are deposited; and that, near the bridge of Sevres, are seen prodigious blocks of pudding-stones, of nearly thirty-six cubic feet, and which have been torn from strata indubitably very remote, since nothing similar is known in the environs of Paris. Almost the whole of the bodies thus deposited are either quartzose or siliceous; the calcareous fragments have been bruised. Some fossil shells and limestones have been found there, dependent on the formation of coarse shelly limestone, and which are foreign to the strata of the environs of Paris; but they are worn and mutilated. Fragments of siliceous wood are also found there, and many bodies carried away from the chalk strata.

The waters which swept off the enormous masses of rock in this valley, and which deposited rolled flints as far as the elevation of Montrouge and that of the forest of St. Germain, must have been considerably elevated above those places to have been capable of making such deposition upon them.

We cannot doubt but that a violent current, which admits neither of appreciation nor comparison, deposited this stratum. Regarding the direction of this current, some doubts indeed may be raised, but there is every reason to believe that it proceeded in the direction of the actual course of the Seine. The volume of water was so considerable, that the declivity of the soil which causes that river to flow in the direction which it at present takes, might not perhaps be sufficient to establish this conjecture; but the pieces of red granite which have been found at Issy and the Bois de Boulogne, which are also found in the basins of the Oise and Marne, and which are thought to have been detached from those of Burgundy, countenance the opinion that the torrent came from that side rather than from the side of Normandy, where no similar granite is to be found.

The rounded flints, which we find on the shores of the sea, have been forced to take this form by the periodical return of the flux and reflux, which rolls them for a considerable time in the same place. But this is not the case with stones of the stratum of which we have been speaking, which have been worn and rounded by rolling together in the direction of the torrent, and constantly removing to a greater distance from the place where they were first seized by the inundation.

The sand with which the bottom of the Seine is spread at the present day is composed, like that of the plain of Grenelle, of small fragments of granite or of quartz, which have remained angular in consequence of their hardness, and of rounded debris of calcareous substances, which render it probable that this sand also is a dependence of the stratum carried into the basin by the torrent. It is very likely that this sand always descends more and more, for fresh supplies of it may be drawn, in those places where it appeared to have been exhausted.

It is indubitable that when the water was above the elevation of Montrouge that it covered a very considerable extent of soil, both to the right and left of the course of the Seine, and especially in the valleys where those streams and rivers flowed which are received by the Seine. But the total absence of any deposition of rolled fragments beyond the limit of Montrouge seems a convincing proof that the torrent did not pass that limit, and that the waters spread through the valleys pretty nearly in a tranquil state. It is, doubtless, to those tranquil waters, which deposited the most tenuous particles of the earths, and other bodies, carried down by the torrent, and which they held in suspension, that are to be attributed the considerable strata of argillaceous earth which cover the environs of Sceaux, of Bagneux, of Arcueil, of Chatenay, and probably of all the places where the tranquillity of those waters permitted a similar deposition.

Had the waters of the torrent been so elevated as to cover all the heights in the neighbourhood of Paris, a current must have been established above them, which would have transported the rolled flints beyond Montrouge, and even into the valley of the river of Bievre, on that side. It would have carried off the entire of the marine depositions of fine and quartzose sand which cover the summits of the little hills of the neighbourhood, and of which they are even sometimes composed; neither would it have permitted the deposition of fine substances composing the argillaceous earth. We may believe, however, that they were high enough to have formed, in retiring, the ravines which are seen in those depositions of sand.

From observations on the tract to the south of Fontenai-aux-Roses, we may believe that a great part of certain steep and rugged hills have been carried away by waters, the current of which was not above them. The bottom of the valley is covered with a thick stratum of argillaceous earth. Ascending these to the fountain of the mills, on each side of the way, you may see the oyster bed, which follows the movement of the stratum, and which was already in a state of inclination when those animals existed in this place. On proceeding to the summit of the hill, the quartzose sand is found under the superficial soil, disposed in horizontal strata, which could not have had this disposition if the valley had not been filled with them, at least in part, when the deposition took place.

As to the disappearance of these sands, whose origin must be attributed to the waters which covered those places when the valley in which the Seine flows was filled, it may be likewise referred to the same waters which deposited the sands, and which, on return, must have carried them away in part.

The deposition of tenuous substances, which compose the argillaceous earth, appears to be one of the last in the neighbourhood of Paris; and it would seem that, from the period of its deposition, this substance has not been in circumstances favourable to crystallization; at all events, it has not been found in that state in any part of those environs. All its strata seem to be owing to some torrent, whose waters, though troubled, were turned aside, and became comparatively tranquil. It is yet to be desired, that, by ulterior observations on the position of the strata of rolled flints found in so great a number of places, we may ascertain if all the depositions of argillaceous earth have been furnished by torrents which have deposited rolled flints.

The marine sandstone being found at the summit of all the heights of the environs of Paris, we may be assured that the waters of the sea have covered all those heights. These waters could not be there without having extended to very considerable distances, as well in France as in other countries. They must also have retired either slowly, or with rapidity. If they retired slowly, all the parts, which are dry at present, must have been shores in succession. Everywhere we should find the traces of cliffs and escarpments, such as we now find on the shores of the sea—and everywhere we should find flints rounded by the waves. But nothing of the kind is to be seen. There is every reason to believe that the retreat of the waters was rapid, and this is the general opinion. In the last case, when the level of the waters had come to that at the bottom of the sea, and even before they must have furrowed it in retiring, and gaining, on different sides, the lower grounds, they must have found, in various directions, the long valleys, at the bottom of which the streams and rivers flow at the present day, and the principal of which are covered with rolled flints.

The bottom of that sea, which once covered the tracts of land now inhabited, may have been unequal, like that which the sea leaves at the present day. But the correspondence of the strata on each side of the basins leads to the belief that they have, in general, been formed by the breaking up of the latter; and the bodies which they contain, and which are foreign to the places in which they are found, prove clearly that the strata of more elevated countries must have been broken up to furnish them.

Before we quite dismiss this subject of petrifactions in general, it may not be unamusing to our readers briefly to notice certain speculations, which have been put forth, relative to the process of petrifaction itself, or the change of an organized body into stony matter.

This subject was meditated very long and profoundly by the ancient naturalists, and even by theologians, because it is connected with the history of the creation of the world, and consequently with the traditions which constitute the basis of our modern religious creeds. A crowd of authors have dilated upon it with a spirit of rivalship in absurdity never to be sufficiently admired. Some, as we have observed before, would have it, that these fossils were mere sports of nature, the result of the *corruption* of stones; others, of a loftier vein, attributed their production to the stars, and, more especially, to the rays of the moon, which were in the habit of *eating the stones*,—an odd species of banquet enough for such a subject, and which reminds us of Nat. Lee's project,—

"To fatten padlocks with Antarctic food."

These and such like errors were for a long time accredited, and propagated even down to the middle of the last century, at which time, also, some explications of the formation and depositions of fossils were presented to the world, scarcely less remarkable for singularity than the foregoing, and equally false. Voltaire, for instance, who, with all his genius, was exceedingly superficial in scientific subjects, has speculated on this as erroneously as he has done on many others. But such reveries were speedily dissipated by the sudden light thrown upon the natural sciences towards the close of the eighteenth century. Patient and judicious observation has evinced, that the buried or petrified remains of organized beings are the result of a series of revolutions of the globe, more or less numerous. Some of them were, to all appearance, general, or nearly so, since the marine debris have been deposited as far as an elevation of more than 7200 perpendicular feet above the natural elevation of the sea; others were partial, such as those from which originated the fossils which we attributed to the fresh water; others, we find, resulted from the eruptions of volcanos, the lava of which had caused considerable spaces of soil where organized bodies, animal or vegetable, were found.

A question has been raised on this subject, involving some degree of interest. It is this:—"Are new fossils still being found in our present days?" Voltaire cites the example of a certain M. le Royer de la Sauvagère, who, assisted by his vassals and his neighbours, had twice witnessed, in the space of four-and-twenty years, a part of the soil of the environs of his estate of Desplaces, in Touraine, metamorphosed into a bed of tender stone. The shells seen there were at first so small, that they could not be discerned without the assistance of a microscope; but afterwards they increased with the stone, so as to assume, by invisible degrees, ten lines in thickness. If, however, we set aside this observation, and pay attention to those of more practised naturalists than M. le Royer de la Sauvagère, we may say that the present formation of petrifactions, at least in the interior of the earth, is not established on any positive fact, and that simple reason does not warrant the affirmative of this part of the question. In the bosom of the waters, however, such an operation may take place; for it appears, that the immersion of bodies in a fluid, dissolving the matter which petrifies, is a necessary condition to the petrifaction.

A great number of petrifactions being siliceous, the fluid in which they have been found must have had the property of dissolving the silex. Now, we are acquainted with no fluid, at present in abundance in nature, which possesses this property. The same may be said of solvents of carbonates, fluates, or phosphates, which are also equally unknown *.

Two facts only seem to prove the possibility of the recent formation of fossils. The first is, that many travellers, who have examined the coasts of New Holland, have seen, in different parts, some very curious and singular petrifactions. Riche, in the bay which the French call De l'Espérance, having penetrated into a valley sunk between downs of sand, found it covered with calcareous trunks of trees, broken towards the root, and the stumps of which, standing upright, were not

* We have merely heard of certain hot waters containing potash in solution, which had the property of dissolving silex, and which deposited stalactites of chalcedony.

more than a foot in height. On a level with the soil might be distinguished the knots, the ligneous layers, and all the other durable accidents of vegetation. Some stems were more than a foot in diameter. Lesueur, Péron, and Bailly, found similar petrifactions in the Island of Decrés, in that of Josephine, and on some points of the lands of Leuwin, of Edels, of Endracht, and of White. They were leaves, fruits, branches, and roots of vegetables; bones of quadrupeds, and even their excrements. They have endeavoured to explain this formation by saying, that the very fine calcareous and siliceous sand, which borders the coasts of this new continent, is raised by the wind, deposits itself on bodies, and becomes incrusted there, and subsequently assumes so much solidity, that if the branches of this sort of lithophytes be broken while the incrustation is recent, the ligneous tissue may be observed engaged in a solid case, and without any remarkable alteration. But, in proportion as the calcareous envelope increases, the wood becomes disorganized, and changes by invisible degrees into an arid and blackish detritus. Then, the interior of the tube is as yet empty, and preserves a diameter pretty nearly equal to that of the branch which served it for a mould.—Finally, this tube finishes by becoming obstructed, and filled with quartzose and all calcareous particles. Some years elapse, and all is converted into a mass of sandstone. At this last period, the arborescent form alone indicates the ancient state of vegetation.

In certain parts of New Holland there exist elevated downs formed of very fine sand, susceptible of a solidification more or less rapid—at the reverse of these mobile hills grow various species of arbustæ, and even large trees, such as *Banksia* and *Eucalyptus*. In such a position, all the sand which the rains, winds, and storms precipitate from the summit of the downs, deposits itself at the foot of these trees. it rises insensibly along their stem, reaches their first branches, and concludes, in the long run, by burying them under its ever-increasing masses—the vegetable tissue becomes altered in the trunks and branches; the substance of the ligneous layers, being much more solid than that which fills their intervals, is decomposed much more slowly than the latter. From this proceed those concentric circles which give to those extraordinary incrustations the appearance of genuine petrifactions. But, however, on examining them carefully, it is easy to see that those pretended petrified trees are nothing but masses of a sandstone, more or less hard, which preserved only the form of the vegetables which have served them for moulds.

Such are the observations of MM. Péron and Lesueur. In adopting the views of those naturalists, then, we cannot consider these trees as true petrifactions. We find in them, however, a mode of incrustation altogether singular, and to which we know of nothing analogous except what is remarked on the roots of trees, which penetrate into a sandy and ferruginous soil. An example of this is seen in the roots of oaks, in the environs of the Sablonnière, of the marsh or pool of Auteuil, in the Bois de Boulogne, near Paris. These roots are changed into hollow tubes, which are tolerably thick, and it is observed that the oxide of iron serves as a cement to unite the quartzose particles.

The second fact in favour of recent petrifactions is to be found in the *Bibliothèque Universelle* for July, 1818. Mr. Mackenzie describes there a petrified tree, which was observed near Pennycuilk, ten miles from Edinburgh. The trunk alone exists, which springs vertically from the ground some feet. It is almost four feet in diameter from its base; its roots sink into the earth in different directions; and, in a word, it appears to have grown in the place in which it was found. Its substance is now a genuine sandstone, and what remains of the bark is in the state of pit-coal, as is often the case with fossilized wood.

In this fact there is an analogy with the one first cited, in the circumstance of the substance of the tree itself being changed into sandstone: as to the root, the means of explaining this transformation are entirely wanting.

Thus, having no examples, under our inspection, of the mode in which true petrifaction is effected, we are obliged to form hypotheses respecting its formation. In a field, therefore, of this kind, where imagination is allowed unbounded latitude, we need not wonder that system after system has been built up, and pulled down, with unparalleled industry and rapidity,

The hypothesis most usually admitted consists in supposing that the stony matter is substituted for the animal or vegetable in proportion as the latter is decomposed, and (according to the opinion of a celebrated mineralogist, M. Haüy) because the substitution takes place successively, and, as it were, molecule by molecule, the stony parts arranging themselves in the places left void by the retreat of the ligneous or animal parts, and, moulding themselves in the same cavities, take the impression of the vegetable or animal organization, and copy its forms exactly. According to the same writer, in the petrified wood, the organization is destroyed, and the appearance of it alone remains.

M. Patrin, allowing the ingenuity and plausibility of this theory, states certain facts, which, according to him, render its admission impossible. Among others, he instances the trunks of trees converted into silex, which are found in the midst of mobile sands; and he is astonished that the fluid which held the stony matter, which has taken the place of the molecules of the wood, has not agglutinated and converted into quartzose sandstone, the sand of which touched this petrified wood—this consequence appearing to him to be inevitable. He denies that the organization is destroyed, because the fibres of the petrified wood, scarcely discernible by the microscope, have completely preserved the form and situation which they had in the most perfect state of the wood, and that, moreover, the colours have not changed. Now, continues this author, if the stony molecules had taken the place of the ligneous molecules, all the petrified mass would be of an uniform colour, since the same stony matter would have successively filled all the places left empty by the retirement of the ligneous particles.

To this observation we may answer, that it is easy to conceive that each fibre of the wood has had its parts replaced successively, in such a manner that its general form and direction have been in nowise altered; and that the number of fibres of a piece of petrified wood may be found equal to that of the fibres of a piece of living wood of the same volume. As to the colours, it appears to us that the observation respecting them is more just; nevertheless, there are many specimens of petrified woods, all of which have the same tint.

M. Patrin also refuses to admit the preliminary decomposition of the petrified wood; and, on this subject, he instances many specimens of petrified woods which contain holes made by worms, and even the worms themselves, and their eggs, changed into agate. From this he concludes that, if the fluid, charged with siliceous matter, had deposited this matter, it would also have filled the vacant spaces which we have just mentioned. To this, again, may be plausibly opposed, that the siliceous matter may have been deposited by an affinity near the molecules of wood, without being deposited elsewhere; and thus the vacancies would have been preserved.

According to M. Patrin, the decomposition took place suddenly; for, from the moment in which substances so soft as worms had experienced putrefaction, they would have been so much deformed that not the least distinguishable appearance of them would have remained.

To this, once more, we may reply, that the fluid, which dissolved the siliceous matter, might have possessed a conservative property for the worms; and that, also, the change of the latter into silex may have been more prompt than that of the wood. Moreover, it would be desirable that the existence of these worms were fully authenticated by zoologists; for, with the assistance of fancy, or preconceived theory, foreign bodies might very easily be taken for them, though, certainly, there would be nothing very extraordinary in the fact of their existence.

The entire system of M. Patrin is briefly this:—He thinks that petrifaction is a genuine transmutation of the parts themselves—of the organized body—into siliceous matter: so that a body was by so much the less susceptible of petrifaction, in proportion as it was more decomposed at the period in which it was buried. The petrifaction took place in an almost sudden manner. It must be regarded as a chemical operation, and a combination of gaseous fluids with the constituent principles of organized bodies: an operation which very rapidly changes the latter into stony substances, without touching, in any way, the arrangement of their molecules; so that neither the forms nor the colours are at all altered by this modification.

We may form a just idea of petrifaction by comparing it to congelation—with this difference, that ordinary congelation takes place by the simple abstraction of caloric, whereas petrifaction is a coagulation occasioned by the introduction of a foreign fluid. With respect to the weight which petrified bodies acquire, there is nothing in that in opposition to this theory; for we know how much density the most subtile gaseous fluids may acquire when they come to be solidified;—such as oxygen, for example, when combined with metallic substances. A striking instance of this may be observed in vitreous tin, which is an oxide of tin without mixture of any other matter; and there is only wanting 3 to 4-100 to make its specific weight equal to that of the pure metal, although the oxygen alone makes more than 21-100 of the mass. It is, therefore, susceptible of condensation to the degree of acquiring a much greater weight than that of any stone; and it is infinitely probable that it is the oxygen which plays the principal part in the phenomenon of petrifaction, by its combination with the phosphoric principle which is developed in all organized bodies. It is well known that the most celebrated chemists have regarded earthy substances as oxides ; and every thing leads us to believe that this conjecture is highly probable.

Small rock crystals, with two points, were found by Demeste between the fibres of the heart of a petrified tree ; which fibres were ligneous and combustible, while those of the circumference were entirely petrified. According to Patrin, it would appear that these were owing to the elementary principles of the wood, which were disengaged, under a gaseous form, by the effect of *putrefaction*, and which, finding themselves free in these interstices, had formed there crystals by the effect of the same chemical combinations which had converted into silex the ligneous parts which were not altered.

It is probable that, among these principles of the wood, and of all organized bodies in general, we should reckon as essential a phosphoric principle—as their phosphorescence, at the time of decomposition, more than seems to prove. Now, it appears certain that phosphorus is equally a constituent principle of quartz, according to the observations of Dolomieu.

It is easy to perceive that, of the two hypotheses we have now explained, that of Patrin is by no means the most simple. He also employs it to refute the explanation, given by M. Haüy, of the formation of a nucleus of pure silex, which often takes place in the interior of shells and of fossil echini, according to this philosopher, by the intromission of a liquid, charged with stony molecules, into the cavity of these shells, and ursini. With Patrin, the theory of the gases suffices for all this, and is contradicted by no fact in nature. He says, that we may well suppose that a gaseous fluid penetrates the entire mass of a substance so porous as chalk; and as this fluid cannot produce the matter of silex but by its combination with the fluids contained in organized bodies, it converts into silex only the substance itself of the molluscous animal contained in the shell. When the interior part of this body, which is the most exposed to the operations of external agents, is altered by putrefaction, or otherwise destroyed, it is only the remaining part which forms the siliceous nodule found towards the point of the shell. When the animal is totally decomposed, the shell remains empty, or has been filled only by the chalk itself, when the latter was of a pasty consistence. The testæ of echini, and the scales of shells, are most frequently found in their natural state, or have been converted only into calcareous spath (as the belemnites, for instance); because these bodies contain too little animal matter, and it is too much screened by the calcareous earth of which they are composed, to allow of the siliceous petrifaction. But they may be easily changed into calcareous spath by a water charged with carbonic acid, which operates insensibly the crystallization of their molecules.

Without pretending to decide at all between the different hypotheses now laid before the reader, we may, without adopting, be permitted to say, that the least complicated is preferable to all others,—that which does not admit, *à priori*, imaginary phenomena of which no clear idea can be formed, but presents marked relations with all that is already well known concerning the laws of crystallization.

We shall take the liberty of indulging in a few more general observations, connected with the invertebrated fossils, before we proceed to any notice of their principal families, as we are anxious to collect as much interesting matter as we can, and to weary our readers as little as possible by more specific details.

The fossils, as may have already been seen, which are incomparably more multiplied than all the rest, are shells, and other marine productions. They form of themselves a considerable portion of the calcareous matter of which the most recent strata are composed, which caused Buffon, and other writers, to imagine that all calcareous substances were derived from the debris of marine bodies. But this hypothesis is completely destroyed by observation: for, independently of the primitive limestone rocks, which are evidently anterior to all species of organization, whether animal or vegetable, and the existence of which we may suppose to have been contemporaneous with the formation of the terrestrial globe itself, we observe that the most ancient secondary calcareous strata contain extremely few vestiges of calcareous bodies, whose existence had scarcely commenced when these early strata were formed.

From this point, that is, from the transition strata, the number of marine bodies gradually increases, in an immense proportion to the antiquity of the strata which contain them. We have already noticed the diversity of species, which, on the contrary, is in a direct proportion with the antiquity of the strata. The older the strata, the more different the species from any now existing; and even where there is an approximation of form between them, there is generally an immense superiority of size on the side of the fossils.

It would appear that the first living beings which were found in the ocean were some small shell-animals,—such, at least, are the only animals which have left any certain traces of their existence in the most ancient secondary strata.

On the assumption that this globe was originally submersed in water, we may suppose that when the surface of the ocean was sufficiently lowered to permit the light to arrive at the summit of the mountains, some zoophytes were formed there with solid body and fixed habitation, and that these multiplied progressively, as well as the shells, in proportion as the solar rays could exercise their vivifying action on more extended spaces in the bottom of the seas.

After the retreat of the sea from above the continents of the globe, and before the valleys were completely formed, the rain waters must have collected in numbers of places, and formed vast lakes, the depositions of which gave rise to those strata, little remarked before our own times, and to which the name of fresh-water formations has been given, because the fossils which they contain are very similar to the bodies with which we are acquainted in a living state in collections of fresh water. This leads us to the presumption that they have existed in a fluid of the same nature.

It has been said, and frequently repeated, that the majority of the fossils have their living analogues, but that they are either in the deepest abysses of the ocean, or in very remote seas. But the strictest and most extensive observation establishes a multitude of differences between the fossil bodies and those which have been regarded as their living analogues.

We shall now proceed to the notice of a few of the principal families and genera to which invertebrated fossil remains belong. We must content ourselves with very brief observations on this part of our subject; for to treat it at large, or enter at all into specific details, would occupy a space totally inconsistent with our limits. There are many admirable works, which those who are disposed to study this subject minutely may have recourse to with the greatest advantage. Among them, for the fossils of our own island, the Messrs. Sowerby's work on "Mineral Conchology" stands pre-eminent.

The Ammonites are a genus of shells of the class of univalves, the characters of which are a discoïd spiral, with contiguous turbinations all apparent, and the internal parietes articulated by sinuous sutures; they have also transverse partitions, lobated through their centre, and pierced by a marginal tube.

The fossil shells, which compose this genus, derive their name from their resemblance to rams' horns, the symbol of Jupiter Ammon. They are the *cornua Ammonis* of oryctologists. They exhibit very great relations with the *nautili;* but they differ from them essentially by having all the turbinations of their spires visible, while in the nautili they are concealed in the last one. They also differ by having their partitions always more sinuous, and their tube or siphon always placed under the keel of the back.

The ammonites are regarded as pelagian shells, that is, as

having lived in the ancient sea; because, hitherto, no living analogues have been recognized of them, and they are only found in mountains of tolerably ancient formation. They are to be seen of a monstrous bulk, nearly six feet in diameter: sometimes they are accumulated to such a degree as to form entire rocks. The oryctologists have engraved many species, but their works are, in general, so little methodical, that great difficulty is experienced in studying them. Fossil ammonites are sometimes to be met with, under their testaceous form, without any stony concretion in their interior. Their structure is then very visible; and it was on species of this kind that Bruguières established the genus, which, before his time, had been but simply indicated.

Lamarck has separated from the ammonites the species which were not articulated, to form a new genus, under the name of PLANULITES.

The ammonites have at all times formed a very striking object of human contemplation, whether we consider their bulk, their abundance, or the places in which they have been found. In India they constitute, or rather their moulds, an object of veneration to the people, under the name of *Salagraman*, because it is believed that one of their gods is concealed therein.

Bruguières, in the *Encyclopédie Méthodique*, mentions twenty-two species of ammonites, most part of which have been figured by Bourguet and Langius; but this number might easily be tripled with the species found in France alone. The chain of secondary mountains, which extend from Langres as far as the environs of Autun, that one near which the town of Caen is built, and many others, contain such immense quantities, that the roads are paved with them. It is usually in very argillaceous schists, in very calcareous and ferruginous argillæ, and in the lower chalk, that they are to be met. They are also found frequently in calcareous rocks, and are often adherent thereon by one of their sides; a fact which is not explained in a very satisfactory manner. Some of them are occasionally pyritous, or have been so, and have become iron ore. Some have the surface smooth—others are marked with striæ, or ribs, —others with tubercles, &c.

The animals, which produced the fossil shells of the genus NAUTILUS, were contemporaneous with the ammonites in the earlier ages of the world; but they have been enabled to resist the causes which proved destructive to the latter, the most recent debris of which are found only in the lower chalk. But the debris of the nautili are to be met with, even up to the era in which the most ancient strata of that substance were deposited, in the crag-limestone of later origin, and also, in the living state, in the equatorial regions. The early species were, however, incomparably more numerous than the present, since scarcely two are known in the living state, while, perhaps, more than thirty have been recognized in the fossil. This genus is of the very small number of those which are found in the ancient as well as the most recent formations; and if nothing of it has been found in the strata of the upper chalk, it is probably owing to the soluble nature of those shells which have disappeared in that formation.

The species of this genus are rather difficult to determine; because, the testa having disappeared in so many instances, there only remains the internal mould, which does not present all the characters of the shell. Those found in our own country are described and figured by Messrs. Sowerby, with their usual accuracy—one of which, the *N. zigzag*, is a very singular shell, and has not precisely either the characters of the nautili or those of the ammonites. It might constitute, says M. de France, a genus approximating to the latter, because the tube, or siphunculus, appears to be marginal; but Mr. Sowerby places it nearest the inside. Denys de Montfort makes a peculiar genus of it, under the name of aganide; but in his figure he places the siphunculus in the centre.

M. de France considers that varieties of this same shell have been regarded as ammonites by Mr. Sowerby, under the names of *A. sphæricus* and *A. striatus*.

A very considerable number of species of fossil nautili, found in England, and described and figured in the work of Messrs. Sowerby, have been met with in the strata anterior to the chalk, or in the most ancient strata of that substance. One of these, the *N. imperialis*, from the Highgate clay, we have figured. The coarse shell-limestone of the environs of Paris contains some pearly specimens; some of which have been referred by M. Lamarck to *Nautilus pompilius*, which is found, in the living state, in the great Indian ocean. M. de France has other specimens in his possession, which he thinks may belong to *Nautilus umbilicatus*, found at present in the same habitat.

At Haudan, in the department of the Seine and Oise, and at Dax, are the remains of a large species, (*Nautilus Deshayesii*, De France,) which is very remarkable, because, independently of its siphunculus being very near the last whorl, and formed by kinds of funnels which enter one into the other, each of its septæ presents on either side a conical depression, which ends in the direction of the siphunculus against the interior paries of the sides of the shell, but which does not communicate with the chamber which precedes.

It is probably something similar to this which is found in certain nautili above the siphunculus, near the last whorl, which has given rise to the belief, that some species had two siphunculi; and on this character Denys de Montfort established the genus Bisiphite. But an examination of those shells which appeared to have two siphunculi seems to have proved that this was a mistake.

Nautili have been found in the fossil state in Italy, various parts of France, Germany, and the Low Countries, &c. Some of these shells are a foot and a half in diameter.

In Eiffel, a canton of the duchy of Juliers, in the environs of Chimay, and in some parts of Ireland, in very ancient formations, fossil shells of several species have been found, which De Montfort had formed into a genus called by him *Bellerophon*. One of the characters which he assigns to this genus is the having of smooth septæ, pierced by a siphunculus. This is one of the many errors published by this author. Two of these species, one from De Montfort's own collection, were in possession of M. de France. Suspecting that they were single-chambered, this gentleman sawed one of them transversely, and found, in fact, that it was rolled on itself like a nautilus, but had no septæ; so that instead of entering into the division of the polythalamous cephalopodes, this genus should be placed in that of the monothalamous, near the argonautæ. Our figure, from Mr. Sowerby's work, was found in Westmorland.

Several species of ORTHOCERATITES are found in England, as may be seen by the work of Mr. Sowerby. The following are the characters assigned to this genus in the "Mineral Conchology"—shell straight, or but little curved, fusiform, with septæ traversed by a siphunculus; the edge of the septæ smooth, with one or two slight undulations.

In his voyage to the North Pole, M. de Buch found, near Drontheim, a black limestone filled with orthoceratites, in the neighbourhood of Christiana. Some of them are several feet in length. They are plain on one side, convex on the other, and traversed by a siphunculus throughout their entire length. There is reason to believe that these tubes were cylindrical or oval, and that the plain part has been destroyed, which has happened to many ammonites, &c. M. de Buch says, that, at Konigsberg, he has seen some which were three or four feet long.

Orthoceratites have been found in a variety of other localities through Europe, and even in Siberia.

By the word BELEMNITES is designated a petrified shell, of a conical, straight, and elongated form, whose structure appears to have much analogy with that of the ammonites. It is composed of two distinct parts, the case and the alveolus. The first presents, in general, the form which we have just mentioned. Sometimes it is rounded towards its extremity, and then terminates in a small point. At other times it is swelled in two-thirds of its length, and resembles a spindle. It may also be flatted on the sides, and present lateral ridges. In general it is hollowed at its surface, which is smooth, with a longitudinal furrow. When it approaches the cylindrical form, the interior is marked with circular striæ, or ribs. On being broken transversely, it presents a sort of crystallization composed of needles, which proceed radiating from the axis of the cone towards its circumference. On its being divided longitudinally, these needles incline towards the summit a little. But we observe, at the same time, in this section, chiefly in the cases of belemnites which have been polished, a series of longitudinal lines which set out, two by two, from the axis under a very acute angle, and direct themselves towards the base. They indicate a series of layers or cornets of stony matter, inclosed, as it were, one in the other. The case has, in its interior, a conical cavity, which extends, in the most entire belemnites, as far as the moiety of its length. Sometimes it has them but for a third. This cavity is exactly filled by the alveolus, formed by an union of cells, the number and extent of which vary much, according as the partitions or septæ which intercept them are more or less distant one from the other. There are also four small kind of caps, the convexity of which is directed towards the summit of the cone; their centre is sometimes pierced by a siphon, or canal, which at other times is on one side. It traverses all the small cells, and communicates only with that of the base or the last, as is the case with the nautili, or ammonites.

It is rare to find belemnites with their alveoli. But they are extremely common without this part, which is also found separately, but more seldom. The majority of naturalists who have spoken of these bodies have not seen them together —from this has arisen the variety of opinions advanced upon their origin. Some have termed the alveolus, *orthoceratite*, and have observed in this part of the structure an evident analogy with that of the ammonites. But we must not, however, conclude that all that has been called orthoceratites is referable to the alveoli in question. The case of belemnites has been successively taken for a stem of echinus, or for the tooth or dorsal spine of a fish. It is not so long since M. de Luc advanced that it was the tooth of a soft fish.

The belemnites are known only in the state of petrifaction. They are usually of a calcareous nature; and it is very common to find them encrusted in marble with ammonites. Those of Meudon, near Paris, are semitransparent, and in the state of yellowish calcareous alabaster:—their size varies considerably. Some of large dimensions are found in Sweden and Norway: the ancients discovered them on Mount Ida.

The living analogues of the Belemnites are utterly unknown, and the genus appears to have been destroyed; or, if it do exist, it is only in the profound abysses of the main ocean.

Passing over an immense number of the genera of fossil shells, on which nothing general can be advanced of interest, we come to the TEREBRATULITES, or fossil *terebratulæ*. This genus, like that of the ammonites, in which the species are so numerous that each locality would seem to furnish some peculiar to itself, presents exceeding difficulties in the way of determination; for the forms of the shell, when the animal is young, differ from those which it exhibits when it has acquired its full dimensions. There are few species of terebratulæ in which the under edge is not terminated by folds more or less great, or more or less numerous; but if we examine one of the shells which has these folds, we shall find that they do not yet make their appearance, while the shell is only half of its proper length. To prevent the mistaking of species for what, in fact, are only the differences of age, it would be right to consider, as species, those shells only which appear to have acquired their full size; which may be judged of, either from their thickness, or where the locality presents a good number of individuals of the same size and form.

The same locality often presents us with one species smooth and another folded; but a great number of species are not found mingled together.

The terebratulæ shew themselves in the strata anterior to the chalk, in those of that substance where they have almost always preserved their testa, and in the coarse limestone. But it does not appear that they are found in more recent strata than the last; and, indeed, they seem less frequent, according as the strata where they are found are less ancient. Some ancient strata appear to be composed of nothing else.

M. Lamarck has stated that this genus may be divided into some others; and, in fact, some which are pierced with a round hole at the summit of the largest valve, appear to have been attached by a tendinous pedicle, like those which are known in the living state. But there are others in which this hole is entirely wanting; others appear to have had a triangular hole above the beak.

There are, in the species of the terebratulæ, as in those of other genera, individual differences, and others which may arise from locality, either in magnitude or form; so that, doubtless, varieties have often been taken for species.

The genus of the OYSTERS, which, in a fossil state, are termed *ostracites*, presents a very considerable number of species. They are found in all the shelly strata, from those of the ammonites inclusive to the most recent. They are so common in certain localities, that many of these, to the extent of many miles, are entirely composed of them; as, for example, Mount Audona in Piedmont.

As the testa of oysters does not disappear, when they are found in localities where that of other soluble shells has disappeared, it frequently happens that we find them alone in strata, where they were accompanied by shells of other genera no longer found there.

The form of individuals of the same species is often so varied, that it is very difficult to distinguish the different species, especially when they have some relation between each other, which often happens.

Some species of oysters, not foliated, and which have the faculty of attaching themselves on other bodies, by a considerable portion of their lower valve, not only modify this valve, which, at its external part, represents the hollow mould of the body in which it has been applied, but copy this body exactly, in relief, on the upper part of the upper valve. Many may be seen which have adhered on pectines, turritellæ, polyparia, and other bodies, which are figured on the upper valves. This is a property which the oysters have in common with some species of gryphææ, of anomiæ, and balanæ. Some of these mollusca, in the living state, possess the faculty of giving to their shells the colour of those in which they have adhered.

Some fossil oysters exhibit, in a very obvious manner, the kind of mechanism by which they destroy the hinder part of their adductor muscle, while they advance that which is anterior. There are certain species, in which this muscle has been displaced several inches, to be removed from the heel against which it was situated at the birth of the animal, and carried forward, in proportion as the latter increases in growth. In forming each of the calcareous layers which have given thickness and extent to the shell, the molluscum displaces itself to get forward; thus abandoning the heel, where it sometimes leaves vacant septæ. The adductor muscle being attached to the valves by each of its ends, it cannot make this displacement except in consequence of the augmentation of the anterior part of this muscle, and its attachment to the new layer; while, at the same time, that which is opposite to it is destroyed, in almost an equal proportion, by a trenchant calcareous plate, which follows it in its displacement, and frequently does not touch the bottom of the shell. Certain species have the faculty of detaching the posterior part of their muscle, without having need of this plate, for it is not found in their shell.

The species in this genus of oysters are so numerous, and their form so varied, that it is very difficult to make exact divisions of them, for the intermediate species blend, as it were, these divisions one into another. M. de France, for the convenience of studying them, divides them—1st, into those whose valves are simple or waved—2nd, those in which the lower valve is folded and toothed on its edges, and the upper valve plane, with the edges smooth—3rd, those which are covered with folds, and have both edges of the valves toothed. The shells of this last division are found in the strata of the chalk, and those which are anterior to that formation, but never in those which are more recent.

GRYPHITES is the name under which are designated the different species of fossil *Gryphææ*. This bivalve shell is remarkable for the inequality and dissimilitude of its two portions. The lower valve is very concave, and terminated by a recurved crook. The upper valve, on the contrary, is flat, and seems destined to cover the other, and serve it as a kind of opercle. Each of these has but a single muscular impression, and the hinge has but one cardinal fosset, which is arched and without teeth.

The gryphites are found in the argillaceous limestone which borders the red and variegated sandstone, and which sometimes immediately covers the primordial stratum. This particular limestone, which often receives its denomination from the gryphites, is very frequently seen on the confines of the coal strata, and if it does not precisely belong to the same epoch, it is probably but little subsequent to it. It precedes the red sandstone from Metz to Sarrebruck;—it is found in Burgundy, in the environs of Couches, just before we come to the coal strata of Creusat. It shews itself in Normandy, in the village of Pont-Rond, near the pit-coal of Litry, and in other localities neighbouring to the same strata. The gryphites are excessively numerous in some of these places, and sometimes so crowded and so little adherent, that they have been used to pave the roads in the neighbourhood of Chavagnac, near Terrasson, in the department of Dordogne. These shells, which approximate to the oysters, like them have existed in large families, for they form entire banks of three, four, and sometimes six feet in thickness. They are often accompanied by other shells, the living analogues of which are not known, such as belemnites, ammonites, &c. Pectines, terebratulæ, and some other shells, are also equally associated with them in the different places just mentioned, and probably in all those which are analogous to them, in England, in Germany, in Belgium, &c. A sort of argillaceous limestone, very much employed at Lyons, contains an immense quantity of gryphites, and it is well known how near to this great city is the pit-coal formation. From all this we are led to believe, that the gryphites, which are so rare in the living strata, formerly existed in great abundance, towards the period when the pit-coal strata were formed. A great number of varieties are known.

This genus presents considerable difficulty in the characterizing of species, because it blends invariably with other genera which approximate to it.

Of the univalve shells, the genus CERITHIUM is that which presents by far the greatest number of species in the fossil state. M. de France, in 1817, had already more than one hundred in his splendid collection. Almost the whole of these species are found in the most recent strata, and the naturalist just mentioned has never seen any properly characterized which were found in the most ancient strata. M. de Gerville, however, in the "Journal de Physique," mentions, that in the formation with ammonites and belemnites of the environs of Bayeux, there were found four species of cerithia.

Though contrary to our plan, we shall notice here one species which is too remarkable to be passed over: this is the *cerithium gigas*. This is a very long and turriculated shell, composed of from thirty to thirty-five whorls. The aperture is oblong and a little oblique, terminated at the base by a canal, whose extremity is somewhat recurved. It is sometimes from fifteen to sixteen inches or more in length, on a diameter of four or five inches at the last whorl. A thread, turned over its suture from the summit to the base, was found to be more than eight feet in length. In proportion as the animal grows, it abandons the summit of the shell for the aperture. It forms concave septæ as it retires. This shell, being extremely heavy, and at the same time verymuch pointed at its summit, is exposed to be broken in that part during the life of the animal. The attrition which some of them have experienced in being transported from one place to another, has so much shortened and worn them on one side, that the columella of the first whorls becomes visible. If the animal had not retired, and in retiring had not formed septæ, it would have been exposed to be mutilated or attacked in this part by its enemies. Nature, to prevent this, bestowed upon it the faculty of withdrawing and forming septæ. The same phenomenon is observable in all univalve shells, and more particularly in those which are turriculated. This species is remarkable, not only for its giant size in the abstract, but also for the sudden advance of that size beyond the other species of its genus. It is found in all the strata of shelly limestone. They are so common in the falunières of Hauteville, that, in some places in the neighbourhood, they use them in making the high roads. This shell is also found in the existing state in the South Seas. There are in the Placentine territory, and many other localities, an immense number of cerithian shells in the fossil state. We are indebted to Mr. Sowerby's valuable work on Mineral Conchology for most of our figures of fossil shells, to which we have added a few from our own cabinet.

The remains of impressions of CRUSTACEA found in the bowels of the earth have been designated under a variety of

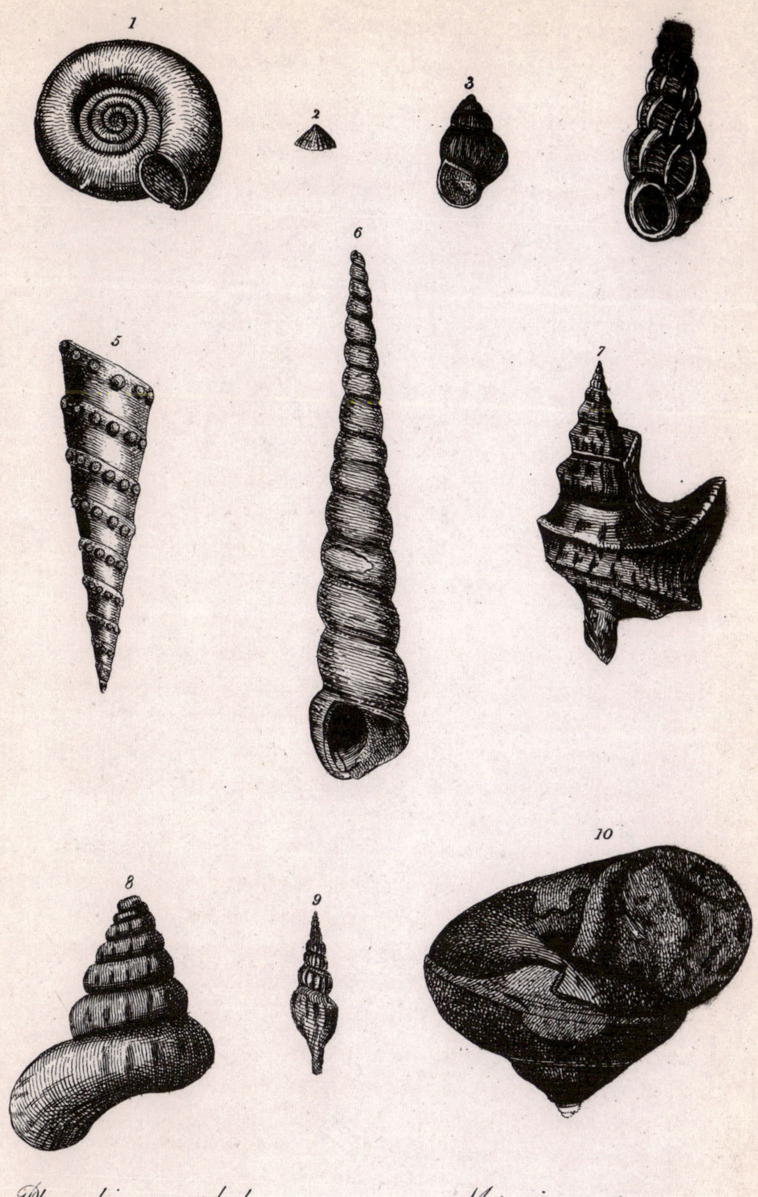

1 . Planorbis euomphalus.
2 . Pileolus plicatus.
3 . Paludina carinifera.
4 . Scalaria similis.
5 . Nerina tubercula.
6 . Melania sulcata.
7 . Rostellaria Pes Pelicani.
8 . Cirrus nodosus.
9 . Pleurotoma brevirostrum
10 . Helix carinatus.

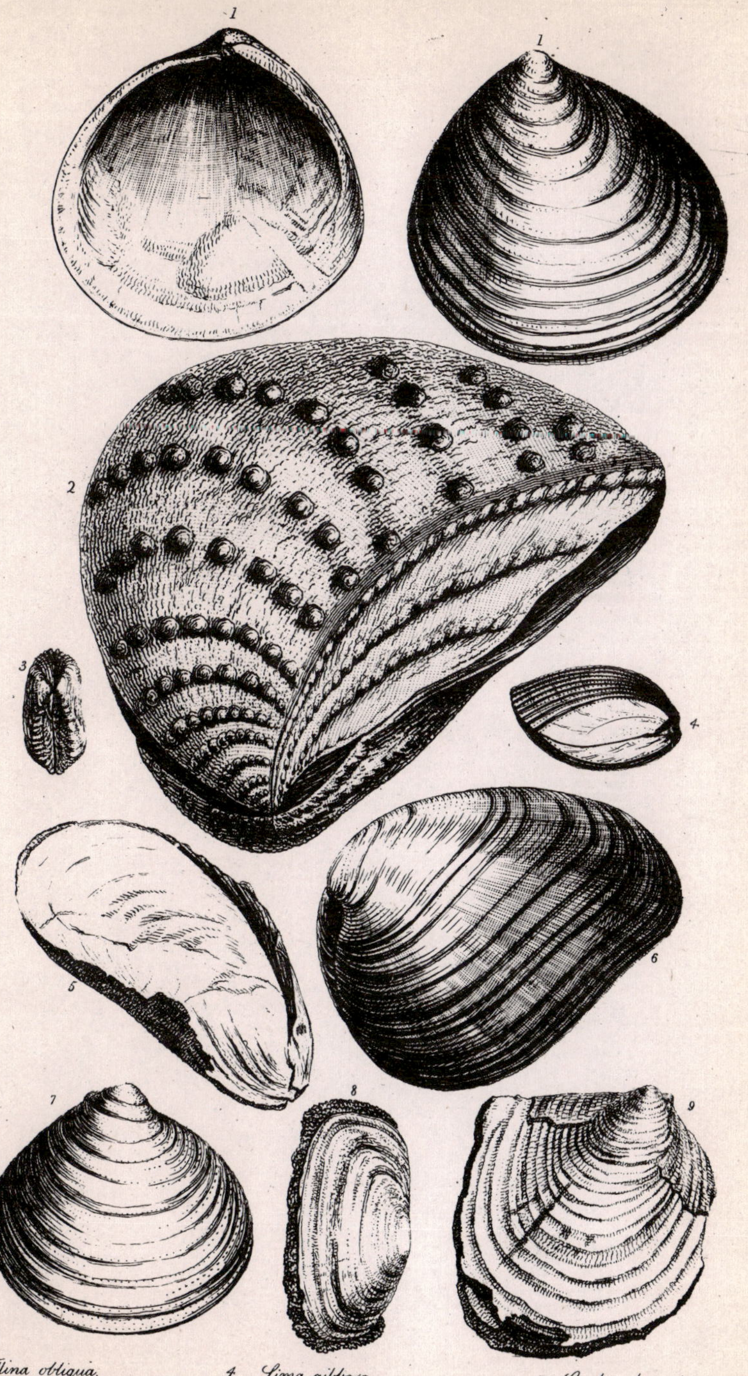

1. *Tellina obliqua.*
2 *Trigonia clavellata*
3 *Saxicava rugosa.*

4 . *Lima gibbosa.*
5 *Modiola depressa.*
6 *Unio Listeri.*

7 . *Cyclas deperdita.*
8 *Sanguinolaria compressa.*
9 . *Inoceramus latus.*

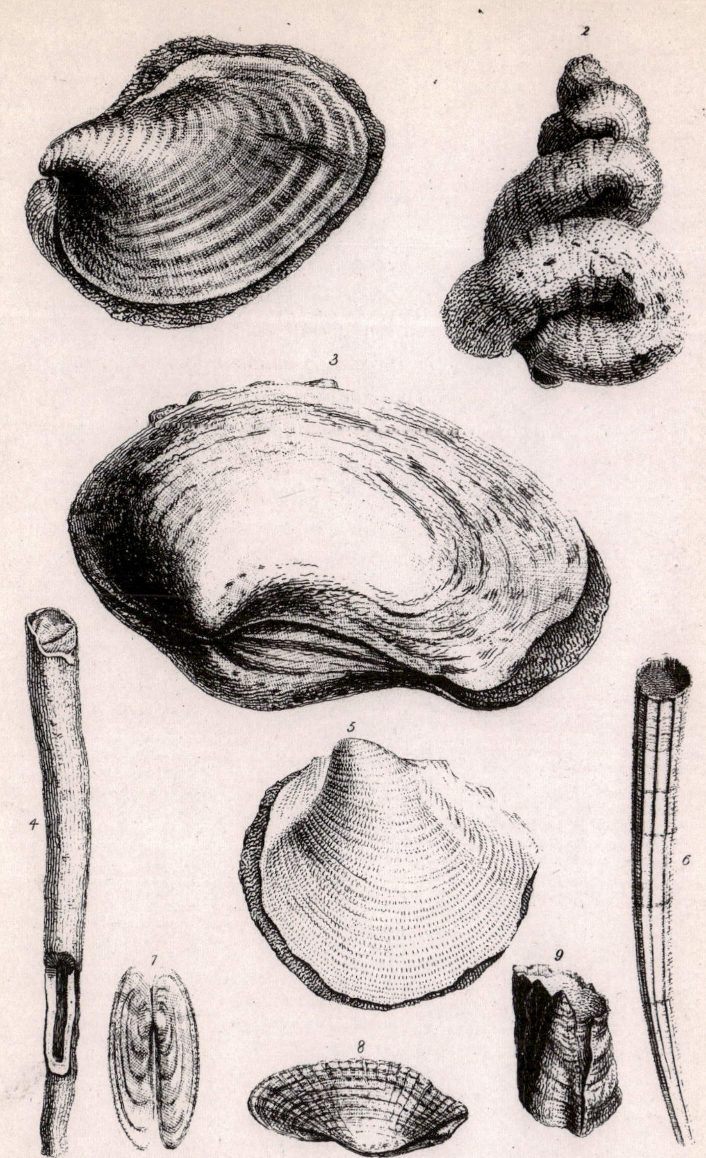

1. *Mya mandibula.*
2. *Serpula dentifera.*
3. *Lutraria gibbosa.*
4. *Tube of Teredo antenautæ.*
5. *Corbula globosa.*
6. *Dentalium elephantinum.*
7. *Solen affinis.*
8. *Pholas cylindricus.*
9. *Balanus concavus.*

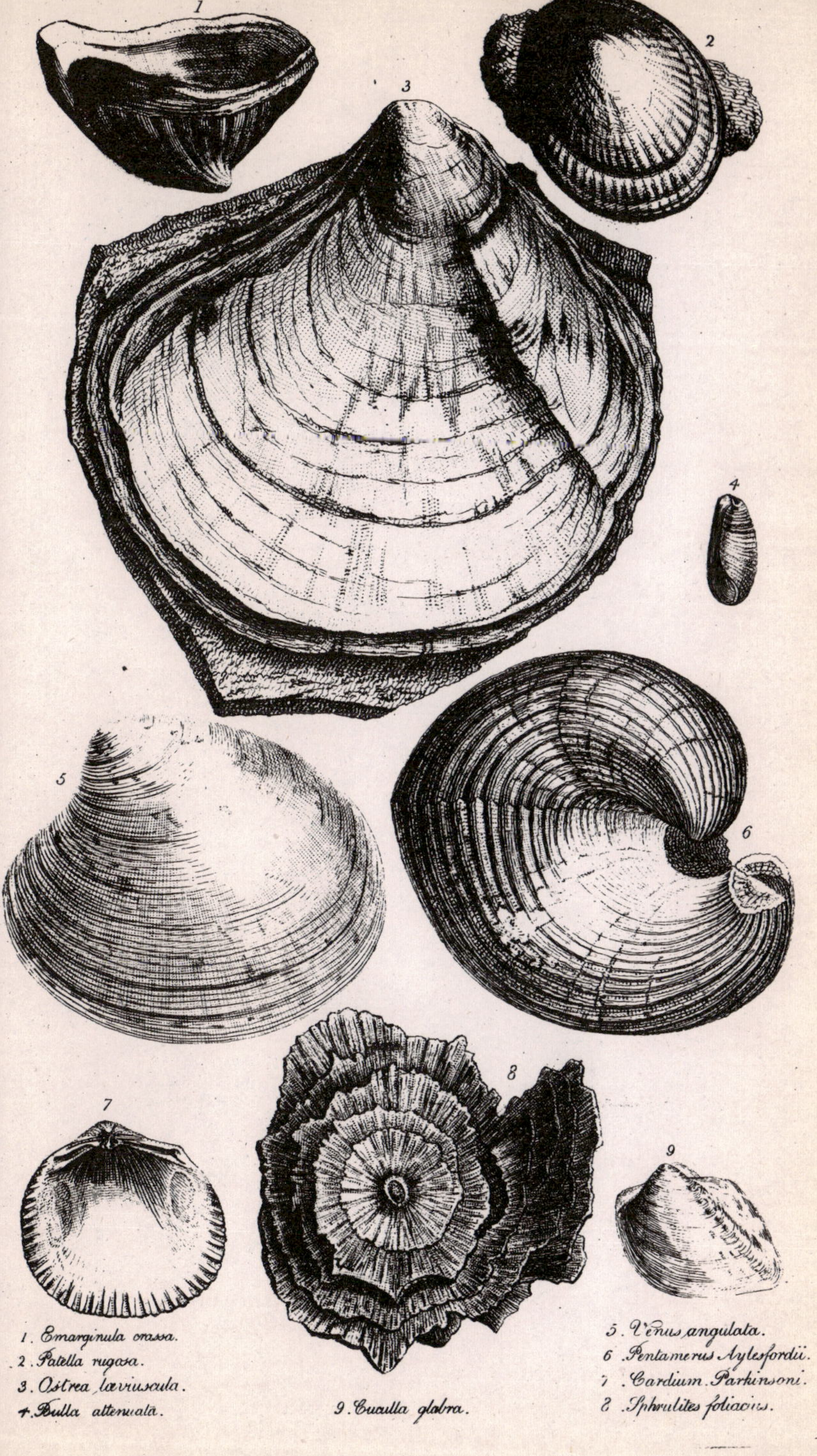

1. Emarginula crassa.
2. Patella rugosa.
3. Ostrea læviuscula.
4. Bulla attenuata.

9. Cucullæa glabra.

5. Venus angulata.
6. Pentamerus Aylesfordii.
7. Cardium Parkinsoni.
8. Sphrulites foliacius.

. Ovula Leathesi.

. Bellerophon cornu-arietis.

. Belemnites lanceolatus.

. Mitra Scabra.

5. Ammonites elegans.

6. Rotalites trochidiformis enlarged.

7. Cassis carinata.

8. Nautilus imperialis.

names, such as *crustacite, carcinite, &c.* These fossils, for the most part, are referable to known genera, but not to species yet admitted into our systems of natural history. All that has been written about them to the present day is very vague, and the figures of them which have been given are, in general, far from exact. The majority of naturalists who have mentioned them, wrote at the period when philosophers were only beginning to perceive that the parts of the globe which are now dry, must formerly have been submerged; and all that they proposed to prove at that time was, that the different bodies which were found buried, had some analogy with the actual productions of our seas. Thus they did not attach themselves to details, but confined their labours to descriptions and figures, which, bad as they were, always designated marine bodies.

At present, when, in consequence of the great progress of the natural sciences, a system so much more rigorous is pursued respecting the distinction of species and genera, the major part of the labours of preceding naturalists has become almost totally useless, unless very detailed and minute figures are added to them done by able artists. In fact, with the exception of the figures of Knorr, all that has been published on petrified crustacea, by a crowd of authors, is too incomplete to furnish anything like an adequate acquaintance with these sorts of fossils.

It is not rare to find fossil crustacea, and their debris more especially, in the most recent as well as in the most ancient strata; but it is not common to find any in a state of complete preservation. The fragility of their testa has not often permitted it to be preserved entire in those depositions where so great a number of the relics of other marine bodies are found. These animals appear to be much in the same predicament as fish. Those which die naturally become the prey of the different marine animals, of which they constitute the ordinary aliment; and it is only by chance that some of them escape destruction.

It is even extremely probable that the depositions in which they are found abundantly, and in good preservation, have been so circumstanced, that these animals were surprised and enveloped there at once, as it were by the effect of a volcano, or some other sudden cause. It is remarked, indeed, that the greater part of the places in which they are found have been volcanized, as, for example, the Philippine Islands, the coast of Coromandel, and various other situations.

These fossils have been found in many different strata. Those which were most anciently buried are anterior to the formation of the crystallized rocks, which, for a long time, were confounded with the granite, and which differ from them only in not containing quartz. Some have been found in calcareous strata anterior to the chalk. The former are the asaphi, the paradoxites, and probably the ogygiæ ; the latter are the calymenes, the cancer Leachi, the atelecyclus rugosus, &c. Others belong to the formation of the chalk itself—such are the paguri of Maestricht— and others to the formation of the cerithian limestone. A great quantity has been found in the deposition of foliated limestone in the margraviate of Anspach, whose geological position does not appear to be well determined. They are principally the limula Walchii, the eryon of Cuvier, &c. Some are found fossil, and yet anomalous, in the limestone formation of the environs of Verona. A great number are also found in the pottery earth in the East Indies, respecting which there is no positive information. A lobster and a crab, both unknown in the living state, are found commonly in the lias at Lyme.

With respect to their state of preservation, we may finally remark, that most of them have but their claws and antennæ. Some are totally changed into calcareous stone ; of others, the internal mould alone is visible, or the external impressions are very clear. There are some whose testa has assumed a deep brown or black tint, and others are covered with a slight pellicle of sulphur of iron, &c.

Geol. Tr.

1. Fossil Crab found in the green sand at Lyme?

2. Another Fossil Crustaceous Animal from the same locality.

Crustaceous Fossils

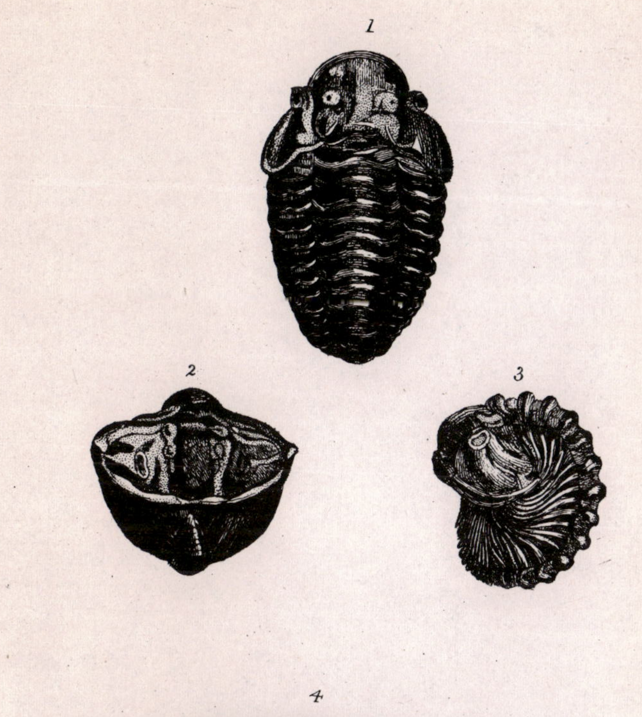

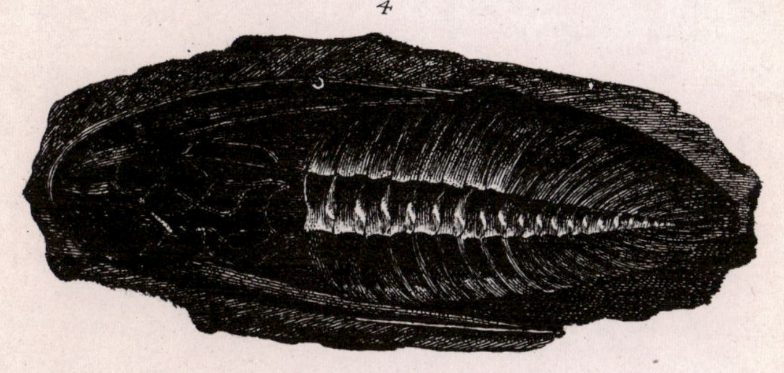

1. Calymene Blumenbachii Brong: from Dudley.

2, 3. The same in different positions.

4. Ogygia Guettardi.

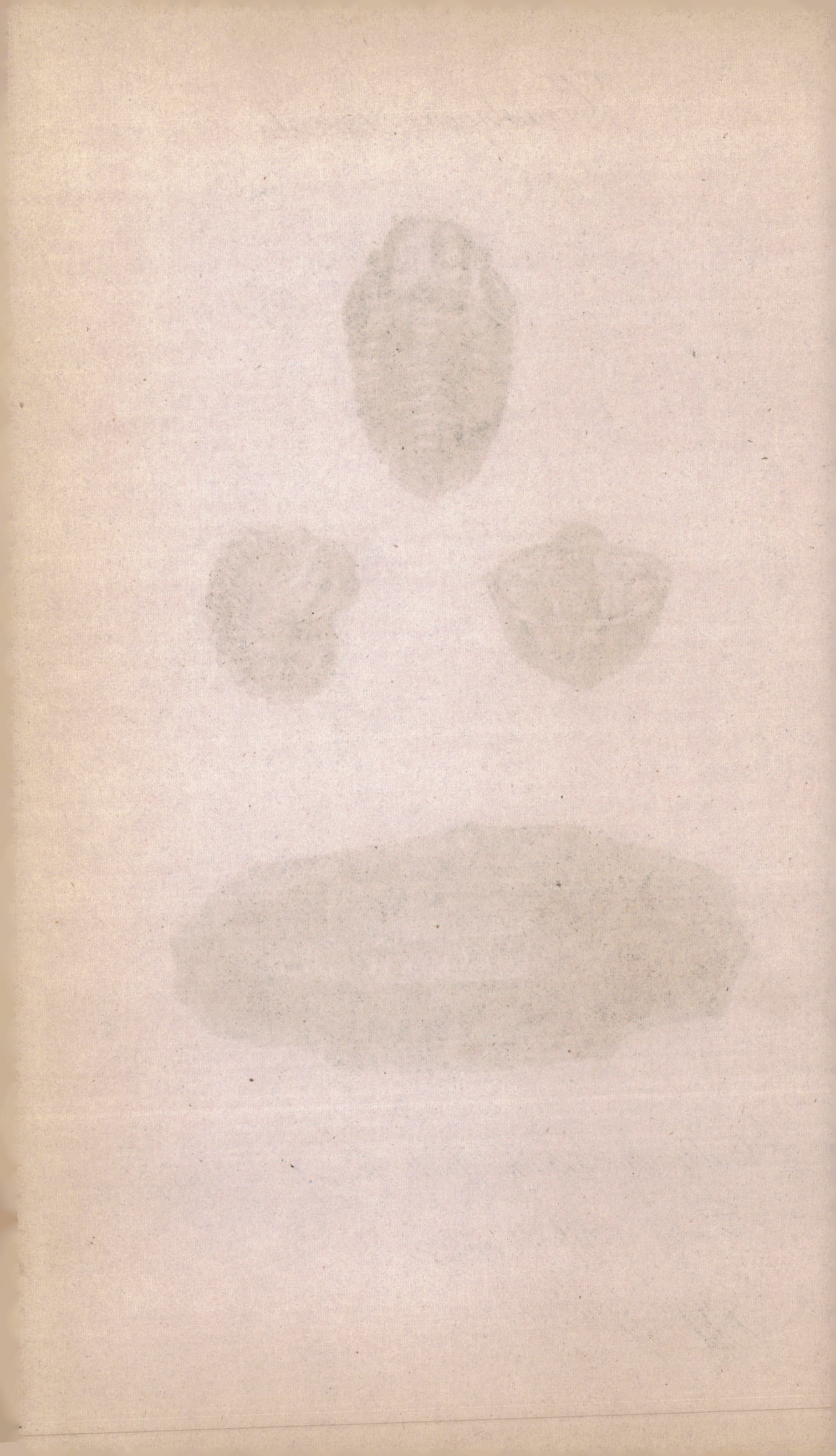

The fossil INSECTS, to which the name of entomolites has been given, are found either in amber or in fissile stones. The first are perfectly preserved in all their parts, and even the species may be recognised. In this substance flies have been found, tipulæ, ichneumons, ants, &c. M. de France mentions a piece of amber, about the thickness of one's thumb, and somewhat flattish, in which twenty-eight insects were distinctly to be seen, such as ants, tipulæ, small coleoptera, and a curculio, not of a species now existing in Europe.

After storms, the amber, and the insects which it contains, are found on the coasts of the Baltic Sea, and chiefly on those of Pomerania and Prussia, and also on some shores of the Mediterranean, such as those of the Marches of Ancona, of Genoa, and of Sicily.

This fossil resin is also found in the interior of the earth, in Lithuania, in Poland, in Italy, and in Provence, near Sisteron. It is usually in blackish sands, among fossilized wood, which is either pyritous or bituminous.

The insects which are met with in stones are not nearly so well preserved as those in amber. The head, however, and corslet may be distinctly observed, and often the body, divided by rings. But it is difficult to ascertain whether these be perfect insects, or merely larvæ, or chrysalids of necroptera, which live in the fresh water until their entire development takes place.

Some of these fossils are found accompanied by debris of small shells; and there is reason to believe that the catastrophe which involved them took place in waters which had been tranquil, and where those larvæ or chrysalids could have continued to exist.

Figures of these fossil insects may be seen in Knorr's work on petrifactions, and in that of Scheuchzer. This last author tells us of a libellula found with its wings in Monte Bolca, a large scarabeus in a stone from Œningen, and a scolopendra in a grey stone of Lubeck.

Aldrovandus mentions an insect of this last genus and some petrified grubs on a black stone from the canton of Glaris.

Vallerius tells us, that in the stones of Œningen flying insects have been found, such as scarabei, flies, libellulæ, and butterflies.

Bromel announces that vestiges of insects, of the wings of butterflies, and scarabei, have been found in the aluminous slates of the quarries of Andra-Rumen, in the province of Scania, in Sweden. He also mentions the wings of flies in the stones of Frankenberg, and large insects, accompanied by brilliant pyrites, in those of Wurtzburg.

In the quarries of Vestena Nuova, with the skeletons of fish, a marine insect has been found, which is referred to the genus pygnogonum of Fabricius.

Different authors, such as Buttner, Richter, Vogel, Longinus, Lippi, and Bruckmann, have stated, that in the schists of Œningen have been found ichneumons, dipterous insects, envelopes of larvæ, and nymphæ; and that in Ethiopia, the cells of bees, and the eggs of insects, have been seen in the fossil state.

We may remark here, that superficial observation may have represented matters to certain writers somewhat differently from the truth. It is not easy to believe, for instance, that grubs could have passed into the fossil state, and it is extremely probable that oolites have been mistaken for the eggs of insects. As to the supposed combs of bees, they seem evidently to have been fossil astrææ, in which the laminæ that filled each cell have been destroyed—by no means an uncommon case. We may certainly believe that scarabei can pass into the fossil state; but there is also reason to think that pyritous paradoxites and their debris have often been taken for them.

On the schists of Solenhoffen, of Pappenheim, and of Eichstadt, impressions have been seen which have been taken for earth-worms, and to which the name of helmintholites has been given. But, from the figures given by Knorr, it is probable that these petrifactions have a very different origin ; at least, this must be the case with some, which are four or five times the length of the earth-worms of our present day.

In the possession of M. de France is a stem from Solenhoffen, which contains small asteriæ, and on which a kind of tube is visible, which might be taken for a portion of a fossil worm, but to which bivalve shells appear to be attached in many places, with their two valves striated circularly and open. But there is little reason to believe that this body, any more than the others above-mentioned, is a relic of the earth-worm tribe.

The generic name of ECHINUS was formerly given to the different genera of the family of *Echinides*, which are now known under the names of *scutella, clypeastra, fibularia, &c. &c.* Those in the fossil state were called *echinitis, echinometra, echinodermata, ovarium, &c. &c.* Rumphius believed that those bodies fell from the sky, as well as the belemnites, and has called them *bronita, tonitru, ombrias.* Nonnius imagined that they were the petrified eggs of serpents.

The Romans believed that these bodies fell upon the earth with the heavy rains or thunder, or that they were the eggs of toads, or petrified toads themselves.—See Pliny, lib. xxxvii. and xxix.

The authors of the fifteenth century believed every thing that was said about them by Pliny. Agricola was the first who rejected those fables, but he did not point out the true origin of these bodies. Mercatus took them for figured stones, to which nature was pleased to assign their peculiar form; and we are informed by this writer that they were formerly used in enchantment. Gesner fell into the error of those who believed that they fell from the sky, and was ignorant that the *Judaic* stones, which are the points of echini, of which he has spoken, had any relation with the echinites. Ferrand Imperati appears to have been the first who, in the commencement of the seventeenth century, referred these stones to the echinus or sea-urchin, and who demonstrated that the Judaic stones were nothing but the petrified points of these echini. But, notwithstanding all that was said by this naturalist, the ancient errors respecting the origin of echinites subsisted until the time of Aldrovandus, who demonstrated the true origin of these fossil bodies.

Luid was the last author who doubted that the fossil echinides were true sea-urchins, because they were never found provided with their points; but the analogy of these fossil bodies with those which are living, should have been quite sufficient to prove this, even though no examples had been seen, as there have been, of fossil echini found with their points.

The fossil echini, properly so called, are found in the strata anterior to the chalk, and in those posterior to that substance; but it is more rare to find them in the chalk itself, in which, however, some others of the same family are common; and those perhaps which have been found there depend upon the cidarites as well as the echini, between which the line of demarcation does not appear to be very clearly traced as to the individuals in the fossil state. The localities of the fossil echini are considerably various and extended.

On the rest of the numerous genera of this family of zoophites, as the cidarites, ananchites, &c., it is impossible to add anything in a general way. We must, for the same reason, pass entirely over the family of STELLERIDES, and conclude with noticing whatever there may be of general interest in the great order POLYPARIA.

In employing the classic denomination of *Polypi* for the animals constituting the genera of madrepore, tubipore, eschare, flustrum, &c., naturalists have necessarily been led to give the name *polyparium* to the bodies more or less solid with which these animals are found, without any respect to the nature of these bodies, and still less to their form, and the manner in which the polypi produce them, and are situated in them. This word is synonymous with the words *coral*, *coral-*

lium, and *stirps*, employed by Pallas and Linnæus, the first for the stony polyparia, or *lithophytes ;* the second, for the flexible or corneous polyparia, or *ceratophites*.

M. Lamarck thus defines a *polyparium*—a fixed envelope, more or less solid, calcareous, or corneous, in which a polypus inhabits, and which is the evident result of a transudation from its body of an excretion through certain pores of its skin of substances sufficiently compounded to form by their approximation a concrete body, more or less solid, and altogether inorganic. This definition, in reality, is only applicable to the madrepores of Linnæus, and also to his escharæ. It is scarcely so to the sertulariæ, cellariæ, and not at all to the isis, coral, and gorgon. Accordingly we find M. Lamarck himself obliged to return to the use of the words *stirps*, *axis*, *rachis*, *stem*, *&c.*, to signify the body, at once fleshy and solid, which forms the common part of the pennatulæ, although there are the greatest affinities between this body and that which constitutes the solid part of the isis, gorgons, and coral. If we persist, then, in the generalization of the word *polyparium*, it should be defined as a solid body, calcareous or corneous, the result of one or many polypi, without troubling ourselves with the mode of its formation or the manner in which the polypi are placed in it. In this sense, the fibrous mass of a true alcyone, the fleshy mass of a pennatula, the corneo-calcareous plate of an eschara, the tubes of the tubulariæ, the phytoïd stems of the cellariæ and sertulariæ, the calcareous, arborescent, frondescent, and phytoïd masses of the madrepores, will all be equally polyparia. We may even range under this name the handsome tufts of corallines, if we admit that they carry polypi, which, however, is not the case.

In contemplating the nature of polyparia, we find them of many kinds, according as they are calcareous or stony, corneous, fibrous, corticiferous, or pasty.

The first one, called *lithophytes*, a denomination very anciently employed, and which is derived from the idea that these were a kind of stones which vegetated, or plants whose tissue was solid, like that of stone.

The second, or the corneous polyparia, have been named, for a similar reason, *ceratophytes* (κεϱας, a horn), supposing that they were also plants, whose tissue had, more or less, relation with that of horn.

The corticiferous polyparia are those which, either calcareous or corneous, are clothed, as their name implies, with a sort of bark in which there are polypiferous compartments, as in the isis, the corals, and the gorgons.

As to the fibrous polyparia, which are also evidently corneous, they have not received any particular denomination; they are, with little difference, the pasty or glued polyparia of M. Lamarck.

Finally, in contemplating the fleshy and contractile mass, which constitutes the common part of the pennatulæ, as a true polyparium analogous to that which forms the alcyone, we would have another species of polyparia, which might be designated under the name of fleshy polyparia.

Taking into consideration the solidity of the polyparia, they have been again divided into those which are solid, not flexible, and those which are flexible. By polyparia not flexible are understood all those which, whether calcareous or corneous, present a resistance more or less considerable; and by flexible polyparia, the different species into the texture of which enters a greater or less quantity of fibrous tissue, which permits this flexibility.

But it is especially on the consideration of the forms of the polyparia, that a greater number of subdivisions has been established, such as arborescent or phytoïd, frondescent, lamellated, or lamelliferous, for animated and vaginiform polyparia, &c.

The composition of polyparia has likewise given rise to their division into simple and compounded, or complex, polyparia.

Finally, admitting the possible existence of polypi with free polyparia, a division of the latter has been established into free and adherent,—denominations which require no definition.

As some organized fluviatile bodies exist which have been compared to sponges, and which, for a long time, have even been ranged in that genus, another division of polyparia has been necessitated into fluviatile and marine.

The great quantity of calcareous or stony polyparia, which have been found in the living state in the seas of warm climates, and the considerable masses which have been met with in the fossil state in the composition of calcareous rocks, have caused these productions of polypi to be considered as forming a very notable constituent part of our continents, and as capable of modifying, in an exceedingly rapid and powerful manner, the surface of our globe which exists under the waters of the sea. Nothing is more common than considerations of this kind in treatises of geology, and especially with the authors who flourished towards the close of the last century. Until very lately, these speculations were founded, in general, only on the observations of travellers, of mariners, and particularly on those of Captain Cook, and other navigators, who have traversed the South Seas and explored Australasia. Since that period they have been corroborated in an especial manner, first by Forster, and afterwards by Peron during his voyage to New Holland, in the expedition of Captain Baudin; and they have been adopted, in the sequel, by all zoologists and geologists. It was easy, in fact, to perceive, that, if we could admit that the animals which produce these polyparia, designated in a general manner under the names of madrepores or corals by mariners, and even by some geologists, could germinate with as much rapidity as those which form the escharæ do, as we have learned from Spallanzani, the stony polyparia must really produce, at the end of half a century, or less, by means of strata superposed almost indefinitely, calcareous masses of enormous extent and depth. But, in the first place, this preliminary position is more than doubtful;—it is more than doubtful, we say, that the astrææ, the caryophylli, produce with the same rapidity as the escharæ. Moreover, it is indubitable that these stationary animals, not being able to live, either at depths where the action of the solar light and heat is not exerted, which we know positively to be the case with the true coral, nor sufficiently near the surface of the sea, to experience the violent movements with which it is often agitated, and, still less, out of this surface, even supposing all the other conditions to be most favourable, nothing greater can result than some strata a few fathoms in thickness. It is, then, most palpably evident, that the islands, archipelagos, and reefs, with which the Indian and Southern oceans are so thickly sown, cannot be entirely madreporic, as has been for a long time supposed, but merely portions of a soil analogous to that of the nearest continents, and for the most part entirely volcanic, which have been encrusted with madreporic depositions of greater or less thickness. Such is the opinion of MM. Quoy and Gaimard, naturalists on the expedition of Capt. Freycinet, who, having visited the same points as Peron, and, among others, Timor and the Isle of France, have endeavoured to demonstrate, in a memoir on the growth of lithophyte polypi, considered geologically, that all that has been asserted or believed to the present day, relative to the immense labours which the saxigenous polypi are capable of executing, is inexact, excessively exaggerated, and most frequently erroneous. It may be, however, that these observers have underrated the influence of the saxigenous polypi in the composition of the islands and reefs of the seas of those torrid climates. They begin by premising that the encrusting polypi, such as the astræa, the caryophilli, &c., which are evidently those whose limits of growth are the most extended, or the least confined, are also those which live at the shallowest depths; for they say that they have never met with them below a few fathoms. Nevertheless, by their own avowal, the branching polypi can live at much greater depth; for these authors mention an instance of their having obtained, by sounding, some small branching madrepores at the depth of twenty-four fathoms, in 56° south latitude; and we know that, in the Mediterranean, even the coral exists at a depth of ten or twelve hundred feet. Now, as this is the case, may it not also happen that the astrææ live at a greater depth than these gentlemen have stated, although they did not happen to find them so? for there is the greatest analogy between this genus of animals and certain of the madrepores. Moreover, may not these reefs and islands, which must certainly have had for their basis some portion or projection of primitive, secondary, tertiary, or volcanic stratum, which constitutes the bottom of the sea, be at first increased to a certain height, by the assistance of the numerous ramifications of the branching polypi, united and solidified by the shelly animals which seek out these anfractuosities, and the rest be then formed by the layers of astræan and other encrusting polypi, whose action must be so much the more lively and rapid, as the animals become subjected to the more favourable influences of light and heat? If there be any probability in this conjecture, which has been made by M. de Blainville, and we think that there is a great deal, we may adopt a middle theory between that of Forster and Peron, with whom the polyparia are the principal cause of the growth and formation of the South Sea Islands, and that of MM. Quoy and Gaimard, who will attribute to them only a slight incrustation of a few feet. As to the support which the opinion of Forster and Peron may derive from the fact of madrepores being found on certain islands at very considerable elevations, we must be careful to ascertain if the nature of these islands be not volcanic; for, in that case, these madrepores might have reposed upon the soil at very great depths, and have been raised only with the volcanic substance itself, or with any other which may have been thrown up by the eruption.

It is not very long since it has been asserted that we have

no proof that any change has been effected in the temperature of the climates which we inhabit. But the study of organic fossil has clearly established that this temperature has been lowered, or that beings, which cannot exist at the present day but in regions much hotter than ours, were formerly capable of existing in those where the winter would now destroy them.

On this change of temperature we have already dilated, and adduced what we conceive to be very adequate proofs in favour of this opinion. Additional proofs are, however, to be gathered from the study of fossil polyparia.

We know that the seas of the northern regions support very few of the genera of this family or order, and that they contain none which are remarkable for size. It is not so with those found in the fossil state in the same regions. Not only are many very large specimens to be found there in the fossil state, whose genera are no longer living but in the equinoctial climates; but an era has existed when the bottom of the sea, in our part of the world, was literally covered with them. In the department of Calvados, says M. de France, strata are observed to a very great extent, which are scarcely composed of anything but the debris of polyparia. In other regions, still further north, there exist, in like manner, numerous remains of polyparia in a fossil state, whose genera are at present confined to the warmest latitudes of our globe.

With the exception of the strata of lacustral formation, which contain the debris of beings which belong evidently to genera, and perhaps to species which live in our climates at the present day, the others present us in general only the remains of vegetables and animals, whose genera exist no longer, but under a temperature much more elevated than ours.

The fossil astreææ, to which the name of Astraïtes has commonly been given, are found in strata anterior and posterior to the chalk, and in the chalk itself.

Although the fossil species of this genus are numerous, it

is difficult, especially for those of the ancient strata, to seize all their characters, or distinguish them all, in consequence of their imparted state, or the destruction of the lamillæ of their stars; or, finally, from the change of their substance into silex or crystals.

M. de France mentions two very curious specimens of this genus in his possession. In one, the entire substance, as well as a part of the lamellæ contained in the stars, is converted into a crystallization, which has preserved the form and contexture of the astræa; in the other, a net-work of crystals, with their points, but without any form of organization, has replaced the matter which constituted the stars, which have remained empty. It is difficult to conceive how a crystallization, which appears to have been very tranquil, could have replaced the matter of the cells of the polyparium, which are three or four lines in diameter.

M. de Lamarck, in his work on Invertebrated Animals, has divided the species of this genus into two sections: in the one, he has placed those whose stars are separated even from their base; and, in the other, those whose stars are contiguous. But there are some intermediate species, which render this division rather difficult of application.

A considerable mass of the polyparium of the *Astræa dendroïdea* was recently discovered in the cliffs of Berneville, department of Calvados, the description of which, in the Memoirs of the Royal Academy of Sciences of Caen, by its discoverer, M. Lesauvage, is worth inserting here. "This singular production is formed of a considerable bundle of branching stems, simply contiguous, from ten to fifteen lines in diameter, and presenting, through their entire length, a tolerably regular series of rounded dilatations and of circular contractions. The branches terminate in blunt points at unequal heights; and all their surface is covered with lamellated stars, rounded, contiguous, and almost superficial. If we exâmine the transversal section of a stem, we shall find that its interior is formed of numerous laminæ, which leave between them angular spaces, and assume somewhat of the starred form. In the longitudinal section, we can perceive a series of cavities, sometimes regularly comparted, which seems to indicate that the interior of the branches was partitioned; but these cavities seem owing, at least in a great measure, to a confirmed crystallization of the calcareous part of the organized body. The colour of this calcareous substance is a dull red, contrasting strongly with the whiteness of that by which the mass is encrusted. Considering the fine preservation of a polyparium of this prodigious size, we may be led to believe that it has undergone no sudden displacement; and that it was engaged, by the calcareous matter which surrounds it, in the place where it was originally produced.

"The disposition of its stars, spread over the entire external surface, removes it from the order of astræa, which is thus characterized by M. Lamouroux:—stars, or cellules, circumscribed, placed at the upper surface of the polyparium."

M. Lesauvage thinks that this polyparium ought to be placed among the madreporic animals, as a new genus.

Encrinus is a name, first given by Ellis, to a very curious and little-known animal, whose exact situation in the natural series may even yet be pronounced rather uncertain. It has been variously allocated, by zoologists, in the radiated class—being sometimes made a genus of the family of Asterias, and sometimes placed near the Iris. This animal has, perhaps, not yet been examined with all the attention which it merits; for, if it do not exist in our present seas, it is an undoubted fact that it must have been very common there formerly, since nothing can be more multiplied, in certain calcareous strata, than those fossil remains, known under the names of entrochi, encrinites, &c.

The characters of this genus, in the present state of science, may be thus given:—A stelliform or radiated body, composed of five principal rays, subdivided into three or four articulated

Foſsil Polyparia.

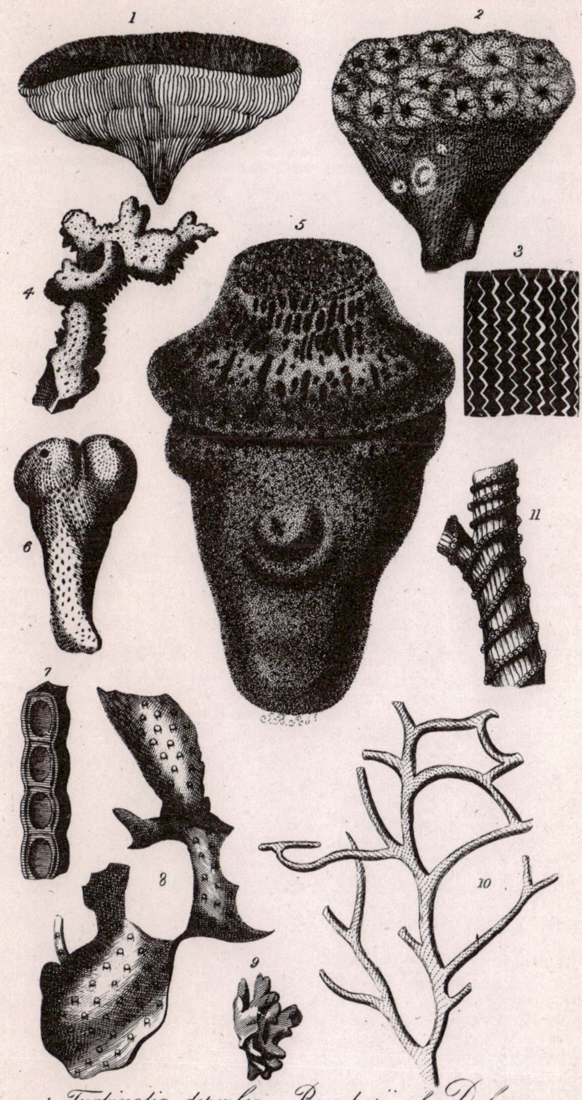

1. Turbinolia depreſsa. Bosachesii of Defrance.
2. Microsolena porosa. Lamou.
3. Portion of lateral surface of same magnified.
4. Tilesia distorta. Lamou.
5. Jerea pyriformis. Lamou.
6. Eudea clavriculata.
7. Vincularia fragilis. magnified.
8. Diastopora foliacea.
9. Chrysaor cornea.
10. Spiroporus elegans.
11. Portion of same. magnified.

Fossil Polyparia.

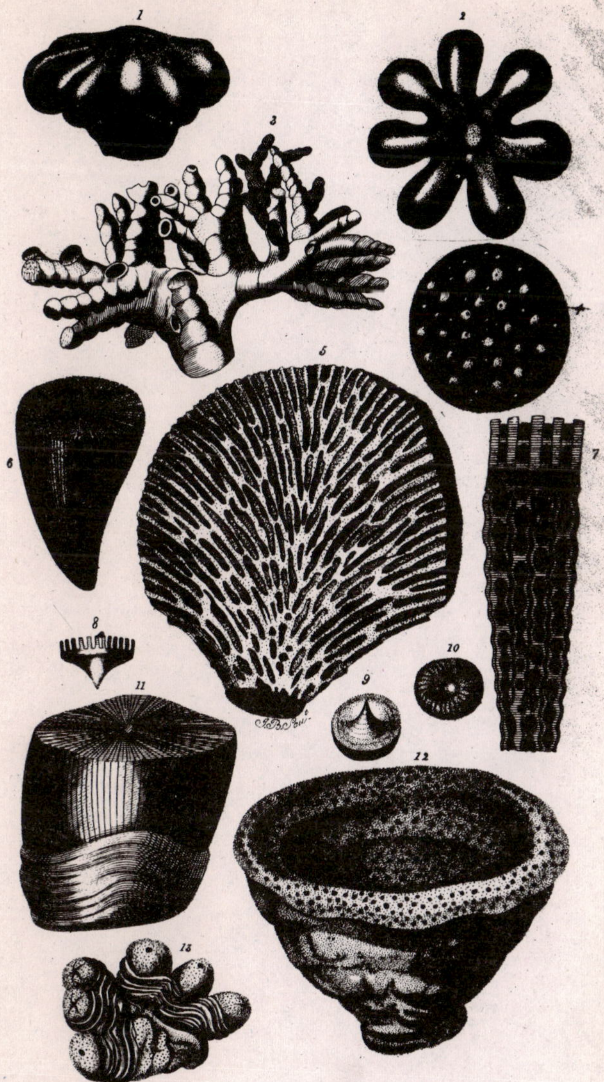

1. *Kallirhoa lateralis.*
2. *Another vien of same.*
3. *Terebellaria ramosissima.*
4. *Lycophris lenticularis.*
5. *Eunomia radiata.*
6. *Turbinolopsis ochracea.*
7. *Portion of same magnified.*
8. *Pelagia clypeata.*

9 & 10. *The same in different positions.*

11. *Montlivaltia caryophillata.*
12. *Chenendopora fungiformis.*
13. *Symnorea mamellata.*

branches, pinnated in their entire length, presenting, at the upper concave face, a series of pores. This body is supported at its extremity by a long stem, vertical, polygonous, and articulated; and furnished, in its length, with a variable number of verticillæ, composed of five small simple branches, equally articulated, and very probably adherent to submarine bodies. We know, in reality, little more of this animal than what is here defined; and which is limited, if we may use the phrase, simply to its skeleton. Ellis, who, it would appear, has seen it in a better state of preservation, adds, that, in the midst of a kind of funnel or rose, formed by the rays of the individual which he observed, there remained a sort of section of crustaceous substance, of an oval form, about an inch long, three-quarters of an inch wide, and a quarter of an inch high, in the centre of which there was a little hole which appeared to communicate with the internal part of the articulations of the stem. But he says nothing respecting a membrane which covered it altogether, and still less of the polypiferous tribes, which zoologists speak of who have adopted the opinion of Linnæus. The stem, or peduncle, the termination of which is not known, is composed of a variable number of small calcareous pieces, somewhat unequal in height and diameter; in general, by so much the less long as they are more highly situated, and with five angles, so much the more marked as the pieces are equally approximated more to the body of the animal; so much so, that, superiorly, these pieces have five radii, separated by as many flutings. They are articulated together by a plane surface, which exhibits a star with five rays, from whence proceed the fibres which serve as a means of union to all this stem, and render it, without doubt, somewhat flexible. In the middle of each piece is a rather small hole which, continued in all the others, forms a canal which terminates in the centre of the body, or *umbel*. It appears that there is, moreover, another between the vertebræ, in the middle of each furrow. On the length of the stem, and at different distances, kinds of verticillæ spring from the indentations of the corresponding vertebræ—which are formed of fine cylindrical branches of equal length, and composed of small pieces, all cylindrical, pierced with a little hole, and whose succession forms a canal which communicates with one of the lateral holes of the vertebra, or joint of insertion. On the lower side of the last articulations are found four small crustaceous tubercles, two at each extremity, the latter of which has the form of a crook. As to the body, or umbel, it is composed of five rays, or arms; the base of which supports the sort of testaceous section of which we have spoken above: this base is short, and formed of three articulations only. Each ray is then subdivided into two branches, and those into two or three others, but without much regularity. These branches, secondary and tertiary, are formed of a vast number of small articulations, rather decreasing in diameter than in height; on each side of which springs a barble of half an inch in length, and the twentieth of an inch in breadth, and which is similarly formed of a great number of little pieces. All the articulations, small and great, are round or convex below, but flat above, with a longitudinal groove, deep in the middle, and furnished with two ranks of suckers, as in the asterias; and we shall be less astonished at the great quantity of articulations of the encrinus found in the bosom of the earth, when we are told that Guettard, who amused himself by reckoning the number of articulations in an individual scarcely twenty inches long, found them to amount to twenty-five thousand seven hundred and thirty-five! As to the very varied formation exhibited by the entrochi found in the earth, we must also observe that, on the same living individual, some are found almost round, so very little are the angles of the pentagon marked; some are evidently pentagonal; others are in the form of stars, with fine rays, more or less distinct; and, finally, there are some smaller ones, which are channelled in one half of their diameter.

These animals, in the living state, very probably inhabit the bottom of the sea, at very considerable depths; but we know not if they are fixed there, though it appears very probable that they are. It is to chance the discovery of the few specimens existing in the collections of Europe is owing; and they came from the American seas.

The fossil debris, belonging to this genus, are very common in certain ancient strata. Their singular and varied forms have caused them to be remarked by the ancient oryctographers, who have given them the names of *pentacrinos*, *lapis pentagonus*, *volvolæ*, *stellaria columellæ*, *asteriæ*, *cylindrita*, &c. They have also been called petrified sea-stars. The name of *entrochi* has been given to inconsiderable portions of the stem of these polyparia; and that of *trochi*, to the articulations separated from the entrochi. These stems are composed of pieces placed one above the other, whose thickness varies, either according to their position, or according to the species. The great variety in these pieces, or articulations, seems to prove the existence of a great number of species of this genus. They may generally be divided into two parts—the one round, and the other polygon: the first, furnish the thickest and the longest pieces; some are two inches in diameter, and others less thick and more than three inches in length. It is clearly demonstrated, by these specimens, that these stems have been adherent on other bodies; as those of the gorgons, or other branched polyparia; and that several stems were often raised on the same basis. The round stems are pierced, in the direction of their axis, with a round hole, more or less wide, according to the different species: that of some of them is smooth; in others, it is channelled circularly. Each articulation is marked, on its two flat surfaces, with rays or striæ, diverging from the centre to the circumference.

It is observable that the thickest stems have been clothed successively, at their base, with layers which have enveloped the primitive stem which supported the diverging rays. Although on certain pieces of these thick stems we still find, externally, the traces of articulations, the rays, nevertheless, do not proceed to the circumference. These rays, which catch in with those of the contiguous articulation, appear to have been intended to hinder the stem from turning on itself; but, considering the great number of articulations of which this stem is composed, it is extremely probable that these polyparia had a certain movement of rotation and of libration on their axis.

There are some specimens, with very thick stems, found in the island of Gothland, and at Pffeffingen; but their species is undetermined. That which is called *lilium lapideum* by Ellis, and *encrinus liliiformis* by Lamarck, is supported on a round stem, of from ten to fifteen inches in length. The head, or corona, of this species, is formed of ten bifurcated rays, which have the form of the flower of a closed lily when contracted. These rays are supported on a piece, which has been named the root of the rays; so that two of these last have always a common root. This root is attached to a part of a pentagonal form, which has been named the base, and which is composed of five pieces, often joined in one, by which the corolla is united to the stem. This last is composed of articulations of different forms: in the part most remote from the corolla, after six or eight cylindrical articulations, which are smooth externally, one is found more thick, and resembling a compressed musket-ball. On the whole, there are many varieties in those stems.—There are some which are channelled circularly; some round and smooth; some with swellings of different thicknesses, which alternate one with another; some with gibbous or convex articulations; and others, with three slender articulations placed between two more thick, &c. Encrini of these kinds are found at Wessembourg, Lower Rhine; in Saxony; in Thuringia; in Silesia; in the environs of Frankfort on the Oder; in Switzerland; at Dudley, in this country; and in many other places. Figures of them may be seen in Ellis's work on Corallines, and in that of Kruezz on Fossils, &c.

At La Haye-du-Puits, department of the Manche, are found

detached pieces, changed into calcareous spath, the largest of which are seven or eight lines in diameter. They are pierced at their centre by a circular hole, in which some slight striæ are to be found. Some are traversed by one axis, or tube, channelled circularly; and this axis is itself hollow at one of its ends.

Many articulations are to be met with, which have depended on round stems, in the Vosges; in the environs of Besançon; near Dijon; in the ancient strata of the environs of Valognes; in the Vicentine; at Bradfort, in this country; and even in the island of Timor. They constitute the greatest part of the elements of Flanders marble, which has been termed the lesser granite. The corolla, however, which should terminate the stems, on which these numerous articulations depend, is never to be found: Pallas has met with these sorts of articulations, of different sizes, in the schistose strata near Konstantinovo, and in other parts of Russia. In the principality of Salm, they constitute, of themselves alone, an iron mine.

In the schists, on the summits of the Alleghany Mountains, in Pennsylvania, and in the sandstone at Buttenrode, are found impressions and internal moulds of articulations, and of portions of the round stems of encrini. Some of these moulds prove that the axis of some species was pierced with a very wide hole, and channelled circularly in its interior.

The encrini, with heads, or corollæ, as the *lilium lapideum*, are not exclusively supported on round stems—at least, in their entire length. To the base of one of these heads, found in the environs of Dijon, are still attached the two final articulations, which have five corners or flaps. The basis is not formed of pieces similar to those just described: a figure of a similar one may be seen in Parkinson.

Some portions of round stems, channelled circularly, and with striæ divergent from the centre to the circumference, present, on an articulation more projecting than the others, five appendages, which indicate that they have been furnished with five lateral articulated branches. These portions of stems are to be found at Valognes; and on one of them, whose articulations are equal between themselves, we see a small isolated appendage, which may have served to sustain a lateral branch, or which, perhaps, is a sort of bud from which a new polypus might spring.

A portion of a round stem, in the collection of M. Brogniart, and on which we find, irregularly disposed, certain little round places, having radiated striæ, might lead us to suspect that these encrini possessed the faculty of reproducing by shoots, like some other species of polypi.

There are stems which seem to occupy a medium between those which are round and those which have several corners. Some have their axis pierced with a pentagonal hole; and their striæ are not radiated from the centre to the circumference: some are smooth, and others carry verticillated asperities on each of their articulations. These last are found in the environs of Besançon.

The five-cornered stems, whose articulations have a star well marked on each of their flat surfaces, present, also, many varieties. Some are almost round; others have five rounded projections; and others have very sharp angles. In all these species, on some articulations, may be remarked the traces of five lateral branches, which must have been attached in each of the re-entering angles. We may believe that these species have some relation to the encrinus called *Pentacrinites briareus*, which is known in the living state, and which has been found in the neighbourhood of Martinique. This fossil is very common about Charmouth, and we have engraved a specimen from that locality.

Articulations of this sort are to be met with in the environs of Valognes and of Caen; at Nevers; in Mount Jura; at Dijon, &c.

It was believed, until recently, that these articulations were to be met with only in the most ancient strata of the globe; but M. de Gerville, a clever naturalist, has found them in the falunière of Nehau, department of the Manche, which is analo-

1. Pentacrinites briareus.

2. Portion of stem distinct.

3. Transverse section of same.

gous to the stratum of Grignon, which we have mentioned before in this sketch. These articulations, however, seem to differ in nothing from those of the ancient strata, except in their white colour, and a certain degree of transparency.

There are articulations, with five corners or points, of all sizes, but the diameter of the largest scarcely ever exceeds five lines; and although, from analogy, we might be led to believe that the stems, on which they depended, were adherent on other bodies, like the round ones, still there appears no decided fact in proof of this opinion.

At Charmouth, and at Dudley, has been found a species of encrinus (*ramosus*). On a five-fronted stem is found a head, which is divided into a prodigious number of branches, composed of an innumerable quantity of calcareous articulations, of a different form from that of the stem. In the flattened state in which these heads are found, they are sometimes nine inches long and of an equal diameter.

Hiemer has given the description of an encrinus of this species, found in the duchy of Wirtemburg, near Dombde, on a slate stone, about four feet high and more than three in width. This piece is composed of a great number of calcareous or articulated stems, of a considerable length, which grow in different directions, and have, each at their summit, a sort of pencil, formed by an assemblage of articulated branches.

Guettard has stated that the stem of encrini commences by being round in the lower part, and terminates in five fronts or corners; but there seems to be no proof whatever of the generality of a fact of this kind. The same author adds, that encrini are to be found with four rays and several corollæ on the same stem. Hærembert tells us that certain species have five or six rays, or little branches. Schulz describes some with eight rays and a square basis; and others with twenty rays. Rosinus mentions some with twelve rays and an hexagonal basis; and Scheuchzer has figured encrinites whose rays are not divided at the base—which is composed of an assemblage of small and smooth pannels.

In one of the ancient strata of the environs of Nevers is found a stem, which is composed of small articulations of a rhomboïdal form. They are changed into calcareous spath, and pierced with a small hole in the direction of their axes. In the collection of M. de France is a specimen of this stem, almost as thick as one's fist, which is absolutely composed of nothing but those little articulations. Some stems have also been found of the same form. There is reason to believe that these stems and articulations belonged to a species of encrinus whose entire form is not yet known.

Parkinson has figured a body, which he has called *encrinites testudinarius*. This ovoïd body, of the thickness of a large nut, appears to have been adherent at its base; and is composed of slender pieces, irregularly pentagonal, of five or six lines in diameter, which rest one against the other, and form a kind of roof: similar ones are found at Valognes, in the department of La Manche. They are somewhat convex on one side; the concave side is smooth: the convex side is furrowed with strongly-marked striæ, some of which proceed from the centre, and others originate from these last and proceed to the edges.

In the same strata is found a body which may belong to the genus or family of encrinus. It is often of an irregular hexagonal form: its breadth is six or seven lines, and its thickness three lines. One of the sides, which may be called the under side, is smooth; the other presents a centre, from which proceed from ten to fourteen articulated arms, which go down and appear fixed underneath.

Some handsome specimens, from North America, of bodies found in Genesee, may perhaps be referred to this family. They are about the size and form of a large nut. At their base may be seen traces of a stem on which they must have been supported. They are divided into five parts. Each of these parts is covered with transverse and very regular striæ, cut in their middle by a longitudinal line, a little sunken, and which proceeds from the summit. These regular bodies are changed into a kind of chalcedony. There is reason to believe that the head had the faculty of opening like that of the other encrini.

The ancient strata of Ranville, near Caen, present very singular cylindrical bodies, to which, in Germany, the name of astropodes have been given, and sometimes that of scyphoïds. These bodies, changed into calcareous spath, and composed of slices applied one above the other, are pierced in the direction of their axis. They have some relations with the encrini, by their division into five parts; but they differ from them in other respects. Their form is greatly varied: some have that of a vase; others that of a tun, or little cask. Their sizes vary from that of a middling hen's-egg to that of a small nut. In some the base is concave, and covered with striæ, radiating from the centre to the circumference. On the edge of the upper part are five small elevations, on each side of which are seen two small indentations, which must have carried other pieces not now found there. The inside of this body is concave as far as the base; others have the upper part a little convex, and covered with radiating striæ. On this kind of platform are seen five carinæ, star-wise; and between each of them is a cleft, which extends from the centre to the circumference. There are some of these bodies, however, in which these apertures are not seen.

Another species, or variety, very remarkable, which has the form of a vase, and to which M. de France has given the name of *astropodium elegans*, has, on the edges of its upper part, five indentations, or sinkings, agreeably striated, which may be compared to the external impressions left by five bivalve shells, covered with striæ, proceeding from the summits. This upper part is almost fourteen lines in diameter, and has a hole in the centre. It is very probable that the indentations served to support certain pieces which are not found—or organs, not calcareous, and which could not be preserved.

It appears that the layers, or slices, which composed these bodies, may have been detached one from the other after the destruction of the polypus ; for they are sometimes found isolated. Many figures of these kinds of fossil polyparia may be seen in the works of Parkinson and Knorr.

The *Caryophylli*, being divided into five parts at their head, and being changed into calcareous spath, appear to have many relations with the encrinites.

The generic name of Pentacrinus was given by M. Oken to a species of living encrinus, inhabiting the sea of the Antilles, and which he thus characterises :—Stem articulated ; very long, pentagonal, bearing verticillated branches in different points of its length, and a similar one at the summit. The name of *Pentacrinite* had been anciently given by Knorr and other writers to the species of encrinites whose stem has five faces. In the Natural History of the *Crinoïdea*, by Mr. Miller, this name has been given to a genus, in which that gentleman recognizes the following characters :—An animal, with a column formed of numerous junctures, with five sides or faces ; articulated in their surface by marks, with five semi-striated petals. The upper juncture of the column, supporting a pelvis with five junctures, on which are five supports of the first ribs, followed by five of the second ribs, and by five of the shoulders, from which proceed ten arms, having each two hands, composed of many tentaculated toes.

It is not easy to suppose that any specimens could have contained, with exactitude, all the characters here detailed. The figures, however, of the species referred to this genus by Mr. Miller, are to be found in Parkinson's Organic Remains. These species are four in number, viz.—*P. caput Medusæ*, *P. briareus*, *P. subangularis*, *P. basaltiformis*, and belong to the lias of this country.

It is important to observe that, though this genus, until the other day, was considered to exist in the fossil state only, two specimens of living species, from the American seas, are now in England: one in the British Museum, the other in that of the Geological Society.

It does not appear that either *corals* or *corallines* are found in the fossil state. Fortis speaks of the latter; but it would appear that he has mistaken for them certain small ramous polyparia.

Imperati appears to be the first author who has employed the word MADREPORA to signify a species of stony polyparia. Marsigli extended it to all polyparia of this nature: Boerhaave and Tournefort applied it to those which are very porous; and Linnæus, in fine, restrained it to the species which present excavations at their surface, in the form of lamellated stars; of which he has constituted a distinct genus in his class of zoophytes. It is not necessary here to trouble the reader with the reformations of subsequent naturalists on this genus. Suffice it to say, that the best of them, that of M. Lamarck, who has made the madrepores of Linnæus and Pallas a section of the polyparia, which he names *lamelliferous*, and has reserved the name *madrepore* for the lamelliferous and dendroïd polyparia, is unfortunately based only on the knowledge of the polyparium, or cretaceous mass, produced by the animals. Though science is in possession of some facts respecting certain of these animals themselves, there is far from being sufficient information respecting them all; yet such information is indispensable for the proper formation of genera, founded on the consideration of the polyparium only. Imperati was the first who had the least idea that the madrepores of Linnæus belonged to animals. This fact was subsequently confirmed by other naturalists; but still we have no information concerning any animal of the genus madrepore, as it is restricted by M. Lamarck. The definition of that genus must, therefore, stand thus:—Unknown polypi, contained in cellules, or lodges, more or less deep; more or less projecting; hardly stelliform; scattered on the surface of an arbusculum, entirely calcareous, which is fixed to its base; ramified in an irregular manner, and pierced with a great number of pores. This genus corresponds to the division of the *madreporæ ramosæ* of Gmelin.

Such madrepores are not found in our seas, and have been met with hitherto only in those of America and the East Indies, where they especially abound. Fixed by their basis at very considerable depths, they appear to develop themselves by elevating, more or less, the foliaceous expansions, or caulescent ramifications, which constitute them. We are totally ignorant of their modes of growth, of multiplication, and of death. We only know that the polyparium, which is entirely calcareous, is of a tissue so much the closer as it approaches the parts which constitute its base; and that, on the contrary, the extremity of the ramifications is always more porous. The lower cellules are always more defaced, while the upper, and the extremity of the branches, are often terminated by an excavation tolerably profound.

To the very rapid growth of the madrepores proper, and especially that of the *muricata*, is attributed the formation of the numerous reefs which exist in the Southern and Indian Oceans, and in the Red Sea. Be this as it may, it is certain that the majority of the islands, in those parts of the world, rest on a calcareous soil, entirely composed of stony polyparia, and that their most elevated mountains are formed of the same material; but it would be difficult to ascertain that the madrepores are the species found there in the greatest number.

The polypi of the madrepores, which are extremely common in the seas of the hot climates, and principally in those of the torrid zone, can no longer exist in the climate which we inhabit. We find, however, the polyparia of these species in the ancient strata, as well as in the more modern, of our countries, where they have been formed; but it appears that they were not so abundant as in the seas where these polypi exist at the present day: in proof of which we may cite the reefs and immense masses aforesaid, which they form in the Southern Ocean.

Linnæus was the first naturalist who deemed it proper to separate from the madrepores, under which name, before his time, all the stony polyparia had been confounded, a considerable number of species, which are distinguishable, at the first glance, by the smallness of the pores, or polypiferous cellules. M. Lamarck has made two or three alterations in the arrangement and situation of the MILLEPORÆ, not necessary to be mentioned here.

Donati, and especially Cavolini, have given us some details on the animals of the true millepores ; but we must here confine ourselves to their habitation. This consists in lodges, or simple cellules, oval, with a very small rounded aperture; these form, by their accumulation and intimate union from bottom to top, a calcareous polyparium, with branches nearly round, of equal diameter, irregular, and sometimes truncated at the extremity; and, at other times, in the form of subcrustaceous or foliaceous expansions. They exist in all seas, but more especially in those of the hot climates.

It may be generally affirmed that the fossil remains of marine genera, whose species inhabit the existing seas, are more usually found in the tertiary strata than in those which are more ancient ; but this is not the case with the millepores. Although their species are tolerably numerous in the living state, no remains of them appear to have been hitherto found, except in the strata anterior to the chalk, or in the lowest of that substance. A great number have been found in the environs of Caen.

To pursue any further these notices of families and genera of invertebrated fossils would lead us far, indeed, beyond the limits to which this essay must of necessity be confined. For the incompleteness of our sketches, those limits, and the immensity of the subject, must plead our excuse ; at the same time, though none can be more conscious of the general deficiencies of our labours than ourselves, we are inclined to hope that we have, in these pages, performed all that we proposed to perform at the outset of our undertaking—namely, to give an abstract of every thing important in the discoveries in fossil osteology; and to present as accurate an idea as possible of those bodies found in the fossil state, whether by comparison with their living analogues, or by simple description. It was for this purpose that we have thought it right, after our generalities on the invertebrata, to take especial notice of a few of the most remarkable families or genera. To render our work, in this way, as satisfactory as possible, we shall subjoin a tabular view of all the classes, orders, and genera, hitherto found in the fossil state; which, with some little alteration, we have adopted from the Synopsis of M. de France, appended to his Essay on Petrifactions.

Respecting the execution of our task, we have but another word to add.—We claim no pretensions whatsoever to originality, but labour has not been spared; the best authorities have been consulted and collated carefully, and their substance transfused into our pages, as far as the prescribed extent of those pages would permit. We have already amply acknowledged our obligations to the Baron Cuvier: in all that part of our performance, previous to our notice of the fossil fish, his great work on the Ossemens Fossiles has been our chief guide, and the storehouse from which our materials have been principally selected. On this portion of our labours we may be permitted to observe, without vanity, that we have been the first to attempt a transfer, into our own language, of the principal substance of that immortal work. That this might have been done much more completely, and in a style far superior, none can be more ready to admit than the author of this superficial sketch; but he must claim some little merit to himself for having done it at all; and those who are best acquainted with the original, can best appreciate the extreme difficulty of any attempt to condense the matter of five quarto volumes, whose minutest details are pregnant with interest and importance; and, we might add, almost indispensable to the perfect elucidation of the subject.

As to the other sources from which our observations have been derived, we have not been backward in indicating them in the proper places. We are deeply indebted to several other distinguished naturalists and geologists, as well as to M. Cuvier; and the names of Buckland and Conybeare, of De la Bêche, of Mantell, of Sowerby, of De France, and De Blainville, so frequently referred to in our pages, will serve to prove that we are not insensible to the extent of our obligations.

On the whole, we have endeavoured to present to our readers as accurate a review as we could of the state of a science which, even yet, can hardly be said to have passed the limits of its infancy. The spirit of scientific discovery, awakened in every portion of the civilized world, is continually pouring in new materials, which no prescribed space can comprehend, and whose rapid succession no individual industry can overtake. In fact, those who attempt to describe the origin, progress, and existing state of any natural science, are in the predicament of writers of memoirs, not in that of historians. There is much relating to the subjects of both, which, though existing, may not be available, and more that is yet undeveloped in the womb of futurity. There is one chance, however, in favour of the memoir-writer, which is denied to him who traces the footsteps of scientific research—he may sometimes hope to outlive his subject and complete his work; but ages must pass away before human science shall have arrived to its utmost limit of discovery,—even though that limit depend on the finite but still undefinable extent of the faculties of man. He who argues that he can complete the history of any science, is like the rustic, tarrying until the river should pass by—

. . . . "at ille
Labitur et labetur in omne volubilis ævum."

All this is more especially true respecting the subject upon which it has been our fortune to treat. Researches into geology and fossil remains have brought, and are daily bringing, to light a host of interesting facts; but it must be confessed, even by the most ardent cultivators of these studies, that their results are not, in all respects, alike satisfactory. That the external crust of this globe has undergone several changes and modifications, more or less general, and more or less violent, these researches have proved satisfactorily enough: that these changes must have been successive, in point of time, is more than presumable from the character and state of the fossil remains found in the different formations. But, though we are accustomed to talk of the eras of such formations, we are very far removed, indeed, from anything like the precision of certainty on this point. Without the fossils, nothing could have disproved the position, that all the known strata of the globe had been formed simultaneously; but even with them, no man can pretend to assign the time in which such and such formations were produced. Therefore, whatever may be asserted, concerning the immense antiquity of this globe, must be considered, if not false, to be at least conjectural. Not that we are at all inclined to agree with a recent writer on geology, who would comprehend all the revolutions, which are traceable in the crust of this globe, within the short period of sixteen hundred years. The succession of different forms of animal existence, the state of the respective remains, not to mention many other circumstances, strongly militate against this hypothesis; nor does the worthy maintainer of it lend much assistance to the cause which he advocates by his efforts in its support.

> "Non tali auxilio, nec defensoribus istis
> *Liber* eget."

But when the theorists, on the other side of the question, speak indefinitely of myriads of ages, which costs nothing, as Cuvier remarks, but a dash of the pen, or, with greater absurdity, pretend to assign definite epochas, we cannot but admire the fancy from which such reveries proceed, and smile at the credulity which believes them.

It cannot, moreover, have escaped our readers, that equally satisfactory conclusions are not deducible from all the fossil remains found within the bosom of the earth. The extinct genera, especially those of quadrupeds and reptiles, speak pretty clearly in favour of the succession of formations, and of different, though not exactly defined, eras. But it is not so with most of the others. From the fish, for instance, little can be concluded as to the order of succession; and, except a few genera of the ancient strata, as little from the remains of mollusca, crustacea, and radiata. It may be added, too, that there is much doubt respecting distinction from, or identity with, living species, as to many of these aquatic remains; nor is it possible to pronounce with certainty respecting the extinction of genera, even, and much less of species: for who has explored the profound abysses of the ocean—who can classify, and name, and number, all its inhabitants?

And, after all, are we much more enlightened respecting the recesses of the earth itself? What proportion does the degree to which we can penetrate bear to the diameter of the globe? Who can tell whether the impassable granite constitutes a solid nucleus of this planet, or reposes itself on other strata, concealing marvels as great as those with which we are already acquainted; but, unlike them, destined peradventure to remain for ever impenetrable to human investigation? Again, how small a part of the crust of the earth has been examined, and what proportion does that part bear to its entire superficies? And, lastly, have geologists been invariably successful in disentangling the confusion of strata in numbers of localities, and in accounting for the causes of such confusion?

These queries serve to show the extreme folly and presumption of theórizing on a grand scale, so much the fashion with the philosophers of the last century. Knowing next to nothing of the mere superficial crust of this earth, both in extent and depth, yet they presumed not only to speculate, but to dogmatize, respecting the origin and formation of the planet itself. But subsequent geologists have abandoned the *ignis fatuus* of hypothesis for the steady light of observation; and by an industrious accumulation of facts, united with the most salutary cautiousness of deduction, have laid a safe and just foundation for the science. By following in their footsteps, geological speculations cannot fail to assume more and more a character of greater certainty; and as we can set no limits to the sources of power which futurity may disclose, we may indulge the hope that a "theory of the earth" may be yet established, on a firmer basis than the air-built edifices of former world-makers. That such a hope is not altogether chimerical, we are warranted in believing, by the rapid strides made, within but a few years, in the practical application of scientific principles. Time has already realized much that the most enthusiastic votaries of science scarcely dared to dream of; and it may realize so much more, that our posterity may one day have occasion to exclaim—

> . . "Quod optanti Divûm promittere nemo
> Auderet, volvenda dies en attulit ultro."

A TABULAR VIEW

OF ALL THE GENERA OF

ANIMAL BODIES FOUND IN THE FOSSIL STATE,

WHETHER IN STRATA ANTERIOR OR POSTERIOR TO THE CHALK, OR IN THE CHALK ITSELF, &c.

VERTEBRATED FOSSILS.

MAMMALIA.

Ursus	Strata posterior to chalk	Living and fossil.
Martes	" "	"
Canis	" "	"
Hyæna	" "	"
Felis	" "	"
Phoca	" "	"
Didelphis	" "	"
Castor	" "	"
Arvicola	" "	"
Lagomys	" "	"
Lepus	" "	"
Megalonyx	" "	Fossil only.
Megatherium	" "	"
Eliphas	" "	Living and fossil.
Mastodon	" "	Fossil only.
Hippopotamus	" "	Living and fossil.
Sus	" "	"
Anoplotherium	" "	Fossil only.
Xiphodon	" "	"
Dichobune	" "	"
Anthracotherium	" "	"
Adapis	" "	"
Chœropotamus	" "	"
Rhinoceros	" "	Living and fossil.
Palæotherium	" "	Fossil only.
Lophiodon	" "	"
Tapir	" "	Living and fossil.
Elasmotherium	" "	"

Equus . .	Strata posterior to chalk	Living and fossil.
Mus . . .	" " .	"
Cervus . .	" " .	"
Bos . . .	" " .	"
Myoxus . .	" " .	"

CETACEOUS MAMMALIA.

Manati . .	Strata posterior to chalk .	Living and fossil.
Delphinus .	" " .	"
Balæna . .	" " .	"

AVES.

Sturnus . .	Strata posterior to chalk .	Living and fossil.
Pelecanus . .	" " .	"
Pelidna . .	" " .	"

Obs.—It is probable that there may be more genera of birds in the fossil state, but their remains are exceeding difficult of recognition.

REPTILIA.

Testudo . .	Chalk and posterior . .	Living and fossil.
Crocodilus . .	Anterior to the chalk .	"
Plesiosaurus .	" " . .	Fossil only.
Ichthyosaurus .	" " .	"
Mosasaurus .	Chalk	"
Pterodactylus .	Posterior to chalk . .	"
Rana . .	" " . .	Living and fossil.
Salamander . .	" " . .	"

PISCES.

Anenchelum .	Strata anterior to chalk .	Fossil only.
Palæorhyncum .	" " .	"
Palæoniscum .	" " .	"
Palæothrissum .	" " .	"
Anormurus .	Posterior to chalk . .	"
Palæobalistum :	" " . .	"
Clupea . .	Ant. and posterior to chalk	Living and fossil.
Zeus . . .	" "	"
Erox . .	" "	"
Stromateus . .	" "	"
Elops . .	Anterior . . .	"
Labrus . .	Anterior and posterior .	"
Cyprinus . .	" "	"
Pæcilia . .	Posterior to chalk . .	"
Pleuronectes .	Chalk and posterior .	"

Squalus	Chalk and posterior	Living and fossil.
Amia	Posterior to chalk	„
Raia	„ „	„
Ammodytes	„ „	„
Anarrhicas	„ „	„
Apterichthus	„ „	„
Balista	„ „	„
Lophius	„ „	„
Blennius	„ „	„
Callionymus	„ „	„
Cæcilia	„ „	„
Caranxomorus	„ „	„
Centriscus	„ „	„
Chætodon	„ „	„
Coryphæna	„ „	„
Cyprinodon	„ „	„
Diodon ?	„ „	„
Exocetus ?	„ „	„
Fistularia	„ „	„
Gades	„ „	„
Gobius	„ „	„
Holocentrus	„ „	„
Petromyzon	„ „	„
Lochius	„ „	„
Loricaria	„ „	„
Lutjanus	„ „	„
Monopterus	„ „	„
Mugil	„ „	„
Muræna	„ „	„
Ophidius	„ „	„
Perca	„ „	„
Salmo	„ „	„
Sciæna	„ „	„
Scomberoïdes	„ „	„
Scomber	„ „	„
Scorpena	„ „	„
Silurus	„ „	„
Sparus	„ „	„
Syngnathus	„ „	„
Tetrodon	„ „	„
Torpillus	„ „	„

INVERTEBRATED FOSSILS.

POLYPARIA.		
Acetabulum	Posterior to the chalk	Living and fossil.
Flustra	Chalk and posterior	„
Cellepora	„ „	„
Eschara	„ „	„
Retepora	„ „	„
Alveolites	Anterior ? posterior	„
Ocellaria	Posterior ?	Fossil only.
Dactylopora	Posterior	„
Lunulites	Chalk and posterior	„
Orbulites	Posterior	Living and fossil.
Disticopora	„	„
Ovulites	„	„
Millepora	Anterior and chalk	„
Favosites	Anterior	Fossil only.
Catenipora	„	„
Tubipora	„	Living and fossil.
Sarcinula	„	„
Caryophyllus	Anterior, chalk, and posterior	„
Turbinolia	Posterior	Fossil only.
Cyclolite	Anterior and chalk	„
Meandrinus	Posterior	Living and fossil.
Monticularia	Anterior	„
Astrea	Anterior and posterior	„
Pocillopora	Posterior	„
Madrepora	Anterior and posterior	„
Seriatopora	Anterior, chalk, and posterior	„
Oculinus	Posterior	„
Isis	„ ?	„
Gorgon	Anterior	„
Flabellaria	Posterior	„
Spongia	Anterior	„
Alcyone	Chalk	„
Virgularia	„	„
Encrinus	Anterior and posterior	„
Poteriocrinites	Anterior to chalk	Fossil only.
Platycrinites	„ „	„
Apiocrinites	Anterior and chalk	„
Pentacrinites	Anterior to chalk	Living and fossil.
Cyathocrinites	„ „	Fossil only.

Actinocrinites	,, ,,	Fossil only.
Rodocrinites	Anterior to chalk	,,
Caryophyllites	,, ,,	,,
Marsupites	Anterior and chalk	,,
Alecto	,, ,,	,,
Apsendesia	Anterior to chalk	,,
Berenice	,, ,,	,,
Chenendopora	Chalk	,,
Chrysaora	Anterior to chalk	,,
Eudea	,, ,,	,,
Eunomia	,, ,,	,,
Fabularia	Posterior to chalk	,,
Halliroe	Anterior to chalk	,,
Homeria	Posterior	Living and fossil.
Tilesia	Anterior	Fossil only.
Diastopora	,,	,,
Spirobora	,,	,,
Microsolenus	,,	,,
Limnorea	,,	,,
Lichenopora	Chalk and posterior	Living and fossil.
Pelagia	Anterior	Fossil only.
Montlivaltia	,,	,,
Ierea	,,	,,
Idmonea	,,	,,
Intricaria	,,	,,
Entalophora	,,	,,
Theonia	,,	,,
Terebellaria	,,	,,
Turbinolopsus	,,	,,
Nubecularia	Posterior	,,
Orizaria	,,	Living and fossil.
Palmularia	,,	Fossil only.
Polytripus	,,	,,
Saracenaria	,,	,,
Lycophrus	,,	,,
Textularia	Chalk and posterior	,,
Vaginopora	Posterior	,,
Vincularia	,,	,,
Thamnasteria	Anterior	,,
Pagrus	Chalk	,,
Ventriculites	,,	,,
Larvaria	Posterior	,,

STELLERIDES.

Genus	Stratum	Occurrence
Euryalus	Chalk and posterior	Living and fossil.
Asterias	Posterior	"
Comatulus	Chalk and posterior	"
Ophiurus	Posterior	"

ECHNIDES.

Genus	Stratum	Occurrence
Scutella	Anterior	Living and fossil.
Clypeastrum	Anterior, posterior, and chalk	"
Galerites	Chalk and posterior	Fossil only.
Ananchites	Chalk	"
Spatanga	Anterior and chalk	Living and fossil.
Cassidula	Anterior, posterior, and chalk	"
Nucleolites	" "	Fossil only.
Echinus	" "	Living and fossil.
Cidarites	Anterior and chalk	"

CRUSTACEA.

Genus	Stratum	Occurrence
Agnostus	Anterior	Fossil only.
Calymena	"	"
Paradoxides	"	"
Asaphus	"	"
Ozygia	"	"
Portunus	Stratum doubtful	Living and fossil.
Podophthalmus	"	"
Cancer	Posterior	"
Goneplax	" ?	"
Gelasimus	" ?	"
Gecarcinus	" ?	"
Atelecyclus	"	"
Leucosia	Stratum unknown	"
Inachus	"	"
Dorippe	"	"
Raninus	"	"
Pagurus	Chalk	"
Eryon	Stratum unknown	"
Scyllara	"	"
Palinurus	"	"
Palemon	"	"
Astacus	Chalk	"
Galatea	Stratum and fossil, genus doubtful	"
Spheroma	Posterior	"

Limulus	Posterior ?	Living and fossil.
Cypris	”	”
Crust. Macroura	”	”
INSECTA.		
Curculio	In amber and posterior	Living and fossil.
Scorpio	”	”
Musca	”	”
Blatta	”	”
Tipula	”	”
Aranea	”	”
Ichneumon	”	”
Libellula	”	”
Scarabæus	”	”
Scolopendra	”	”
Papilio	”	”
Hemerobion	”	”
Carabus	”	”
ANNELIDES.		
Siliquaria	Posterior	Living and fossil.
Dentalus	Anterior and posterior	”
Entalus	Chalk	”
SERPULEA.		
Spirorbis	Chalk and posterior	Living and fossil
Serpula	Anterior, posterior, and chalk	”
Vermilia	” ”	”
Rotularia	Anterior	Fossil only.
CIRRHIPEDES.		
Balanus	Anterior and posterior	Living and fossil.
Pollicipes	Posterior	”
TUBICOLŒA.		
Penicillus	Posterior	Living and fossil.
Clavagella	”	Fossil only.
Fistulana	”	Living and fossil.
Teredinus	”	Fossil only.
Taretta	”	Living and fossil.
PHOLADARIA.		
Pholas	Posterior	Living and fossil.
Gastrochen	”	”
SOLENACEA.		
Solen	Posterior	Living and fossil.

Gervillia	Chalk	Fossil only.
Panopæa	Posterior	Living and fossil.
Myaria.		
Mye	Anterior and posterior	Living and fossil.
Anatina	Anterior	,,
Mactracea.		
Lutraria	Anterior, chalk, and posterior	Living and fossil.
Mactra	Posterior	,,
Crassatella	,,	,,
Erycina	,,	,,
Corbulea.		
Corbula	Posterior	Living and fossil.
Sphæna	,,	,,
Pandora	,,	,,
Lithophaga.		
Saxicava	Anterior	Living and fossil.
Petricola	Posterior	,,
Venerupa	,,	,,
Nymphacea Solenaria.		
Sanguinolaria	Posterior	Living and fossil.
Nymphacea Tellinaria.		
Tellina	Posterior	Living and fossil.
Corbis	,,	,,
Lucina	,,	,,
Donace	,,	,,
Crassina	,,	,,
Conchæ Fluviatiles.		
Cyclas	Posterior	Living and fossil.
Cyrene	,,	,,
Conchæ Marinæ.		
Cyprinus	Posterior	Living and fossil.
Cytherea	,,	,,
Venus	,,	,,
Venericardia	,,	,,
Cypricardia	Anterior	,,
Cardiaria.		
Bucardia	Anterior and posterior	Living and fossil.
Cardites	Posterior	,,
Isocardia	Anterior and posterior	,,

Arcacea.		
Cucullea	Anterior and posterior	Living and fossil.
Arca	Anterior, posterior, and chalk	,,
Petunculus	Chalk and posterior	,,
Nuculus	Anterior, posterior, and chalk	,,
Trigonea.		
Trigonia	Anterior and chalk	Living and fossil.
Opis	Anterior	Fossil only.
Nayades.		
Unio	Anterior and posterior	Living and fossil.
Anodon	Posterior	,,
Mytilacea.		
Mytiloïdes	Chalk	Living and fossil.
Modiolus	Anterior and posterior	,,
Pholadomye	Anterior and chalk	,,
Mytilus	Anterior, posterior, and chalk	,,
Pinna	Anterior and posterior	,,
Lithodoma	,, ,,	,,
Malleacea.		
Catillus	Chalk	Fossil only.
Perna	Anterior and posterior	Living and fossil
Pulvinites	Chalk	Fossil only.
Inoceramus	,,	,,
Avicula	Anterior and posterior	Living and fossil.
Pintadina	Anterior	,,
Pectidines.		
Lima	Anterior and posterior	Living and fossil.
Dianchora	Chalk	Fossil only.
Plagiastoma	Anterior	,,
Pachytos	,,	,,
Pecten	Anterior, posterior, and chalk	Living and fossil.
Plicatula	Anterior and posterior	,,
Spondylus	Chalk and posterior	,,
Hinnites	Anterior and posterior	,,
Podapsis	Chalk	Fossil only.
Vulsella	Posterior	Living and fossil.
Ostracea.		
Ostræa	Anterior, chalk, and posterior	Living and fossil.
Anomia	Anterior and posterior	,,
Placuna	,, ,,	,,

Genus	Position	Status
Rudistes.		
Caprinus	Anterior	Fossil only.
Spherulites	,,	,,
Radiolites	,,	,,
Calceolus	,,	,,
Birostrites	Chalk	,,
Crania	,,	Living and fossil.
Brachiopoda.		
Pentamerus	Anterior	Fossil only.
Strygocephalus	,,	,,
Orbicula	Posterior	Living and fossil.
Productus	Anterior	Fossil only.
Terebratula	Anterior, chalk, and posterior	Living and fossil.
Uncites	Anterior	Fossil only.
Strophomena	,,	,,
Thecidea	Chalk	Living and fossil.
Spirifer	Anterior	Fossil only.
Lingula	,,	Living and fossil.
Magas	Chalk	Fossil only.
Phyllidiana.		
Oscabrion	Posterior	Living and fossil.
Patella	Anterior and posterior	,,

Obs.—The living species are generally very large, the fossil very small.

Genus	Position	Status
Calyptracea.		
Rimularia	Posterior	Fossil only.
Emarginula	,,	Living and fossil.
Fissurella	,,	,,
Cabochon	,,	,,
Hipponyce	,,	,,
Calyptrea	,,	,,
Crepidula	,,	,,
Ancyle	,,	,,
Bulleana.		
Bullea	Posterior	Living and fossil.
Bulla	,,	,,
Colimacea.		
Helix	Posterior	Living and fossil.
Carocola	,,	,,
Helicina	,,	,,
Bulima	Anterior and posterior	,,

Agathina . .	Posterior . . .	Living and fossil.
Ambretta . .	,, ...	,,
Auricula . .	,, ...	,,
Cyclostoma . .	,, ...	,,
LIMNEANA.		
Planorbis . .	Posterior	Living and fossil.
Physe . . .	,, ...	,,
Limnea . .	. . . .	,,
Rissoa . . .	,, ...	,,
MELANIÆ.		
Melania . .	Anterior and posterior .	Living and fossil.
Melanopsis . .	Posterior	,,
PERISTOMIA.		
Paludininus .	Posterior	Living and fossil.
Ampullaria . .	,, ...	,,
NERITACEA.		
Pileola . .	Anterior and posterior .	Fossil only.
Neritina . .	Posterior	Living and fossil.
Natice . .	,, ...	,,
Nerita . . .	,, ...	,,
PLICACEA.		
Tornatella . .	Posterior . . .	Living and fossil.
Pyramidella .	,, ...	,,
Nerine . . .	Anterior . . .	Fossil only.
SCALARIA.		
Vermetta . .	Anterior and posterior .	Living and fossil.
Scalaria . .	Posterior	,,
Dauphinula .	,, ...	,,
Pleurotomaria .	Anterior	Fossil only.
Euomphalus .	,, ...	,,
TURBINACEA.		
Solarium . .	Chalk and posterior .	Living and fossil.
Maclurites . .	Anterior	Fossil only.
Tupia . .	Anterior, chalk, and posterior	Living and fossil.
Monodon . .	Posterior . . .	,,
Phasianella .	,, ...	,,
Turritella . .	Anterior and posterior .	,,
Cirrus . .	Anterior . . .	Fossil only.

Camalifera.		
Pleurotoma	Posterior	Living and fossil.
Cerithium	Anterior, posterior, and chalk	,,
Cancellaria	Posterior	,,
Nassa	,,	,,
Fasciolaria	,,	,,
Cyclope	,,	,,
Fusus	,,	,,
Pyrula	,,	,,
Potamis	,,	Fossil only.
Struthiolaria	,,	Living and fossil ?
Ranella	,,	,,
Rupes	,,	,,
Triton	,,	,,
Alifera.		
Strombus	Posterior	Living and fossil.
Purpurifera.		
Cassidaria	Posterior	Living and fossil.
Cassis	Chalk and posterior	,,
Purpura	Posterior	,,
Licornis	,,	,,
Lyra	,,	,,
Buccinum	,,	,,
Eburnea	,,	,,
Vis	Anterior and posterior	,,
Tonna	,, ,,	,,
Columellaria.		
Columbella	Posterior	Living and fossil.
Mitra	Anterior and posterior	,,
Voluta	Posterior	,,
Marginella	,,	,,
Volvaria	,,	,,
Convoluta.		
Ovula	Posterior	Living and fossil.
Porcelana	,,	,,
Tariere	,,	,,
Cerapus	,,	Fossil only.
Ancillaria	,,	Living and fossil.
Oliva	,,	,,
Conus	,,	,,

Orthocerea.		
Belemnites	Anterior and chalk	Fossil only.
Orthocera	Anterior	Living and fossil.
Nodosaria	Chalk	„
Hippurites	Anterior and chalk	Fossil only.
Conilites	Anterior	„
Lituolea.		
Spirula	Anterior	Living and fossil.
Spirolina	Posterior	„
Lituola	Chalk	Fossil only.
Cristacea.		
Renulites	Posterior	Fossil only.
Cristellaria	„	Living and fossil.
Miliolus	„	„
Radiola.		
Rotalia	Posterior	Fossil only.
Lenticulina	„	„
Vasculites	„	„
Nautilacea.		
Discorbis	Posterior	Fossil only.
Siderolites	Chalk	Living and fossil.
Nummulites	Posterior	Fossil only.
Nautilus	Anterior, chalk, and posterior	Living and fossil.
Ammonea.		
Ammonites	Anterior and chalk	Fossil only.
Orbulites	„	„
Ammonceratos	Anterior	„
Turrilites	Chalk	„
Baculites	„	„
Hamites	Anterior and chalk	„
Scapulites	Chalk	„
Cephalopodes.		
Bellerophon	Anterior	Fossil only.
Sicca	Posterior	Living and fossil.
Genera little known.		
Amplexus	Anterior	Fossil only.
Planospirites	Chalk	„
Trigonellites	Anterior	„
Receptaculites	„	„

It is evident that the genera Echinus, Balanus, Bucardia, Arca, Nuculus, Modiolus, Pinna, Auricula, Plicatula, Lima, Patella, and Nautilus, which are found in the most ancient and in the recent strata, as well as in our seas, in the living state, must have traversed the chalk, and their debris would have been found there if their soluble testa had not disappeared.

The genera found in all the strata, chalk inclusive, as well as in the living state, are eight in number: viz. Caryophyllus, Seriatopora, Serpula, Mitylus, Pecten, Ostrea, Terebratula, and Tupia. As to the last, however, it has been rarely found in the chalk.

In the foregoing list we have given all the principal genera found fossil, excepting only those of which some doubt appeared to exist; as to the species, we have purposely avoided an enumeration of them, for the obvious reason that the progress of discovery renders it impossible to set limits to their number, any more than to those of the living species.

*** In our tabular view of the fossil genera of invertebrata, the ascending order is pursued, as most convenient.

INDEX.

THE END.